two-sample dependent *t* test
(pages 169 and 177)

$$\text{dependent } t = \frac{\bar{D} - 0}{\dfrac{s_D}{\sqrt{N}}} = \frac{\bar{D}}{\sqrt{\dfrac{\Sigma D^2 - \dfrac{(\Sigma D)^2}{N}}{N(N-1)}}}$$

Mann-Whitney *U* test
(page 195)

$$U = R_1 - \frac{(N_1)(N_1 + 1)}{2}$$

converting *U* to *z* for large samples
(page 200)

$$z = \frac{U - \dfrac{(N_1)(N_2)}{2}}{\sqrt{\dfrac{(N_1)(N_2)(N_1 + N_2 + 1)}{12}}}$$

Chi-square
(page 220)

$$\chi^2 = \Sigma \frac{(O - E)^2}{E}$$

correlation coefficient
(pages 246 and 247)

$$r = \frac{\Sigma z_x z_y}{N} = \frac{N \Sigma XY - \Sigma X \Sigma Y}{\sqrt{[N\Sigma X^2 - (\Sigma X)^2]\,[N \Sigma Y^2 - (\Sigma Y)^2]}}$$

regression line using *z* scores
(page 269)

$$z_{\hat{y}} = r z_x$$

regression line using raw scores
(page 266)

$$\hat{Y} = r \frac{\sigma_y}{\sigma_x}(X - \bar{X}) + \bar{Y}$$

standard error of the estimate
(pages 274 and 275)

$$\sigma_{\hat{y}} = \sqrt{\frac{\Sigma(y - \hat{y})^2}{N}} = \sigma_{\hat{y}} \sqrt{1 - r^2}$$

***F* ratio**
(page 301)

$$F = \frac{\text{between-groups estimate of } \sigma^2}{\text{within-groups estimate of } \sigma^2} = \frac{N s_{\bar{X}}^2}{s_{\text{WG}}^2}$$

sum of squares within-groups
(page 306)

$$SS_{\text{WG}} = \Sigma\Sigma X^2 + \Sigma \frac{(\Sigma X)^2}{n_g}$$

sum of squares between-groups
(page 306)

$$SS_{\text{BG}} = \Sigma \frac{(\Sigma X)^2}{n_g} - \frac{(\Sigma\Sigma X)^2}{N_{\text{total}}}$$

BASIC STATISTICS FOR THE BEHAVIORAL SCIENCES

Kenneth Pfeiffer
The University of California, Los Angeles
Los Angeles, California

and

James N. Olson
The University of Texas, Permian Basin
Odessa, Texas

BASIC STATISTICS FOR THE BEHAVIORAL SCIENCES

Kenneth Pfeiffer

The University of California, Los Angeles
Los Angeles, California

and

James N. Olson

The University of Texas, Permian Basin
Odessa, Texas

HOLT, RINEHART AND WINSTON

New York Chicago San Francisco Dallas
Montreal Toronto London Sydney

Library of Congress Cataloging in Publication Data

Pfeiffer, Kenneth
Basic statistics for the behavioral sciences.

Bibliography: p.
Includes index.
1. Psychometrics. I. Olson, James N., joint author.
II. Title.
BF39.P44 150'.72 80-22778

ISBN 0-03-049866-X

Printed in the United States of America
1 2 3 144 9 8 7 6 4 3 2 1

Publisher: Ray Ashton
Acquiring Editor: Dan Loch
Managing Editor: Jeanette Ninas Johnson
Senior Project Editor: Arlene Katz
Production Manager: Pat Sarcuni
Art Director: Renee Davis

PREFACE

This book is addressed to a wide audience of psychology, education, and other social and behavioral science students who are required to take a course in statistics. It is written for college and university students and stresses practical, useful techniques such students are likely to encounter in readings or in performing required research. This text is not a cookbook. It is based on the premise that statistics will be more meaningful and more easily assimilated if students can gain a basic understanding at the intuitive level. There is an emphasis on useful skills and understanding and away from complex mathematics and derivations. For example, relevant chapters include "Journal Form" to facilitate comprehension of research reports. Some material commonly taught in introductory statistics is omitted in the hope of eliminating areas of potential difficulty for students, particularly with material not usually very useful on a practical basis for the majority of students. Some of the topics not included are interpolation and the complex manipulation of grouped data. These supplementary topics and others can be covered in lectures at the discretion of the individual instructor.

The chapters of the text are unitized and are adaptable to many different types of learning approaches. There is a wealth of pedagogical apparatus including: (1) a concise introductory "Chapter in a Nutshell"; (2) behavioral objectives for each chapter; (3) a "Key Concepts" summary for each chapter; (4) glossaries of symbols, terms, and formulas; and (5) problems and partial solutions for each chapter. The textbook package includes the text, a student workbook, and a comprehensive instructor's manual. The student workbook includes study problems with answers and completion questions useful in learning terminology and principles. The instructor's manual includes suggested teaching techniques, classroom demonstrations, classroom examples, and example test questions.

The authors want to thank Donna Benson, Kim Burroughs, Laura Darke, Nita Jumper, Liz Mack, Earlene Muse, Travis Olson, Patti Phillips, and Glenda Robbins for their help in reading early versions of the manuscript of this book and for correcting errors in the problems solutions. We would also like to thank the staff of the Word Processing Center of the University of Texas-Permian Basin for their contribution to the manuscript.

We are grateful to the Literary Executor of the late Sir Ronald A. Fisher, F.R.S., to Dr. Frank Yates, F.R.S., and to Longman Group Ltd., London, for permission to reprint Tables χ^{2*}, r^*, and t^* from their book *Statistical Tables for Biological, Agricultural and Medical Research* (6th edition, 1974).

Kenneth Pfeiffer
James N. Olson
December 1980

FOREWORD TO STUDENTS

Statistics has traditionally been one of the most difficult and disliked subjects that behavioral science students are required to take. Although statistics *is* a difficult subject, we believe it can be taught in a way that maximizes learning efficiency and minimizes pain. In this book we have incorporated most of the more useful learning aids available today. Each chapter is introduced with a "Chapter in a Nutshell" summary of what is to come in the chapter. Key words, terms, and symbols are located and emphasized in the margin at the place of first occurrence. Each chapter has a "Key Concepts" summary. Finally, there are practice problems and solutions in both the text and in the accompanying workbook.

To make the best use of the learning aids in this book, we suggest you study in the following manner. First, read through the "Chapter in a Nutshell." Do not worry too much if it does not make a whole lot of sense to you. It will make much more sense after you have studied the chapter, and you have become familiar with the terms and concepts. You should read the "Chapter in a Nutshell" to get an overall feeling for what is to come. Second, page through the chapter and look at the concepts, terms, and symbols in the margin, and look at the illustrations and read the captions. Here again, the idea is to get a general overview of what you will be learning. Third, read the "Key Concepts" summary at the end of the chapter. After you have done this, you should have a general idea of the type of material you will be studying, and you will also probably have quite a few questions in your mind about the things you have just looked over.

After completing the three steps above, then start reading the chapter thoroughly from the beginning. Try to understand the things you have seen in the "Key Concepts" summary. As you read, make sure you understand all the terms and concepts emphasized in the margins. You should also be able to answer any questions you have framed in your own mind concerning the contents of the chapter. Many parts of the book will have to be read very slowly to understand the difficult concepts which are presented. The majority of people studying a book like this will have to reread difficult sections over and over to get the point. Statistics is a subject that takes a lot of time to understand because there are many new terms to become familiar with and many complex concepts to comprehend.

After you have read the chapter thoroughly and reread the "Key Concepts" summary, make sure you understand each point in the summary. If you are not sure about some point, then go back over the relevant part of the chapter and reread it until you think you can explain it to someone else. Finally, go back to the beginning of the chapter and reread the "Chapter in a Nutshell." It should make a lot more sense now. It is very important to go back and reread the "Chapter in a Nutshell" because it will help you get an overall understanding of how the various parts of the chapter interrelate.

After you complete all the reading, you then need to test your understanding. Go to the exercises at the end of the chapter and try to work them out for yourself. Try to get the answer before you look at the solution at the back of the book and without referring to the chapter material. The goal is to try to find what you do not know. If you get stuck, then go back to the text and review until you understand where your weakness lies. Then try the troublesome exercises again. Finally, you should further test yourself and practice applying the concepts you have learned by going to the workbook and working out the material for the chapter you have read. We cannot emphasize strongly enough the need to *practice, practice,* and *practice* with problems.

After reading through the chapter you might feel you really understand the material thoroughly and you might not feel you need to try any problems. *This is a mistake!* Unless you try problems, you cannot discover whether there is anything you still do not thoroughly understand. The feeling of understanding after merely reading is deceiving. There is a big difference between feeling you understand something presented verbally and the understanding that comes from actually practicing problems. Solving problems is a distinct skill that cannot be acquired by passive reading but only by solving problems. There is an analogy here with learning to ride a bicycle. You can read about riding a bike, or you can have someone tell you about it, but you cannot learn to ride unless you get out on a bike and try it yourself. You must try, make mistakes, and learn to correct yourself. Likewise, you cannot learn to solve statistics problems unless you practice solving them. That is the reason there is a workbook with this text—to allow you to get more practice.

When studying or reviewing for an exam, the best thing to do is to read the "Chapter in a Nutshell," and then go through the text and make sure you can define and explain the key words and concepts listed in the margin. Then review the key concepts. Try to explain the key words and concepts to a friend, or at least think to yourself (without looking at the book) how you would explain them to someone else. It is important to do this rather than merely passively read, because on an exam you might be asked to explain something. By practicing what you will be required to do, you will be learning in the most efficient way possible. There is another advantage in trying to explain something to someone else. It makes you formulate the concepts in your own words, which helps you understand and remember the material. Finally, trying to explain something will often make you immediately aware of any gaps in your knowledge. This is something that passive reading cannot do. After you have tried explaining all the words and concepts, try some problems. Do not just look at the solutions to problems you have already worked out. Start each problem from scratch and work through it. On an exam you will be asked to solve problems, not read answers or solutions. Therefore you should practice the relevant skill, which is solving problems.

K.P.
J.N.O.

CONTENTS

CHAPTER 6

INTRODUCTION TO HYPOTHESIS TESTING 91

CHAPTER 7

ONE-SAMPLE HYPOTHESIS TESTING 111

CHAPTER 8

TWO-SAMPLE HYPOTHESIS TESTING 147

1 INTRODUCTION

CHAPTER GOALS

After studying this chapter, you should be able to:

1. Describe the difference between descriptive and inferential statistics.
2. Define population and sample.
3. Define parameter and statistic.
4. Describe the characteristics of a nominal, ordinal, and interval scale of measurement, and give examples of each.
5. Differentiate between discrete and continuous data, and give examples of each.

CHAPTER IN A NUTSHELL

Statistics consists of a collection of tools useful for describing, predicting, and interpreting the world. Researchers apply these tools in categorizing objects, events, and processes, and in attempting to discover how objects, events, and processes interrelate. There are three types of categorizing: unordered categorization (such as the categories "male" and "female"); ordered categorization (such as the categories of grades *A, B, C,* etc.); and ordered categorization based on the natural number system (such as the possible automobile braking-time latencies of 1.2, 1.3, 1.4 seconds, etc.). Categorizing is the essence of measurement. Measurement, in turn, allows precise descriptions of observable characteristics of the world, and provides the basis for subsequent inferences about unobserved or unobservable characteristics. The class of statistical tools that aims at supplying precise descriptions of observations is termed descriptive statistics. The class of statistical tools aimed at drawing sound inferences about the unobservable is termed inferential statistics.

Introduction to statistics

The word "statistics" typically elicits thoughts of a boring and useless course riddled with mathematical jargon that must be endured to obtain a college degree. However, after an appropriate study of statistics most students recognize the usefulness of a knowledge of statistics, and find its practice rewarding. This is not to say that the study of statistics is easy or that mastery comes quickly when taught "correctly," but rather that a knowledge of statistics is well worth the effort required to obtain it. There are at least three practical ways in which a knowledge of statistics pays off. First, since you are required to read many books and professional journal articles containing statistical terminology and analyses (such as "standard deviation," "significant," "*t* test," and so on), a knowledge of statistics will enable you to *understand* these articles and thereby maximally benefit from reading them. Second, a knowledge of statistics will enable you to critically assess the manner in which facts and figures are presented and to determine whether or not the conclusions are justified. Third, since you are usually required to participate in one or more laboratory/field research projects, a knowledge of statistics is necessary to collect the desired information efficiently and to analyze the results accurately.

What is statistics?

Statistics is a set of tools that helps researchers to quickly and economically answer a potentially infinite number of questions about the nature of the world. The application of these tools will provide answers to many different kinds of questions, for example:

Do individuals in different socioeconomic classes differ in terms of the amount and type of illegal drugs used?

Assume that you have discovered a new method of muscle relaxation. In an attempt to determine the effectiveness of this method, you found from a group of 20 college students that neck muscle tension was reduced an average of 35% following the use of your method. What is the average reduction that you would expect to observe in *all* college students? Having answered this question with, "based on my results, it is *probably* 35%," within what limits would this estimate be accurate most of the time (for example, give or take 3%)?

Assume that you are a statistics instructor, and you want to determine whether or not "programmed" homework is effective in teaching statistical skills. You randomly divide the class into two groups. You give the students in one group programmed homework, and you give the students in the other group traditional homework problems. Subsequent comparison of the average final exam scores revealed a 97 for the programmed group and a 93 for the traditional group. Is this 4-point difference due to some "facilitating" effect of programmed homework, or due to some "accident" of random selection (for example, better students could have been coincidentally assigned to the programmed group)?

Assume that you have administered a test measuring aggression to 50 males and 50 females and found that the average male score was slightly higher than the average female score. Is this difference representative of a general difference between *all* males and *all* females, or is this difference restricted to the particular males and females you tested?

What is the degree of the relationship between the time an infant spends crawling and his or her subsequent IQ?

descriptive statistics
inferential statistics

The tools of statistics are divided into two major classes, **descriptive statistics** and **inferential statistics.** The tools of descriptive statistics condense groups of scores by providing *descriptive* facts and figures summarizing the scores. For example, one way to summarize the raw scores[1] of a group of 50 students who took a statistics exam is to calculate the average score. Suppose the average score is 75% correct. This one number serves to summarize the performance of all 50 students. Another way to summarize the 50 scores is to draw a graph depicting the number of students obtaining each possible score. Notice that the tools of descriptive statistics supply facts and figures that, if calculated correctly, are not subject to doubt. Assuming no mathematical errors in calculating the above average of 75% correct, then that is an indisputable fact.

The tools of inferential statistics make use of the descriptive facts characterizing a group of scores by attempting to make generalizations (inferences) beyond this group of scores. One goal of inferential statistics is to *estimate* the magnitude of some characteristic in a larger group than the one in which you calculated the descriptive statistics. In this larger group each individual score cannot be obtained for one reason or another, usually because of financial and/or time limitations. This "larger group" is referred to as a **population.** A population consists of all the cases of interest. For example, if you are interested in determining the frequency of marijuana use of U.S. males between the ages of 20 and 30, your population is all the 20- to 30-year-old U.S. males. Since you cannot realistically investigate all 19 million of these cases, you might interview a smaller group of the individuals belonging to the population. This "smaller group" is referred to as a **sample.** A sample consists of a subset[2] of the cases that comprise a population. You could interview a sample of 1000 males shown to be representative of the population and describe their marijuana-using behavior. Since you are really interested in the 19 million 20- to 30-year-old U.S. males, you will have to *infer* that the descriptive facts characterizing your sample are the best estimates of the same characteristics of the population. If you find that the average number of incidences of marijuana usage during the year preceding the interview is 13, you would probably infer that the best estimate of the average incidence of marijuana usage during the preceding year for the population

population

sample

[1]A raw score is the original category/value recorded by the researcher when measuring objects, events, etc. Raw score is generally abbreviated with a capital X.

[2]A subset is a part of the whole.

is 13 as well. Notice that estimates are always subject to doubt. The doubt arises because you have not measured everyone in the population. Different samples from the same population may give you different averages upon which to base your estimate. To constantly remind you that the descriptive facts characterizing samples can be precisely determined (since all members of a sample, by definition, are measured), but that the same descriptive characteristics of a population can usually only be estimated, the descriptive characteristics of samples are termed **statistics**, and the descriptive characteristics of populations are termed **parameters.**

statistics
parameters

Categorization and measurement

Researchers attempt to categorize the world by distinguishing between different objects, events, or processes. Generally, they go a step further by categorizing objects, events, and so on, in terms of the extent to which they possess a particular quantity or quality. Measurement is the expression of the properties of objects, events, or processes in numbers. The three types of categorization used in behavioral research are: (a) unordered categorization, (b) ordered categorization, and (c) ordered categorization based on the natural number system. Each categorization scheme involves a different type of measurement. These types of measurement are referred to as **scales of measurement** or **measurement scales.** The scales of measurement differ in terms of their precision and the information they convey. Each categorization scheme is defined in terms of the scale of measurement it entails.

scales of measurement

Unordered categorization. Objects, events, and the like are merely assigned to categories having no apparent ordering between them. Measurement is by *naming* the category into which an object or event falls. Hence the term **nominal scale** of measurement indicates unordered classification. Since the sole property of a nominal scale is simply naming, to say that assignment to a particular category means one object or event is "better," "bigger," "smaller," "smarter," or whatever, is meaningless. Examples of nominal scales are: assignment of individuals to the categories of "male" or "female"; assignment on the basis of eye color; assignment on the basis of city of birth; and so on.

nominal scale

Ordered categorization. Objects, events, and so on are assigned to categories as in a nominal scale, with the addition of one extra property: there is a specific order among the categories. Measurement is by naming the rank or *order* in which an object or event falls. Hence the term **ordinal scale** of measurement indicates ordered categorization. With an ordinal scale, to say that assignment to a particular category means one object or event is "bigger"/"smaller" than one assigned to another category now makes sense. For example, individuals could be assigned to one of

ordinal scale

the ordered categories "high," "medium," or "low" on the basis of their socioeconomic level; or assignment could be made on the basis of body type (ectomorph, mesomorph, endomorph), college level (freshman, sophomore, junior, senior), and so forth. Anytime you rank, rate, or judge (as "on a scale of one-to-five"), it is an ordinal scale.

Ordered categorization based on the natural number system. This categorization scheme is linked with the natural number system in that the ordered categories used are represented as real numbers. Objects or events are assigned to ordered categories as in an ordinal scale, but there exists an additional property. This property is that the difference between adjacent categories is always the same, just as the difference between adjacent numbers is the same throughout the number scale. For example, the difference between 2 and 3 is 1, just as the difference between 3 and 4 is 1. To say it another way, the *interval* between adjacent categories is the same, and thus the term **interval scale** of measurement indicates ordered categorization based on the natural number system. With an interval scale, not only can you say that an object or event assigned to one category is "bigger"/"smaller" than one assigned to another, you can precisely state "how much bigger" or "how much smaller." A good example of an interval scale is temperature. The difference in energy between the adjacent categories 0°F and 1°F is the same as the difference between other adjacent categories, such as 103°F and 104°F. As a consequence of equal intervals, you can precisely state that 40°F is 30° more than 10°F. It is often difficult to decide whether the measurement is ordinal or interval. For example, measures of psychological traits (IQ, creativity, introversion-extroversion, etc.) and answers to questionnaires are frequently viewed as interval data, but many people argue that the intervals between adjacent categories are not equal throughout the scale. The researcher must decide whether to regard the measurement as ordinal or interval. When in doubt, most researchers assume they have interval measurement because of certain statistical advantages.

interval scale

A still more sophisticated scale of measurement associated with the natural number system is the **ratio scale.** A ratio scale is the same as an interval scale except that it has a nonarbitrary zero point. For example, the zero point for the Fahrenheit temperature scale is purely arbitrary because the inventor of the scale, G. D. Fahrenheit, simply felt like placing the zero point at the lowest level the mercury fell on the coldest day of winter in the year he invented the Fahrenheit thermometer. On the other hand, a true ratio scale will have a zero point located at a place that is intuitively compelling. Length, for example, as measured in meters (or miles, yards, etc.) has a zero point where there is, well, no length! For weight, pounds and ounces form a ratio scale. For temperature, the Kelvin scale would be a ratio scale. A clue to whether or not you have a ratio scale is that negative numbers are not possible if the zero point

ratio scale

is nonarbitrary. Ratio scales are mentioned only in passing, because there is no difference in the treatment of interval and ratio scales in statistics.

Having discussed the types of categorization and their associated scales of measurement, it is important to notice that all ratio scales are interval scales, all interval scales are ordinal scales, and all ordinal scales are nominal scales. The only difference is the increasingly sophisticated properties implicit in ordinal and interval scales. It is also important to note that the raw scores (or data[3]) recorded by the researcher may be either discrete or continuous. **Discrete measurement** categories, or discrete data, generally come from nominal and ordinal scales in which a finite number of raw scores could result, with no "in-betweens." For example, tallying cars in terms of manufacturer (a nominal scale) can never result in the raw score "one-half Ford" or "one and three-fourths Dodges." Counting the number of men and women in the armed services in terms of rank (an ordinal scale) can never result in "captain and two-thirds" being scored. **Continuous measurement** categories, or continuous data, generally come from interval/ratio scales, where it is possible to be infinitely precise in terms of fractions and decimals. Here the limits on precision are constrained only by technology. For example, with temperature it is possible to get readings that vary more or less continuously from, say, 10°F to 11°F, such as 10.1°F, 10.2°F, and so on. With an extremely precise Fahrenheit thermometer, raw scores such as 10.001°F and 10.002°F can be observed. The degree of precision of measurement is limited only by the precision of the thermometer.

discrete measurement

continuous measurement

KEY CONCEPTS

1. Descriptive statistics deals with describing and summarizing groups of raw scores (or data).

2. A population refers to all the scores of interest. Generally you cannot measure all the scores of interest. A sample refers to a subset of scores from a population.

3. Inferential statistics enables you to make inferences about characteristics of populations (parameters) on the basis of observed characteristics of samples (statistics).

4. Scales of measurement are ways of categorizing things, and may be classified as follows:
   a. nominal: unordered categories

[3]Data are a group of scores. The word "data" is plural. The word "datum" is singular and refers to an individual score.

b. ordinal: ordered categories
c. interval/ratio: ordered categories separated by equal intervals

5 Data may be classified as either discrete or continuous. Discrete data can take on only a finite number of values, whereas continuous data can take on a potentially infinite number of values in a limited interval.

EXERCISES

For each of the hypothetical research situations described in Exercises 1 to 11, identify the scale of measurement and indicate whether the data are discrete or continuous.

1 For 16 persons arrested for driving while intoxicated you record whether they live in an urban, suburban, or rural area. Your raw scores are:

suburban	rural	rural	urban
urban	suburban	rural	urban
urban	suburban	urban	suburban
rural	suburban	urban	urban

2 You measure the body lengths (in inches) of 10 full-term infants at birth and record the following:

$17\frac{1}{2}$	$19\frac{1}{4}$	$17\frac{1}{2}$	19	20
21	18	18	$19\frac{1}{2}$	$10\frac{3}{4}$

3 You observe 20 overweight persons and classify them as having high, medium, or low physical-activity levels. Your data are as follows:

low	low	high	medium
medium	low	low	low
low	low	medium	low
high	medium	low	low
low	low	medium	medium

4 You gather a sample of 20 college students to determine the effect of anxiety on memory. In a high anxiety situation, you measure the number of words each person correctly remembers from a list of 20 key words. Your data are:

12	17	18	13	13	12	12
13	14	14	15	16	13	13
16	15	14	16	14	18	

5 You ask a group of eight adults to solve a particular problem. You record the following times (in seconds) required to solve the problem:

12.5	19.0	21.6	16.6
81.6	14.7	18.4	27.9

6 As part of a marketing analysis conducted by a large dairy, you ask ten children to name their favorite ice-cream flavor. The data are:

chocolate	strawberry	peach
vanilla	vanilla	chocolate
vanilla	rocky road	vanilla
	chocolate	

7 You ask 18 individuals to rate the "pleasantness" of a new advertising slogan on a scale from 1 to 5 (using ratings 1, 2, 3, 4, or 5). Their ratings are:

4	3	4	4	3	5
4	4	3	5	3	2
2	5	3	3	4	3

8 You interview 20 couples who have filed for divorce and you record the number of children they have. Your raw scores are:

0	1	5	0	2
3	2	4	2	3
6	0	2	0	2
2	1	2	3	1

9 You ask 15 students to rank order Professors Higgins, Jones, and Smith from 1 to 3 in terms of popularity. Professor Jones received the following ranks and won the popularity contest:

1	1	3	1	1
1	2	1	1	1
3	2	2	2	1

10 You administer a test to 12 delinquent adolescents which is designed to classify them into one of four levels of conformity: extreme nonconformer, nonconformer, conformer, or extreme conformer. You obtain the following data:

conformer	extreme nonconformer
nonconformer	extreme nonconformer
nonconformer	extreme conformer
extreme nonconformer	conformer
extreme nonconformer	extreme nonconformer
nonconformer	nonconformer

11 You administer a personality test to 11 college graduates applying for a managerial position in a large company. The personality test places them in one of five categories: bright, assertive, conscientious, enthusiastic, or controlled. Your raw scores are:

assertive	controlled	conscientious
controlled	enthusiastic	assertive
bright	conscientious	bright
enthusiastic	conscientious	

For each of the hypothetical research situations described in Exercises 12 to 14, identify the population and the sample.

12 A private university desires to estimate the percentage of its students who have full-time jobs. You interview 50 of its students and ask whether they hold full-time jobs.

13 You are interested in the annual incomes of widows over 65 years of age living in Miami, Florida. You select a group of 200 Miami widows over 65 years of age and determine their annual incomes.

14 A memory test was administered to a group of 20 students enrolled in an introductory psychology course. The purpose of this test was to assess the mnemonic aids that introductory psychology students usually use for memorizing key terms.

For each of the hypothetical research situations described in Exercises 15 to 17, identify the population and the sample, and indicate whether the researcher is employing inferential or descriptive statistics.

15 You are interested in hypertension among young male airport flight control tower operators. After examining the systolic blood pressure of 80 male tower operators selected from throughout the United States between the ages of 26 and 30, you analyze the data. You conclude that tower operators between the ages of 26 and 30 tend to have higher than normal blood pressure.

16 You measure the number of times ten albino rats press a lever to obtain food following an injection of amphetamine (an appetite suppressant). The average number of lever presses made by each rat during the session was 187. You state that the average number of lever presses emitted by your ten rats was 187.

17 You want to administer a reading achievement test to all the sixth-grade children in the school district. However, due to the limited number of test materials and personnel, you select one sixth-grade class from each elementary school in the district and administer the tests to only these children. You test 135 out of the total of 674 sixth-graders in the district. After analyzing the reading achievement scores, you conclude that on the average, the sixth-graders in the district are reading at the eighth-grade level.

2

GRAPHICAL REPRESENTATION OF DATA

CHAPTER GOALS

After studying this chapter, you should be able to:

1 Make a frequency distribution using individual or grouped data.

2 Make a cumulative frequency distribution.

3 Make a histogram, bar graph, frequency polygon, and cumulative frequency polygon.

4 Describe the type of data (discrete or continuous) most appropriate for a histogram, a bar graph, or a frequency polygon.

5 Describe the difference between empirical and theoretical distributions.

6 Identify and verbally define: symmetrical distribution; positively skewed distribution; negatively skewed distribution.

CHAPTER IN A NUTSHELL

This chapter describes methods for tallying and graphing groups of scores so that you can see at a glance any "trends," "differences," "idiosyncrasies," and so on. Once the data are tallied and/or graphed, they are often referred to as a distribution of scores. Methods of tallying data include frequency distributions (tables showing how often each specific score occurs) and cumulative frequency distributions (tables showing the number of individuals obtaining scores at or below a specified score). Methods of graphing data include histograms and bar graphs (in which the height of bars indicates how often each specific score occurs), frequency polygons (in which the height of connected dots indicates how often each specific score occurs), and cumulative frequency polygons (in which the height of connected dots indicates the number of individuals obtaining scores at or below each score). There are two types of distributions: empirical distributions constructed from scores gathered in the real world and theoretical distributions constructed from assumptions about the real world and from mathematical formulas. You can "verbally" describe the shapes of distributions of scores as either symmetrical (meaning there is an evenly balanced number of individuals obtaining scores above and below the average score), positively skewed (meaning most individuals obtained low scores, but a few high scores occurred), or negatively skewed (meaning most individuals obtained high scores, but a few low scores occurred).

Introduction to graphical representation

With graphical representation the researcher attempts to present all of the data in such a way that any trends or subtle differences in the scores may be readily observed. The goal of graphical representation is to enable an interested reader to deduce various characteristics of the total mass (mess?) of data at a glance. For example, consider the following study:

Researchers are interested in determining the age at which adolescents show the maximum rate of physical growth. Thirty-five 10-year-olds were measured for height and then remeasured annually until they reached 18. The age of maximum yearly growth was recorded for each subject.[1] The following data represent the age of maximum growth for each subject:

12, 14, 13, 14, 16, 14, 14, 18, 13, 10, 13, 18, 12, 15, 14, 15, 15, 14, 14, 13, 15, 16, 15, 12, 13, 16, 11, 15, 12, 13, 12, 11, 13, 14, 14

As you can see, it is hard to quickly determine trends or idiosyncracies in the data above. It takes more than an instant to find the oldest and youngest maximal growth ages, the most frequent maximal growth age, the middle maximal growth age, and so on.

Before we go on, one important point concerning this example deserves mention. In the age data above, we have rounded off ages to the nearest whole year. Thus an age of "14 years" could represent anything from 13-1/2 years to 14-1/2 years. Age is really a continuous quantity, but to simplify things we have rounded it off to the discrete categories of 10, 11, 12, and so on.

Frequency distributions

One way of organizing the data into a more comprehensible form is with a frequency distribution. A frequency distribution is a table of all the potential raw score values that could possibly occur in the data, along with the number of times each actually occurred.

Constructing a frequency distribution. To construct a frequency distribution, first find the smallest and largest raw scores in the collected data. Then make a columnar table like that in Figure 2-1. Figure 2-1 is a completed frequency distribution of the hypothetical data of the age of maximal growth rate. Label the left column with a capital X, which stands for raw score. Below X list all the potential raw scores from the smallest (put at the top of the column) to the largest (put at the bottom). (Some researchers, however, prefer to put the largest numbers at the top and the smallest at the bottom.) The smallest and largest raw scores actually occurring in the data are used as these extreme values. Label the right-hand column with a lowercase f, which stands for frequency. Below f list the number of times each raw score occurred. For example, using the age

X

f

[1]A subject is one whose behavior is studied, be the subject human, rat, cat, mouse, worm, or so on. A subject is any organism whose responses are measured.

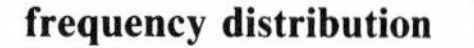
frequency distribution

X	f
10	1
11	2
12	5
13	7
14	9
15	6
16	3
17	0
18	2

FIGURE 2-1
Frequency distribution for age at maximal growth.

data, start with the smallest age and search the data. Count the number of times each specific age occurs and simultaneously cross off each occurrence. The number of occurrences (frequency) is entered under f, adjacent to the corresponding raw score. Notice how easily various characteristics of the data can be determined from Figure 2-1. Ten is the lowest age in which maximal growth occurred, and 18 is the highest age. Fourteen was the most frequent age of maximal growth, followed by the age of 13 and then 15. No one had maximal growth at age 17.

When making a frequency distribution, all potential raw scores must be entered in the X column, regardless of whether or not they actually occurred in the data. For example, in the hypothetical age data, no maximum growth rate occurred at age 17. Nevertheless, 17 was a potential raw score and therefore was included under X. Zero was entered as the corresponding frequency. One way to make a fast check of the accuracy of the f values is to add up all the f values (that is, find the total for the f column). This total should equal the total number of raw scores that were collected and tallied. So that researchers can save words, the total number of raw scores that were collected is abbreviated N. Thus for the age of maximal growth rate data, $N = 35$. Generally, N is the total number of subjects in the study.

N

Grouped frequency distributions

When you are constructing a frequency distribution from a large group of data, it is often desirable to tabulate the raw scores into groups. Figure 2-2 depicts the distribution of ages of the 6588 females interviewed by Kinsey and his associates in their study of sexual behavior. Notice that the ages are grouped in intervals of five years. Each interval into which the scores are grouped is technically called a **class interval** since each score falling within the interval is "classified" as a member of that group. This procedure provides a more compact representation of the data than individually listing each yearly age.

class interval

grouped frequency distribution

X	f
11–15	87
16–20	1860
21–25	1260
26–30	811
31–35	720
36–40	630
41–45	485
46–50	330
51–55	184
56–60	120
61–65	50
66–70	31
71–75	20

FIGURE 2-2
Grouped frequency distribution of the ages of the females interviewed by Kinsey, Pomeroy, Martin, and Gebhard (1953).

Constructing a grouped frequency distribution. Conventional construction of a grouped frequency distribution groups the raw scores into 5 to 20 class intervals of the same width. With fewer than 5 class intervals the data are generally so densely packed that the details of the distribution are lost. With more than 20 class intervals, there are so many intervals in the distribution that it is hard to envision the whole. All class intervals must be equal in width, with class intervals of 2, 3, 5, and multiples of 5 being the most convenient widths. Naturally, the larger the class interval size, the fewer class intervals there will be. The specific number of class intervals and their width is more or less arbitrary and mainly depends on ease of understanding.

Note that the smallest and largest class intervals should contain the respective lowest and highest raw scores. To illustrate with the data in Figure 2-2, it is not necessary to list a class interval of 6–10 since the smallest score is 11. The first class interval should be 11–15. As a rule, you should omit class intervals *at the extremes* of the frequency distribution if no raw scores fall in that interval. You should start the top of the X column with the class interval into which the smallest raw score occurs, and end the bottom of the column with the class interval into which the largest raw score occurs.

Cumulative frequency distributions

You might want to know how many individuals obtained scores *at or below* a particular score. For example, consider the data in Figure 2-1 of the ages of maximal growth rate of boys. If you are interested in the number of boys whose maximal growth rates occur at or before age 14 (or any other age), you can quickly determine this by constructing a *cumulative frequency distribution.*

cf

Constructing a cumulative frequency distribution. A cumulative frequency distribution is simply a frequency distribution with an additional column labeled ***cf,*** for cumulative frequency. Figure 2-3 is a cumulative frequency distribution of the ages of maximal growth rate. Notice that

cumulative frequency distribution

X	f	cf
10	1	1
11	2	3
12	5	8
13	7	15
14	9	24
15	6	30
16	3	33
17	0	33
18	2	35

FIGURE 2-3

Cumulative frequency distribution for age of maximal growth.

the numbers under the X and f columns are the same as in Figure 2-1, because a cumulative frequency distribution includes an ordinary frequency distribution. Look row by row at the numbers under *cf*. These *cf* values tell you the number of scores that are equal to or less than the raw score listed in that row. For example, the *cf* of 1 in the first row means one boy's maximal growth rate was at or below 10 years old. Notice that the number of boys having maximal growth at or below age 10 is simply the frequency of boys having maximal growth at age 10, because there is no boy in the sample younger than 10. In the second row, the *cf* of 3 means three boys' maximal growth rates are at 11 years old or below. This value is calculated by adding the number of boys having maximal growth at age 11 ($f = 2$) to the number having maximal growth below age 11 (which is the *cf* of 1 for age 10). For the third row, the *cf* of 8 is calculated by adding the number of boys having maximal growth at age 12 ($f = 5$) to the number having maximal growth below age 12 (which is the *cf* of 3 for age 11). Thus to calculate the *cf* values for each raw score, start with the row containing the smallest raw score. The *cf* will always equal f for this row. With each successive row, add the value of f at that row to the *cf* value in the row just *before.* One way to make a fast check of the accuracy of the calculated *cf* values is to check that the *cf* value for the last row (corresponding to the largest raw score) is equal to the total number of scores, N.

A cumulative frequency distribution may also be made with "grouped" data, and is called a *grouped cumulative frequency distribution.* A grouped cumulative frequency distribution is constructed by adding a *cf* column to a grouped frequency distribution. Figure 2-4 is an example of a grouped cumulative frequency distribution for Kinsey's data in Figure 2-2.

grouped cumulative frequency distribution

X	f	cf
11–15	87	87
16–20	1860	1947
21–25	1260	3207
26–30	811	4018
31–35	720	4738
36–40	630	5368
41–45	485	5853
46–50	330	6183
51–55	184	6367
56–60	120	6487
61–65	50	6537
66–70	31	6568
71–75	20	6588

FIGURE 2-4
Grouped cumulative frequency distribution of the ages of the females interviewed by Kinsey, Pomeroy, Martin, and Gebhard (1953).

Histograms and bar graphs

Frequency distributions and cumulative frequency distributions are convenient methods for summarizing all of the data in a columnar table. Pictures (graphs) also have the ability to convey large quantities of information at once. Often the reader can "see" key aspects of the data faster with a graph than with a table. Remember the saying, "a picture [graph] is worth a thousand words [numbers!]." One type of graph is a *histogram.*

axes, origin
abscissa
ordinate

Constructing a histogram. The first step in constructing any graph is to make a frequency distribution. Then begin the actual construction of a histogram by drawing two perpendicular lines as in Figure 2-5. These lines are called **axes.** Their intersection, the **origin,** corresponds to zero frequency and a raw score of 0. The horizontal axis, called the **abscissa,** represents all the potential raw scores from the smallest (represented on the left) to the largest (represented on the right). The vertical axis, called the **ordinate,** represents the number of times each raw score actually occurs. The abscissa depicts the values in the X column of the frequency distribution, and the ordinate depicts the values in the f column. These values are listed next to each respective axis and are separated by equal intervals. Tick marks are made on the axes next to each value. Notice that the break symbolized by two slashes is placed along the abscissa to eliminate listing the scores 1 to 9. This convention of two slashes may also be used along the ordinate to make the graph more compact. The next step is to draw adjoining bars above each score listed on the abscissa. The height of each bar corresponds to the frequency with which the score occurs. The higher the bar, the more frequent the score. For example, in Figure 2-5 the bar above age 11 is drawn to the height of 2 on the ordinate to signify that there are two boys having maximal growth at age 11. Notice that the bars are centered above the corresponding scores listed on the abscissa and extend halfway between adjacent scores. For example,

histogram

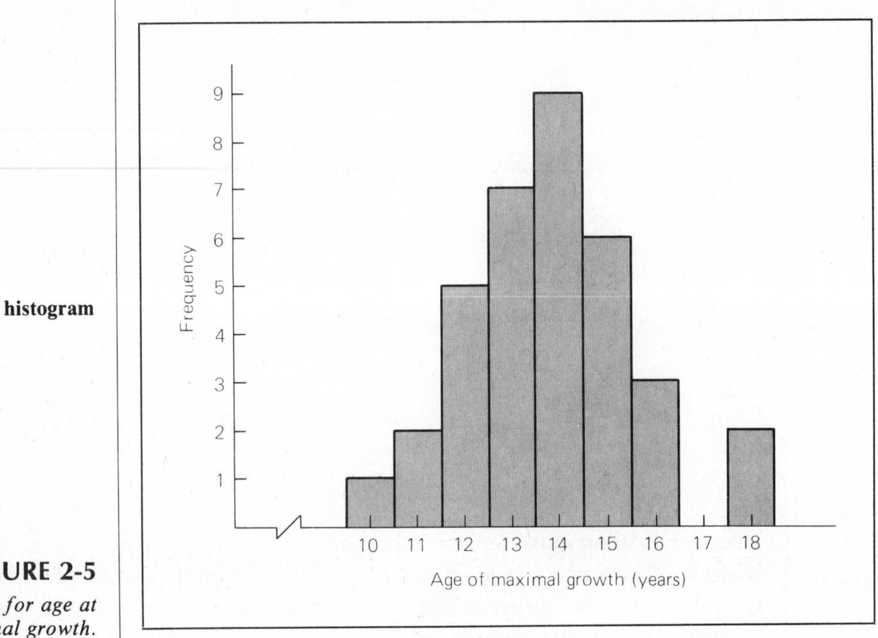

FIGURE 2-5
Histogram for age at maximal growth.

the bar from age 11 is centered above 11 and extends on the abscissa from an imaginary point at 10.5 to 11.5. The point of this "halfway" convention is to show that any boy whose age is between 10.5 and 11.5 is classified as an 11-year-old.

A histogram may also be made with grouped data. You begin by making a grouped frequency distribution. Then follow the procedure for making an ordinary histogram except for the score values listed on the abscissa. For a grouped histogram you can either list the class intervals at each tick mark, as in Figure 2-6, or list the **midpoints of each class interval,** as in Figure 2-7. The midpoint is the midmost value of the class interval. For example, the midpoint of the class interval 11–15 is 13. This midmost value is considered the "best representative" of the scores classified in the interval. You can readily calculate the midpoint by adding the lowest and highest possible scores in the class interval and then dividing this sum by 2:

midpoint of class interval

FORMULA 2.1

$$\text{midpoint of class interval} = \frac{\text{lowest score} + \text{highest score}}{2}$$

Constructing a bar graph. Bar graphs are constructed the same way as histograms except that you leave a space between adjacent bars. The

grouped histogram

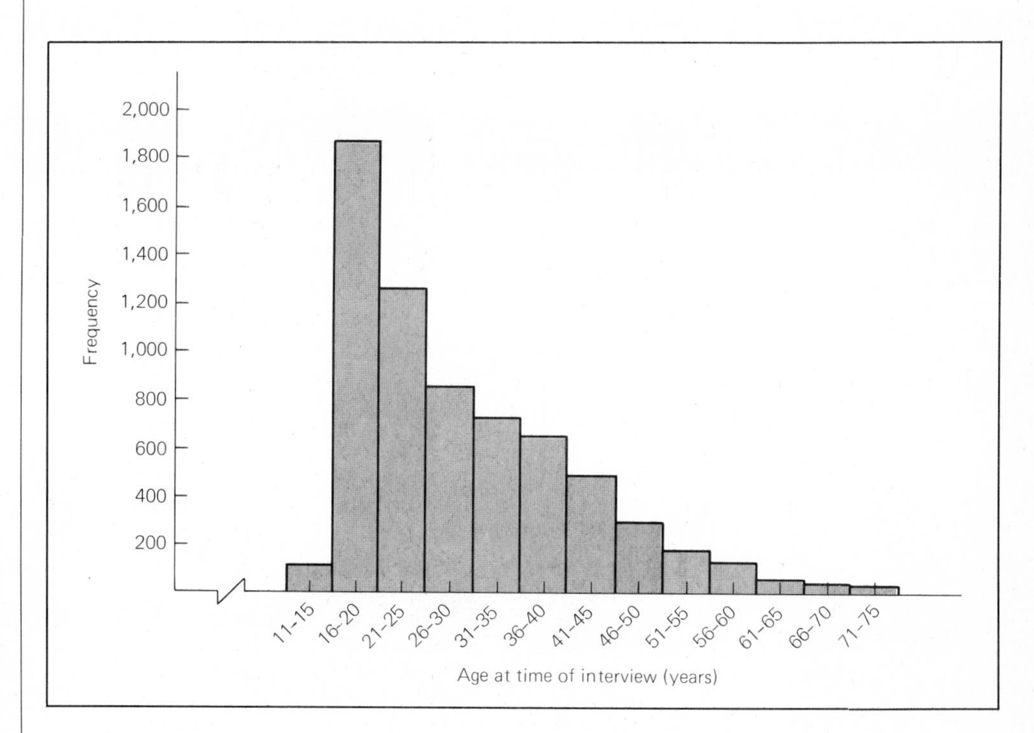

FIGURE 2-6

Grouped histogram of the ages (using intervals) of females interviewed by Kinsey, Pomeroy, Martin, and Gebhard.

grouped histogram

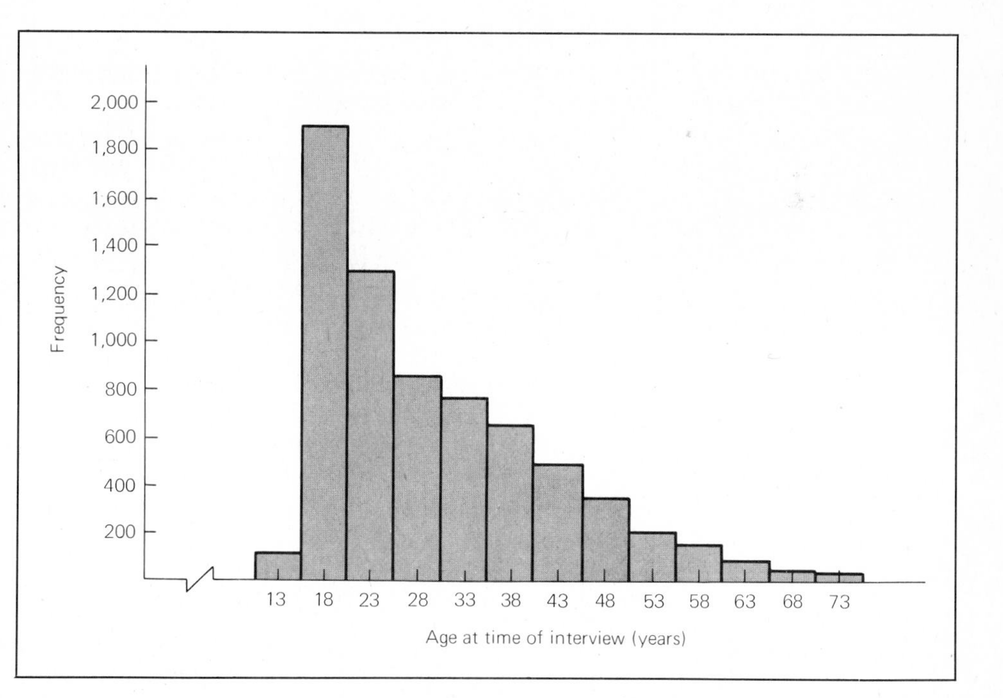

FIGURE 2-7

Grouped histogram of the ages (using midpoints) of females interviewed by Kinsey, Pomeroy, Martin, and Gebhard.

space is usually the width of one bar. For example, suppose there are 30 males and 32 females in your statistics class and you wish to visually depict this information. You might construct a bar graph as in Figure 2-8.

bar graph

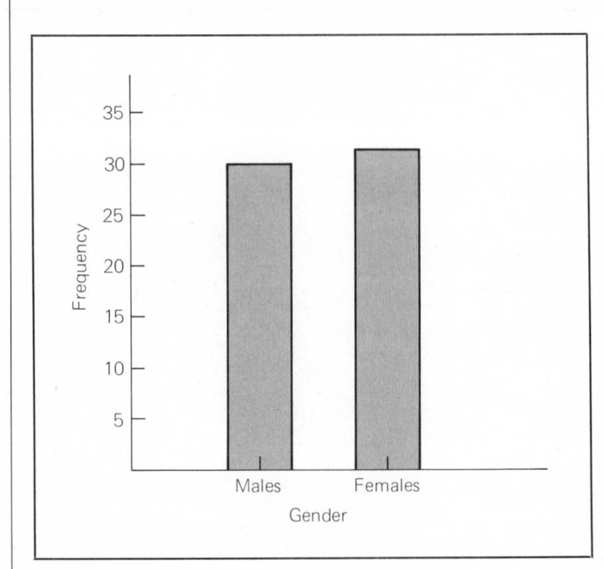

FIGURE 2-8
Bar graph of gender frequency.

Frequency polygons

Frequency polygons are very similar to histograms. The difference is that dots instead of bars are placed directly above each score on the abscissa. The height of each dot corresponds to the frequency with which the score occurred. Notice that if the frequency of a score in the frequency distribution is 0 (the score did not appear in the data) then a dot corresponding to a frequency of 0 must appear over that score in the frequency polygon. The dots are then connected with straight lines as in Figure 2-9. The frequency polygon may also be used for grouped data.

Cumulative frequency polygons

The cumulative frequency polygon is constructed after making a cumulative frequency distribution. The procedure then follows that for making a frequency polygon, but differs in that the ordinate of the cumulative frequency polygon is labeled "Cumulative Frequency" instead of "Frequency." Place dots directly above each score listed on the abscissa at a height corresponding to the *cf* value for that score. Then connect the dots with straight lines as in Figure 2-10. Notice that the line made by connecting the dots in a cumulative frequency polygon will always extend upward to the right because cumulative frequencies can only stay the same or increase.

frequency polygon

cumulative frequency polygon

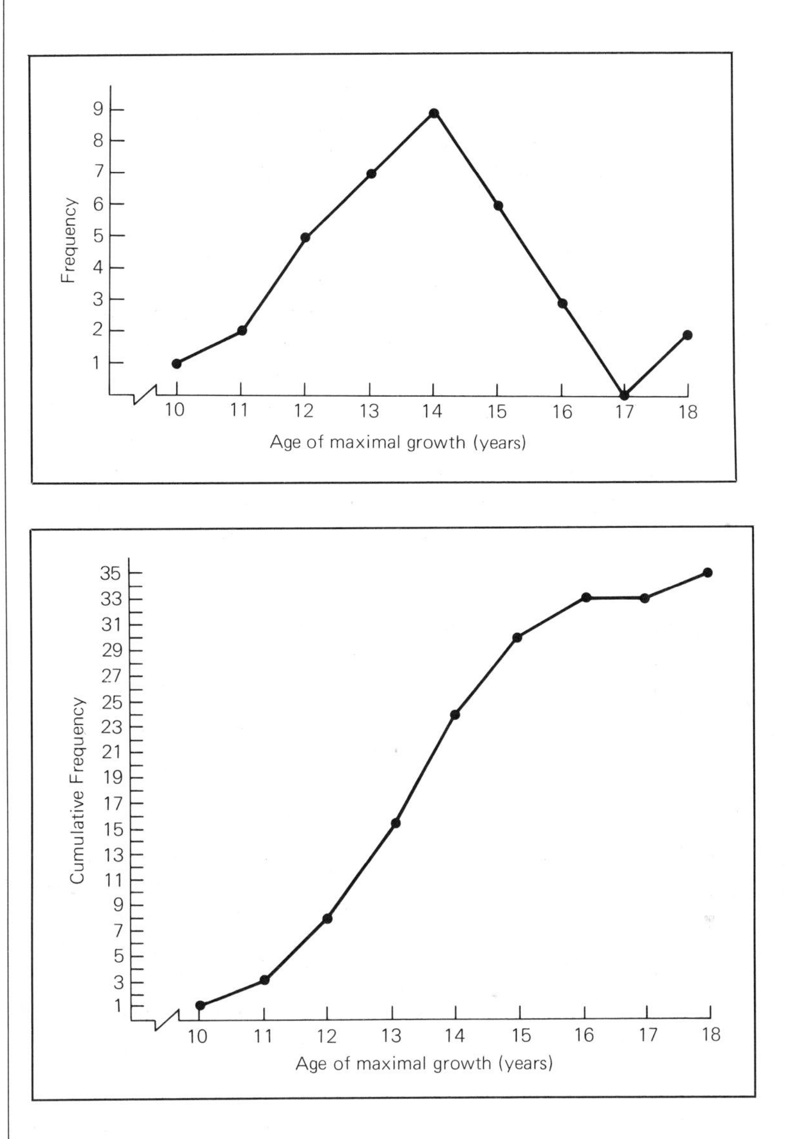

FIGURE 2-9
Frequency polygon for age at maximal growth.

FIGURE 2-10
Cumulative frequency polygon for age at maximal growth.

It is important to notice several conventions that hold for constructing all graphs. First, the axes must have tick marks and must be numbered. Second, the axes must be verbally labeled. Third, the unit of measurement represented on the abscissa must be specified (usually in parentheses after the verbal label). There are few hard-and-fast rules for graphing, but the tick marks, numbers, and labeling conventions are necessary. If you are trying to decide which one of two graphs portraying the same data is better, *use the one that is most aesthetically pleasing and depicts the key aspects of the data in the most quickly understood manner.*

The scale of measurement serves as a useful guide in selecting the method to represent a collection of data graphically. Bar graphs are more appropriate with data measured on nominal and ordinal scales than on data measured on an interval/ratio scale. This is because nominal and ordinal measures render discrete data, and the separateness of the bars in bar graphs projects the idea of a "whole" discrete score, with no in-between values. Histograms and frequency polygons are more appropriate with data measured on an interval/ratio scale. Histograms are often used when the scores have been tabulated into groups, whereas frequency polygons are usually used with ungrouped data. With a large number of potential scores listed on the abscissa, the shape of the polygon made by connecting the dots may look smooth.

Types of distributions

distribution

A **distribution** is the pattern of a group of scores. Frequency distributions and graphs visually display distributions. Distributions are categorized according to whether they are empirical or theoretical and according to shape.

empirical distribution

theoretical distribution

Empirical and theoretical distributions. All of the distributions displayed in Figures 2-1 through 2-10 are **empirical distributions** because they depict actual measurements collected from the real world. There are other types of distributions, called **theoretical distributions,** which are not constructed by taking actual measurements but are derived by making assumptions and applying mathematics to these assumptions. Theoretical distributions are meant to approximate the real world. For example, suppose you are interested in constructing a frequency polygon of the ages at which children start school in Normaltown, U.S.A. If you assume that all children start school at age 5 and there are 400 children of age 5, you could construct the theoretical distribution given in Figure 2-11. This is a theoretical distribution because you have not actually collected data, but instead you have constructed the frequency polygon on the basis of the assumption

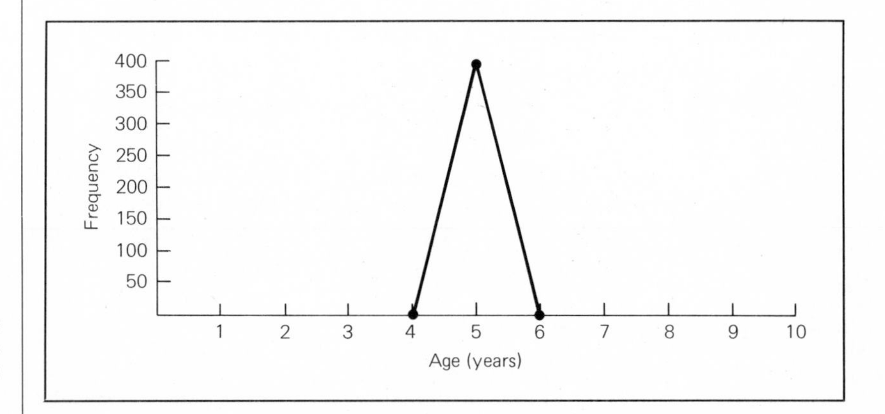

FIGURE 2-11

Theoretical distribution of age at which children start school in Normaltown, USA.

that children begin school at age 5. Now suppose you go out and actually measure the ages of children starting in Normaltown and construct the empirical distribution given in Figure 2-12.

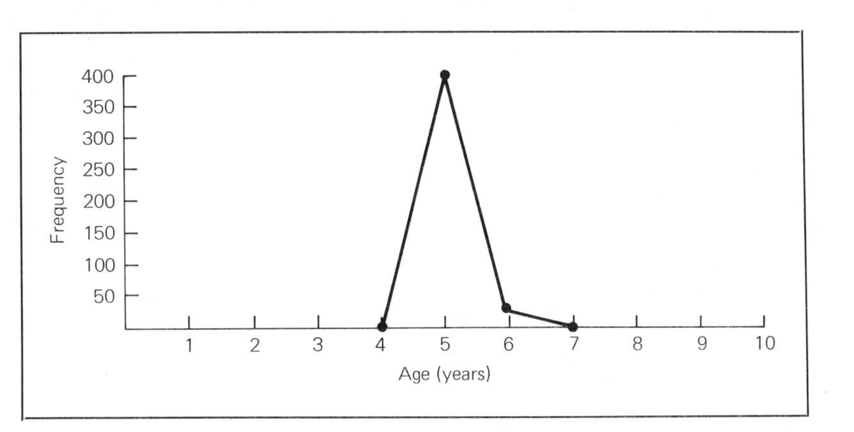

FIGURE 2-12
Empirical distribution of age at which children start school in Normaltown, U.S.A.

The empirical and theoretical distribution of the age at which children start school in Normaltown are very similar because the assumption upon which the theoretical distribution is based bears a good approximation to reality.

Theoretical distributions are often used to approximate reality when it is impossible or impractical to construct empirical distributions. An important theoretical distribution encountered in statistics is the **normal distribution.** The normal distribution is defined by a mathematical formula and is based on the assumption that a large number of random, independent factors are combining to determine some trait. An example of the normal distribution is shown in Figure 2-13. It is a good approximation to many empirical distributions such as height or scores on certain types of tests. The normal distribution is also called a **bell-shaped curve** because it is

normal distribution

bell-shaped curve

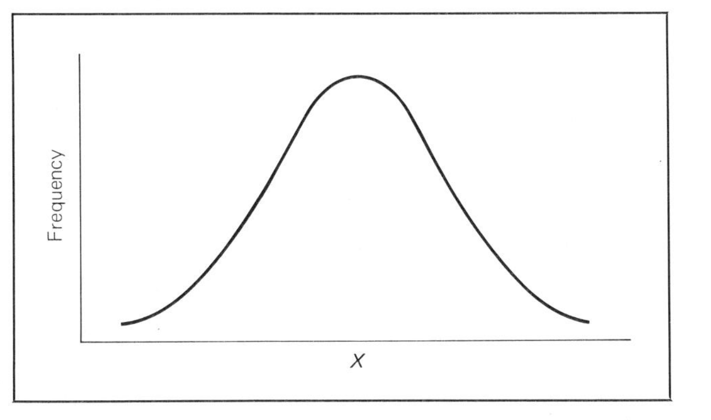

FIGURE 2-13
A normal distribution. The numbers and specific labels have been purposely omitted in this illustrative graph.

moderately peaked in the middle and smoothly tapers downward on both ends like a bell. It is completely smooth because it is defined by a continuous mathematical formula. Empirical distributions depicted by a frequency polygon cannot be so smooth, because discrete points are connected with straight lines.

symmetrical

Shapes of distributions. Suppose you are faced with verbally describing the key aspects of a distribution of scores as that depicted in Figure 2-9, but without the aid of a graph. Just like the word "dachshund" paints a mental picture of a specific type of dog, researchers have special words that paint mental pictures of shapes of distributions. For example, the word **symmetrical** best describes the shape of the distribution in Figure 2-9. A symmetrical distribution is one that could be folded in the middle along a vertical line so that the two halves of the curve coincide. Although the curve in Figure 2-9 may not be exactly symmetrical, it is close enough to be described as "almost symmetrical." Figures 2-14 and 2-15 are also

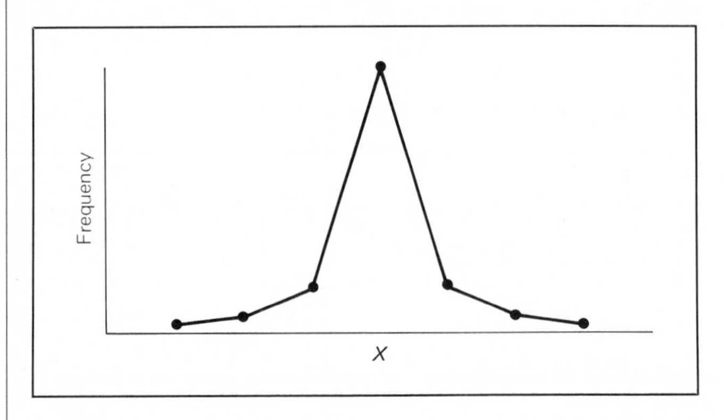

FIGURE 2-14
A symmetrical distribution. The numbers and specific labels have been purposely omitted in this illustrative graph.

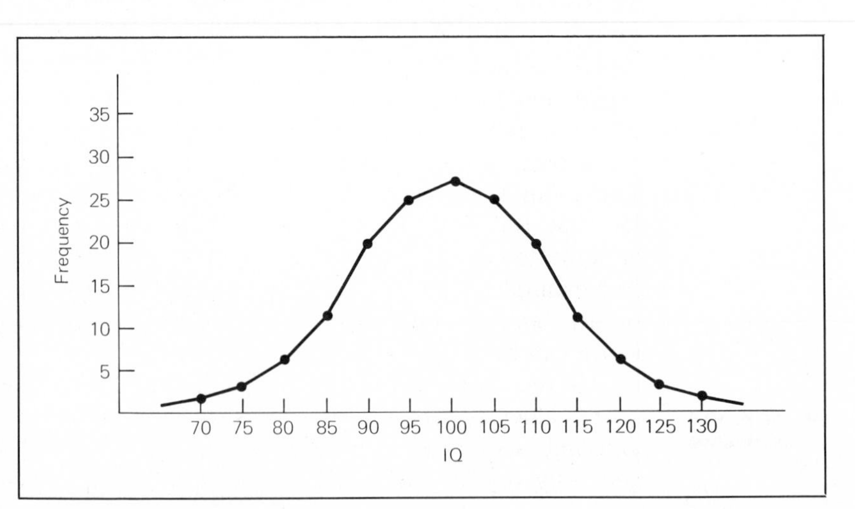

FIGURE 2-15
A symmetrical distribution of IQ scores.

symmetrical distributions. Symmetrical distributions can be very peaked in the middle or can be quite flat. A normal distribution is a symmetrical distribution.

Some distributions are not symmetrical but have a tail-like elongation at one end or the other. These distributions are called *skewed* distributions; the elongation is the skew. If the elongation extends to the right in the direction of increasing score values, as in Figure 2-16, the distribution

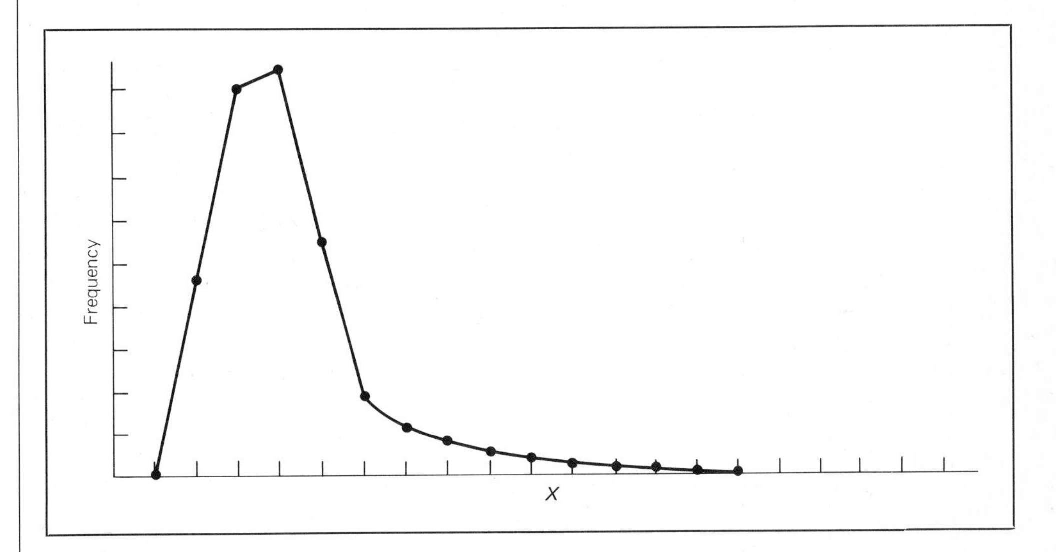

FIGURE 2-16
A positively skewed distribution. The numbers and specific labels have been purposely omitted in this illustrative graph.

positively skewed distribution

is termed **positively skewed.** Low or moderate scores occur most frequently, and just a few high scores occur. Good examples of positively skewed distributions come from "reaction time" studies in which the subject's task is to press a button (or peddle, etc.) as soon as possible after some signal (usually a light) comes on. The delay between the onset of the signal and the press of the button is the reaction time. Reaction time, RT, is usually measured to the nearest hundredth of a second. While most RT scores are low (fast) there are always a few high (slow) RT scores. Lapses in attention, old age, drug state, and so on can all dramatically increase RT. Positively skewed distributions generally occur when there is some lower limit on the value of scores, but no practical upper limit. For example, an RT score cannot possibly be less than zero, or it would be an "anticipation." However, there is no limit as to how long an RT score can be.

negatively skewed distribution

If the elongation extends to the left in the direction of the smaller score values as in Figure 2-17 the distribution is **negatively skewed.** Moderate and high scores occur most often, and just a few low scores occur. This generally occurs when there is some upper limit on score values, but no

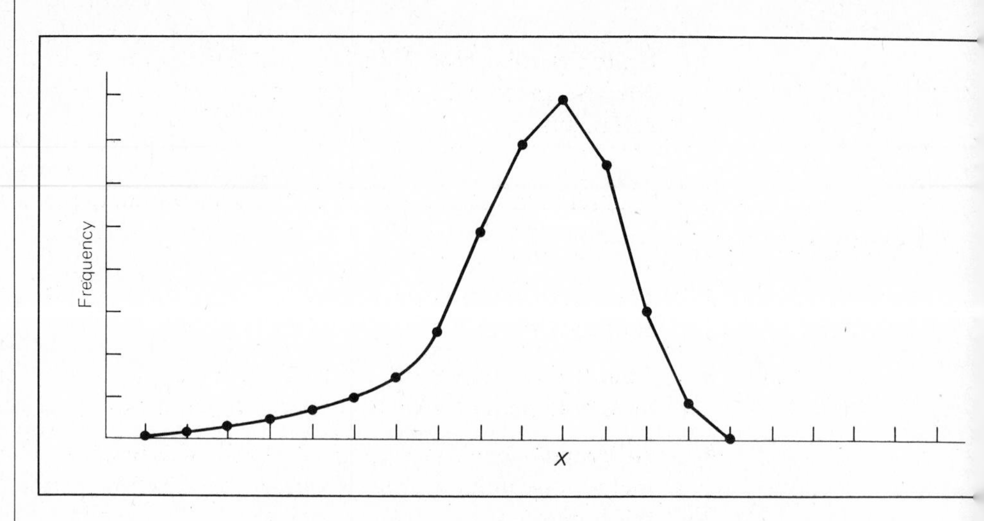

FIGURE 2-17
A negatively skewed distribution. The numbers and specific labels have been purposely omitted in this illustrative graph.

practical lower limit. A good example comes from easy exams, where most of the students score very high, but a few do very poorly compared to the rest of the students.

It is important to realize that skewness is a relative concept. Some distributions are more skewed than others. For example, consider the hypothetical annual incomes of heads of households in the United States and in Mexico, the frequency distributions of which are given in Figure 2-18. The distribution for Mexico is more skewed than the one for the

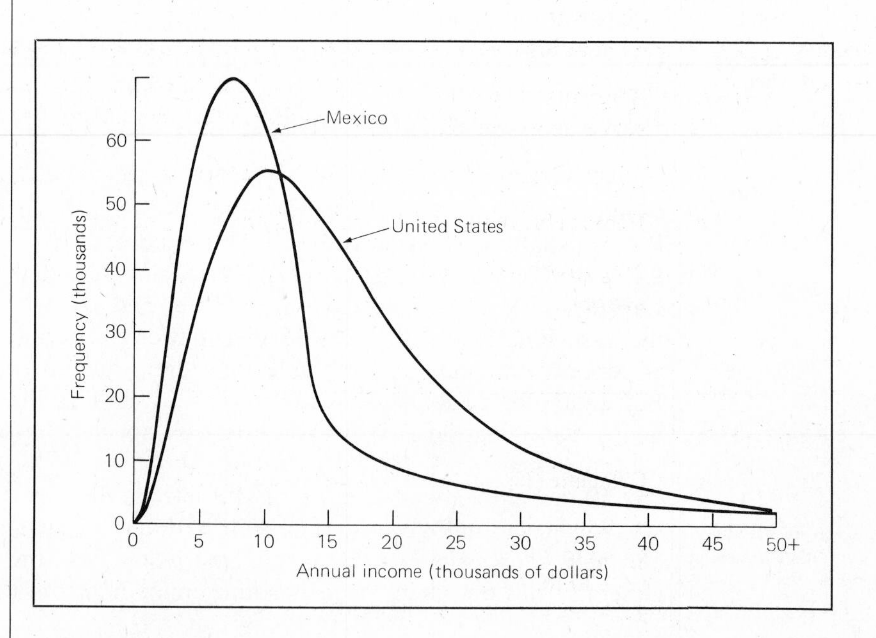

FIGURE 2-18
Hypothetical distributions of the annual income of heads-of-households in the United States and Mexico.

United States. Notice in Figure 2-18 that the distributions appear smooth because income is measured in very small increments and the dots thus run together.

KEY CONCEPTS

1. Frequency distributions are columnar tables of data that show how often each raw score occurs.
2. Cumulative frequency distributions are extensions of frequency distributions and show how often scores occur at or below each score.
3. All graphs should have axes that are calibrated in numbers, labeled, and include the units of measurement. The vertical axis is the ordinate and the horizontal axis is the abscissa.
4. Histograms are graphs in which the heights of adjoining bars correspond to the frequency of the score over which they stand. Bar graphs are identical to histograms except that there are spaces between the bars.
5. Frequency polygons are graphs in which dots represent the frequency of each score. The dots are connected with straight lines.
6. Cumulative frequency polygons are graphs in which dots represent the cumulative frequency of each score.
7. Empirical distributions are constructed from data collected through observations in the real world. Theoretical distributions are constructed by making assumptions about the real world and applying mathematics.
8. A normal distribution is a bell-shaped theoretical distribution that can serve as an approximation to many real-world traits.
9. Symmetrical distributions have vertical axes of symmetry.
10. Positively skewed distributions have elongations toward the larger scores.
11. Negatively skewed distributions have elongations toward the smaller scores.

EXERCISES

1. Calculate the interval width and midpoint for each of the following class intervals:

a. 0 to 2 **b.** 11 to 20
c. 65 to 74 **d.** −5 to 4
e. −10 to −6 **f.** 3 to 5
g. 11 to 15 **h.** 25 to 49

2 You ask 20 students living more than 10 miles from campus what type of transportation they use to get to campus. Their responses are:

car pool	bus	private auto	private auto
bus	motorcycle	motorcycle	bus
private auto	bicycle	bus	bus
private auto	private auto	motorcycle	private auto
bus	bus	car pool	bicycle

a. Construct a frequency distribution for these data.
b. Construct a bar graph for these data.

3 You draw a sample of 30 adolescents and ask them how much time they spent on homework during the past week. You obtain the following times (in hours):

2	8	4	7	4	2	4	3
1	7	5	3	3	2	5	4
1	1	6	3	1	3	9	
5	2	6	2	3	4	3	

a. Construct a frequency distribution for these data.
b. Construct a histogram for these data.
c. Verbally describe the shape of the distribution of time scores depicted in the histogram.
d. Construct a cumulative frequency distribution for these data.
e. Construct a cumulative frequency polygon for these data.

4 You measure the IQs of the students graduating from Pacific High School and Atlantic High School who had been absent 30 or more days during their senior year. Twenty students came from Pacific High and 18 came from Atlantic High. Their IQs are:

Pacific High		*Atlantic High*	
131	99	103	117
123	90	137	130
98	69	132	110
108	99	113	109
93	73	126	121
106	116	78	113
86	81	127	125
121	109	68	131
95	89	136	122
100	103		

a. Construct separate grouped frequency distributions for the IQ scores from each high school using a class interval width of 5 and beginning with the class interval of 68–72.
b. Construct a grouped frequency polygon for each of the grouped frequency distributions obtained in (a) above. Draw these grouped frequency polygons on the same axes.

c. Verbally describe the shape of the distribution of IQ scores from Pacific High and Atlantic High depicted in the grouped frequency polygons.

REFERENCES

Kinsey, A. C., Pomeroy, W. B., Martin, C. E., and Gebhard, P. H. *Sexual Behavior in the Human Female.* Philadelphia: Saunders, 1953.

3

CHAPTER GOALS

After studying this chapter, you should be able to:

1 Calculate the mode, median, and mean.

2 Determine the appropriate measure of central tendency for a given distribution of scores.

3 Calculate the range.

4 Calculate the variance and standard deviation by both the definitional and computational formulas.

5 Determine the appropriate measure of variability for a given distribution of scores.

6 Describe the reason for squaring the deviations from the mean when determining the variance.

7 Describe the reasons for taking the square root of the variance to derive the standard deviation.

8 Describe three properties of the normal distribution in relation to the standard deviation.

CHAPTER IN A NUTSHELL

It is often necessary to reduce groups of data to one or two summary numbers. The most useful summaries are measures of central tendency and measures of variability. Measures of central tendency describe the average size of the scores in the distribution. Three measures of central tendency are: the mode, which is the most frequently occurring score; the median, which is the middle score; and the mean, which is the arithmetic average. Measures of variability are: the range, which is one plus the difference between the smallest and largest score; the variance, which is the mean of the squared deviations from the mean; and the standard deviation, which is the square root of the variance.

Introduction to measures of central tendency

Suppose you were interested in comparing the anxiety levels of freshmen, sophomores, juniors, and seniors enrolled in an introductory statistics course. After administering a test of anxiety you obtain the following data (high scores indicate high anxiety):

Freshmen:	86, 44, 43, 40, 30, 86, 69, 45, 37, 10
Sophomores:	62, 29, 48, 26, 11, 16, 29, 26, 17, 43, 27
Juniors:	39, 42, 48, 31, 39, 20, 39, 31
Seniors:	36, 73, 10, 70, 97, 13, 60, 47, 53, 62

Do freshmen have more anxiety than sophomores? Do sophomores have greater levels of anxiety than juniors? Do juniors have higher anxiety levels than seniors? And so on, and so on. To answer questions like these, you must compare the *group* of 10 freshmen to the *group* of 11 sophomores, the *group* of 11 sophomores to the *group* of 8 juniors, and so forth. Making these comparisons with only the raw scores above (that is, without making further calculations) boggles the mind. Intuitively, questions or statements containing adjectives such as "more . . . than," "less . . . than," "greater . . . than," "smaller . . . than," "higher . . . than," "lower . . . than," etc., involve a comparison between *only two* measurements. Therefore, before you can determine whether the 10 freshmen have more anxiety than the 11 sophomores, you must reduce each group of scores to one number. This one measure should represent the entire group. A **measure of central tendency** is such a measurement; it represents where a group of scores "stacks up." You know the words "measure of central tendency" as the synonym "average." Three different measures of central tendency are the *mode,* the *median,* and the *mean.* The trick is to know which measure of central tendency should be used to "best" represent the group of scores. Unfortunately, as you will see, the measures of central tendency which are easiest to calculate are probably least useful.

measure of central tendency

Mode

mode

The **mode** is the easiest measure of central tendency to calculate. It is simply *that score* that occurs most frequently. For the group of freshman scores above, the score 86 occurred most often (twice), and is thus the mode. You can say, "the modal freshman anxiety level is 86." Looking at the sophomore scores, there is a tie for the most frequently occurring score. The scores 26 and 29 occurred equally often, and both are considered modes. You can say, "the two modes are 26 and 29." For juniors, 39 is the mode. For seniors, all scores occurred equally often. Thus there are ten modes. You can say, "the modes are 10, 13, 36, 47, 53, 60, 62, 70, 73, and 97." In simple formula form, the mode is defined as follows:

FORMULA 3.1

$$\text{mode} = \text{value of score occurring most frequently}$$

Although most distributions of scores have only one mode, distributions having two or three modes are encountered, and it is possible to find distributions having as many modes as there are scores. The word **unimodal** is used to describe a distribution having one mode, and **bimodal** describes a distribution having two modes.

unimodal

bimodal

Many people become confused when the scores are names instead of numbers. This frequently occurs with nominal and ordinal measurement. For example, consider the results of a small survey asking 8 students taking a correspondence course in statistics whether they "recommend," "not recommend," or have "no opinion" as to whether students should take statistics through correspondence. The results are plotted in Figure 3-1. The mode is "not recommend." The mode *is not* the frequency value

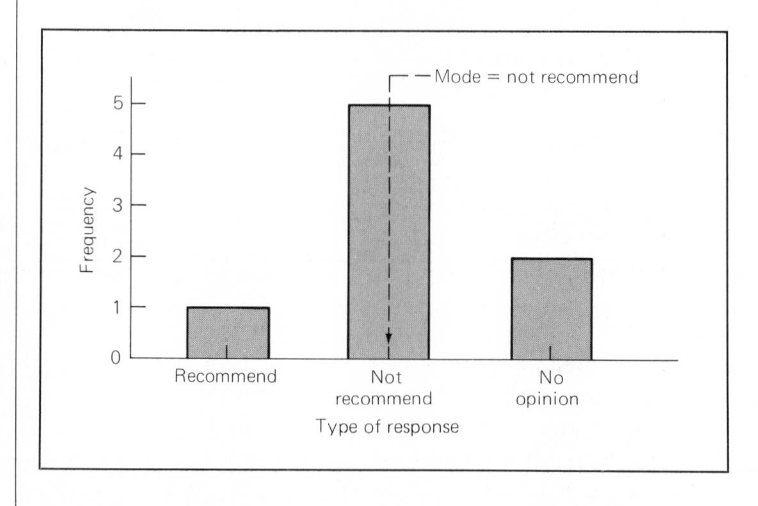

FIGURE 3-1
Frequency of responses to survey question. The type of response represents the categories of this nominal scale of measurement. The mode will always be one (or more) of the categories listed on the abscissa.

5. The mode is always one (or more) of the score categories, regardless of whether they are verbal labels or numbers. If the data are graphed, the mode will be one (or more) of the scores listed on the abscissa, such as "not recommend" in Figure 3-1. Confusion often stems from thinking the mode will always be a number, so people erroneously look to the highest frequency value *as* the mode.

Although the mode has the advantage of quick calculation, its use is limited because of two drawbacks. First, the magnitudes of every score except the most frequent one(s) are completely ignored. For example, the three distributions below have the same mode as the distribution of freshman scores (mode = 86):

Distribution A: 86, 86, 87, 88, 89, 90, 91, 92, 93, 94
Distribution B: 1, 2, 3, 4, 5, 6, 7, 8, 9, 10, 86, 86
Distribution C: 86, 86, 86, 1000, 1001, 1002, 1003, 1004, 1005, 1006, 1007, 1008, 1009, 1010

unstable

You can see the tremendous differences in the magnitude of the scores in each distribution. These differences are not represented by the mode. The second drawback is that the mode is very **unstable.**[1] Changing one score in a distribution can drastically change the mode. For example, suppose you change one of the 86s in Distribution A to an 87. This changes the mode to 87. More drastically, if one of the 86s in Distribution B is changed to a 2, the mode becomes 2. In fact, you can change one of the 86s in Distributions A or B to any other score in the distribution and the mode will shift to that score. The instability of the mode is evident in test-retest situations, such as the one below in which "national grade equivalence" scores are given in the ninth grade and then in the tenth grade for the same group of students:

	Bob	*Pat*	*Jan*	*Tom*	*Jim*	*Sue*
Ninth Grade:	9.5	8.6	8.7	11.0	9.0	11.0
Tenth Grade:	10.5	9.8	9.8	11.9	10.9	11.6

The mode is 11.0 for ninth grade and 9.8 for tenth grade. Since the mode is one type of average, imagine a perplexed parent asking, "Do you mean to tell me that the average grade equivalency decreased from 11.0 to 9.8 in one year of schooling?" The parent has been seriously misled, because each student's grade equivalent actually increased.

Median

median

The **median** is the middle score in a distribution. The median splits the distribution into two equal parts so that there are equal numbers of scores larger than the median and smaller than the median. The easiest way to calculate the median score is to arrange the scores in order of magnitude. Be sure to list every score, even though several scores may be equal! Ordering the freshman anxiety scores gives the following:

10, 30, 37, 40, 43, 44, 45, 69, 86, 86

Then determine whether *N* is an even or an odd number. When *N* is odd, there is a single score in the middle of the distribution. The median is this middle score. This score is found by beginning at the left of the ordered scores and crossing out each successive score while counting up to *N* divided by 2, or $N/2$, rounded up to the next highest number. For the 11 sophomores scores, since $N/2 = 5.5$, round up to 6 and count off six scores as follows:

~~11~~, ~~16~~, ~~17~~, ~~26~~, ~~26~~, ~~27~~, 29, 29, 43, 48, 62

The last score crossed out is the median, 27. You can use this same technique

[1]Unstable means subject to fluctuation. The mode is unstable because different samples from the same population will usually have fairly divergent modes.

if you have constructed a frequency distribution. In simple formula form, when N is odd, the median may be defined as:

FORMULA 3.2

$$\text{median } (N \text{ is odd}) = \text{the middle score}$$

When N is even, there is no single middle score and thus the median is defined as the value halfway between the two middle scores:

FORMULA 3.3

$$\text{median } (N \text{ is even}) = \text{value halfway between the middle scores}$$

You can find the two middle scores by beginning at the left of the ordered scores and crossing off each score while you count up to $N/2$. For the 10 freshman scores, since $N/2 = 5$, cross out five scores:

~~10~~, ~~30~~, ~~37~~, ~~40~~, ~~43~~, 44, 45, 69, 86, 86

The last score crossed out and the score to its immediate right are the two middle scores:

~~10~~, ~~30~~, ~~37~~, ~~40~~, ~~*43*~~, *44*, 45, 69, 86, 86

The median for freshmen is halfway between 43 and 44, which is 43.5. You can find the halfway value by adding the two middle scores and dividing by 2:

$$\frac{\text{sum of two middle scores}}{2} = \frac{43 + 44}{2} = \frac{87}{2} = 43.5$$

When the scores are names, this technique cannot be used. Nevertheless, the median is halfway between the two middle scores. For example, consider the highest degrees held by eight psychologists on the staff at a mental hospital:

M.A., M.A., M.A., *Ph.D., Ph.D.,* Ph.D., Ph.D., Ph.D.

The two middle scores are Ph.D. and Ph.D., and halfway between them is still Ph.D. Thus the median degree held is Ph.D. Now consider the following distribution of degrees held by eight television newswriters:

B.A., B.A., B.A., *B.A., M.A.,* M.A., M.A., Ph.D.

Here, the two middle scores are B.A. and M.A., and halfway between them is what? The median in such cases may be expressed as "halfway between the B.A. and M.A. degrees."

Formulas 3.2 and 3.3 will sometimes give only an approximate value. This will occur when there are ties at the middle score(s), as in the data for juniors:

20, 31, 31, *39, 39,* 39, 42, 48

Using Formula 3.3, the median is 39. However, there are other techniques for calculating the median that may yield slightly different results. These methods make complicated adjustments for ties at the middle of the distribution. Due to the limited usefulness of the median in behaviorial research, we do not believe that the complex derivation of the median warrants the additional effort. We recommend Formulas 3.2 and 3.3 because they are very quick and convenient and yield the same results, or very close to the same results, as the more complex methods.

The median is useful because it conveys information about where the middle of a distribution lies. However, the median has one drawback because it ignores the magnitude of every score except the middle score(s). For example, take the freshman anxiety scores and change the largest score, 86, to any greater value, say, 10087. You will not affect the median of 43.5:

10, 30, 37, 40, *43, 44,* 45, 69, 86, 10087

↑ median = 43.5

Even though the median ignores the magnitude of the scores above and below the middle, it is much more stable than the mode because it takes into account the total number of scores in the distribution (N). N is taken into account because there must be an equal number of scores above and below the median.

Mean

The **mean** is also known as the arithmetic average. The mean is the familiar "average" you probably have been calculating for a long time. But because the word "average" technically refers to either the mean, median, or mode, researchers generally avoid the use of "average." Researchers eliminate ambiguity by using the words "mode," "median," and "mean."

To calculate the mean, add up all the scores in the distribution and divide by N. For the freshman anxiety data, you can find *the sum of* the scores as follows:

$$86 + 44 + 43 + 40 + 30 + 86 + 69 + 45 + 37 + 10 = 490$$

Then divide the sum of the scores by N:

$$\frac{490}{10} = 49$$

Σ

To save words, "the sum of" is abbreviated with the Greek letter **Σ** (sigma). Since "score" is abbreviated *X*, you can abbreviate the phrase, "the sum of the scores" with ΣX. Thus the general formula for the mean is:

FORMULA 3.4

$$\text{mean} = \frac{\Sigma X}{N}$$

deviations

The mean is the balance point of a distribution. The mean balances the **deviations** (distances from the mean) of scores above it with the deviations of scores below it. For example, the balance point of the scores 80 and 2 is the mean of 41. The score of 80 deviates 39 units above the mean, and the score of 2 deviates 39 units below the mean. You can see this balance point concept more readily by graphing the distribution of scores in a frequency polygon. However, instead of drawing it on graph paper, draw it on a piece of wood and cut it out. The cut-out piece will balance at the mean as shown in Figure 3-2.

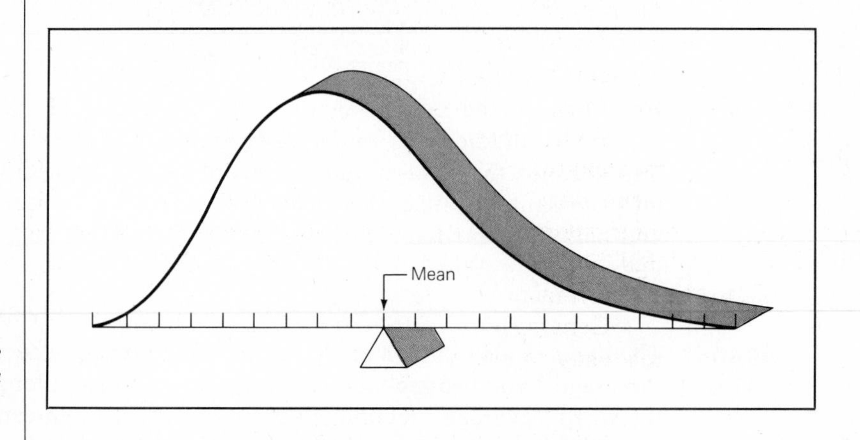

FIGURE 3-2
The mean is the balance point of a distribution.

Unlike the mode and median, the mean takes into account the magnitude of every score. Therefore the mean shifts a little with a change in any score. For example, changing one of the 86s in the freshman anxiety data to 100 changes ΣX to 504, and thus the mean shifts to 50.4. This does not imply that the mean is unstable. Actually, the mean is the most stable measure of central tendency because it shifts relatively little with score changes, especially if you have a large group of data.

In inferential statistics, it becomes important to distinguish whether the mean is derived by measuring every subject you are interested in (the population), or by measuring a subset of the population (a sample).

μ

$\bar{X}$

Although this distinction may seem trivial now, you will see later how important it is. It is important enough to introduce two more abbreviations reflecting this distinction: μ and $\bar{X}$. The Greek letter μ, spelled mu and pronounced "mew," specifies that you derived the mean from the population of scores. Capital $\bar{X}$ indicates that you derived the mean from a sample of scores. There is no difference in calculating μ or $\bar{X}$:

$$\text{population mean} = \mu = \frac{\Sigma X}{N} \qquad \text{sample mean} = \bar{X} = \frac{\Sigma X}{N}$$

When to use the mode, median, and mean

To a large extent, the scale of measurement determines the appropriate measure of central tendency. With a nominal scale of measurement, the mode is the only useful measure of central tendency, since the categories cannot be ordered in a meaningful manner.

With an ordinal scale of measurement, although the mode may be useful at times, the median is the preferred measure of central tendency since it conveys more information than the mode. The mean is not very useful with ordinal data because the mean relies on calculating the sum. A sum is meaningful only when the intervals between successive categories are equal.

With interval/ratio data, the mode and median are both quick and meaningful but do not convey information about the balance point. The mean does. The mean is very useful with interval/ratio data since the intervals between categories are equal. The mean also has the advantage that it is much easier to manipulate and simplify in mathematical derivations and formulas.

General recommendations for the most appropriate measure of central tendency to use with different scales of measurement are:

Scale of measurement	*Preferred measure of central tendency*
nominal	mode
ordinal	median
interval/ratio	mean

These preferences need to be tempered somewhat, however. For example, if you have interval/ratio data and want a quick and easy measure to see how the data are stacking up, you might choose the mode. At other times, with interval/ratio data the mean may give a misleading idea of where a distribution of scores stacks up. This will happen with skewed distributions. For instance, consider the distributions in Figures 3-3 and 3-4. Notice that the mean is closer to the tail of the distribution than

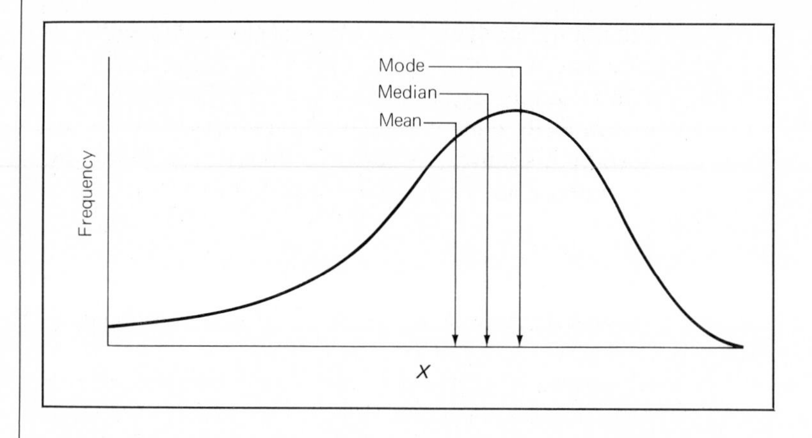

FIGURE 3-3

The location of the mean, median, and mode in a negatively skewed distribution.

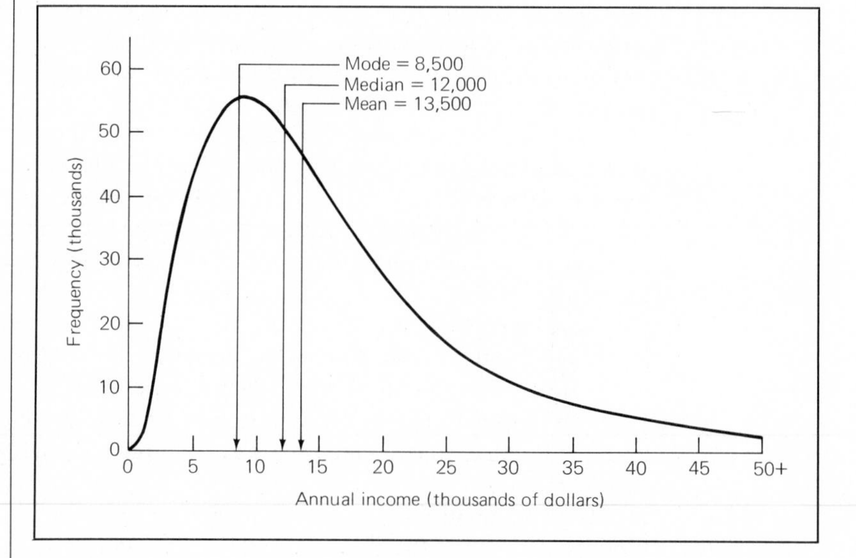

FIGURE 3-4

Hypothetical distribution of the annual incomes of heads of household in the United States. This distribution shows the location of the mean, median, and mode in a positively skewed distribution.

the median for both negative (Figure 3-3) and positive (Figure 3-4) skews. The mean is shifted in the direction of the skew because it takes into account the magnitudes of the extreme but infrequent scores in the tail of the distribution. Since the median is located more with the bulk of the scores, it is a more representative measure of central tendency when a distribution is extremely skewed.[2] Notice that the distribution of the

[2]When a distribution is symmetrical, the mean and median are equal. Thus you might conclude that the most representative descriptive measure of central tendency is always the median. In a sense this is true, because the median is the measure of central tendency closest to all the scores in the group. However, when it is appropriate to calculate the mean, the mean is preferable because of its relative ease of manipulation in mathematical formulas.

annual incomes of heads of household in Figure 3-4 are positively skewed. The choice of whether to use the mean or the median as a measure of central tendency might be guided by the impression of economic prosperity you want to convey. A country desiring Western immigrants and industrialization may use the mean to paint a glowing picture, whereas a small crowded country may choose the median.

Introduction to measures of variability

Measures of variability are as important as measures of central tendency in describing a distribution. When a measure of central tendency and a measure of variability are both found, you can make more elaborate comparisons than with only a measure of central tendency. For example, assume your IQ is 130 and you have just been accepted into an exclusive honor society that has a mean IQ of 120. How do you compare with the rest of the society's members? You are 10 IQ points above the mean, but that is only half the story. Imagine that the society's distribution of IQ scores looks like the one on the left in Figure 3-5. Compare it with

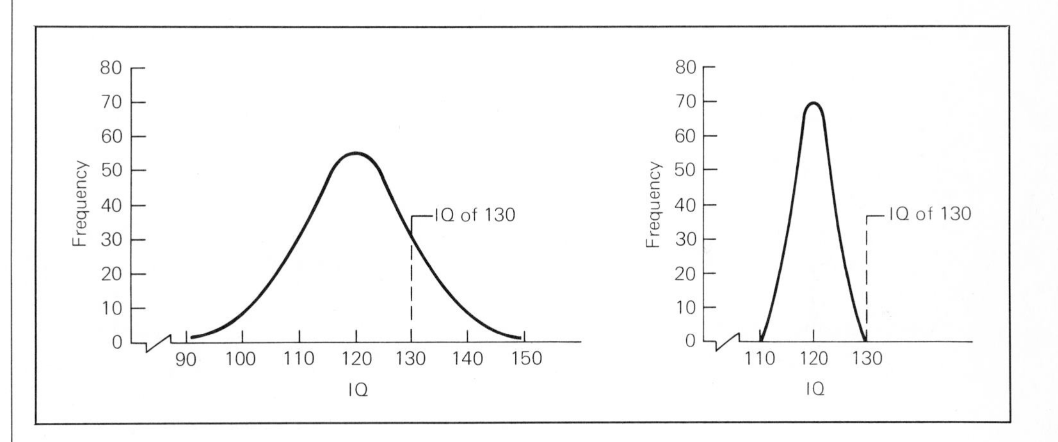

FIGURE 3-5
Comparing an IQ score of 130 in two hypothetical distributions with the means of 120.

the one on the right. Both hypothetical distributions have means of 120. With the distribution on the left, your IQ is higher than more than half of the members, but with distribution on the right, your IQ is the highest. Knowing how spread out the scores are in these two distributions makes a big difference in how you rate with the rest of the members. Whereas a measure of central tendency is the point at which a distribution of scores stacks up, **a measure of variability** indicates how the stack is spread out. Three different measures of variability are the *range,* the *variance,* and the *standard deviation.*

measure of variability

Range

range

The **range** is the easiest measure of variability to calculate. It is simply one plus the difference between the smallest and the largest scores in the distribution. For the freshman anxiety scores, you can calculate the range by subtracting the smallest score, 10, from the biggest score, 86, and then adding 1. The result is 77, which is the range. The reason you add one is to reflect the fact that your scores are rounded off to the nearest whole number. Any scores between 85-1/2 and 86-1/2 would be rounded to 86, and any score from 9-1/2 to 10-1/2 would be rounded to 10. Thus the range would be from the lowest possible score of 9-1/2 to the highest possible score of 86-1/2, which is actually 77 points. You can say "the spread of the scores is 77," or, "the range is 77." In simple formula form, the range is:

FORMULA 3.5

$$\text{range} = \text{highest score} - \text{lowest score} + 1$$

The range is useful because it indicates the span between the lowest and highest scores. However, like the mode, the range is very unstable because all the scores except the the two extremes are ignored. For example, changing one of the 86s in the freshman data to 200 drastically changes the range to 190. This value of 190 is not representative of the spread of the bulk of the scores, and is misleading since most readers would assume that there are more high scores than actually occur.

Variance

deviation from the mean

With interval/ratio data it would be nice to have a measure of variability that takes into account the magnitude of each score in the distribution, not just the two extremes. A reasonable thing to do might be to calculate the distance, or deviation, each score is from the mean and then find the average (mean) of these deviations. The more spread out the distribution, the larger should be the mean deviation from the mean. For example, with the freshman anxiety data, subtract the mean of 49 from each score to find the **deviation from the mean.** This operation is $X - \mu$:

X	$X - \mu$
86	37
44	−5
43	−6
40	−9
30	−19
86	37
69	20
45	−4
37	−12
10	−39

Now, total the $X - \mu$ column to find the sum of all the deviations from the mean and then divide this sum by N. Since the "sum of the deviations from the mean" is symbolized $\Sigma(X - \mu)$, the mean deviation from the mean is:

$$\frac{\Sigma(X - \mu)}{N} = \frac{0}{10} = 0$$

The sum of the deviations is zero, so the mean deviation from the mean is zero.[3] In fact, the sum of the deviations from the mean will always be zero. This **zero-sum problem** reflects the status of the mean as the balance point. The scores above the mean have positive deviations and are counterbalanced with scores below the mean that have negative deviations. Consequently, to generate a measure of variability that does, indeed, increase as the spread of a distribution increases, you must eliminate the zero-sum problem.

zero-sum problem

The best solution to the zero-sum problem is to square each deviation from the mean. This creates positive values. Squaring also has the advantage that it is easy to manipulate mathematically. Now, if you find how much each score deviates from the mean and square these deviations before finding their mean, what will you call your answer? It is called the **mean squared deviation from the mean.** Abbreviating "squared deviation from the mean" with $(X - \mu)^2$, you can calculate the mean squared deviation from the mean as follows:

mean squared deviation from the mean

X	$X - \mu$	$(X - \mu)^2$
86	37	1369
44	−5	25
43	−6	36
40	−9	81
30	−19	361
86	37	1369
69	20	400
45	−4	16
37	−12	144
10	−39	1521

$$\Sigma(X - \mu)^2 = 5322$$

$$\text{mean squared deviation from the mean} = \frac{\Sigma(X - \mu)^2}{N} = \frac{5322}{10} = 532.2$$

[3]For a mathematical proof: $\Sigma(X - \mu) = \Sigma X - \Sigma\mu$. Since μ is a constant, $\Sigma\mu$ says to add up a constant (μ) N times, which is the same as multiplying μ by N. Thus $\Sigma\mu = N\mu$. But $\mu = \Sigma X/N$, so $N\mu = N\Sigma X/N = \Sigma X$. This means $\Sigma(X - \mu) = \Sigma X - \Sigma X = 0$.

However, researchers do not have to go around saying "mean squared deviation from the mean" all the time (try repeating this a few dozen times); they simply say **variance.** The variance of a population is abbreviated σ^2, and is read "sigma squared." The definitional formula for the variance of a population is:

variance

σ^2

FORMULA 3.6
definitional formula for population variance

$$\text{variance of a population} = \sigma^2 = \frac{\Sigma(X - \mu)^2}{N}$$

The variance of a sample can be calculated the same way. We will not give a different abbreviation for the variance of a sample right now because inferential statistics treats sample variances in a special way to be discussed in Chapter 7. If you wish to describe the variance of a sample, treat it as a population and compute the variance using Formula 3.6.

The variance is a useful measure of variability, but it has two slight drawbacks. First, the large deviations of scores farthest from the mean are emphasized more than the small deviations of scores nearest the mean. Squaring a large deviation gives a disproportionately bigger value than squaring a small deviation. Second, the units of measurement in the variance are squared units. For example, if you measure the heights of players on a basketball team (in inches) and calculate the variance, it will be in square-inch units. This is a measure of area, not height!

Standard deviation

standard deviation

The drawbacks of the variance can be somewhat alleviated by taking the square root. This gives a value in terms of the original unit of measurement, and it somewhat deemphasizes large deviations from the mean. The value obtained by taking the square root of the variance is called the **standard deviation.** For example, the standard deviation for the freshman data is 23.07:

$$\text{square root of the variance} = \sqrt{532.2} = 23.07$$

σ

Symbolically, the square root of the variance is $\sqrt{\sigma^2}$, which simplifies to $\boldsymbol{\sigma}$. Thus the symbol for the standard deviation of a population is σ, and is defined by the following formula:

FORMULA 3.7
definitional formula for population standard deviation

$$\text{standard deviation of a population} = \sigma = \sqrt{\frac{\Sigma(X - \mu)^2}{N}}$$

The standard deviation is the "square root of the mean squared deviation from the mean." To get a better understanding of how useful the standard deviation is, consider its role in the normal distribution. A normal distribution is a theoretical distribution that is symmetrical, unimodal, and bell-shaped as shown in Figure 3-6.

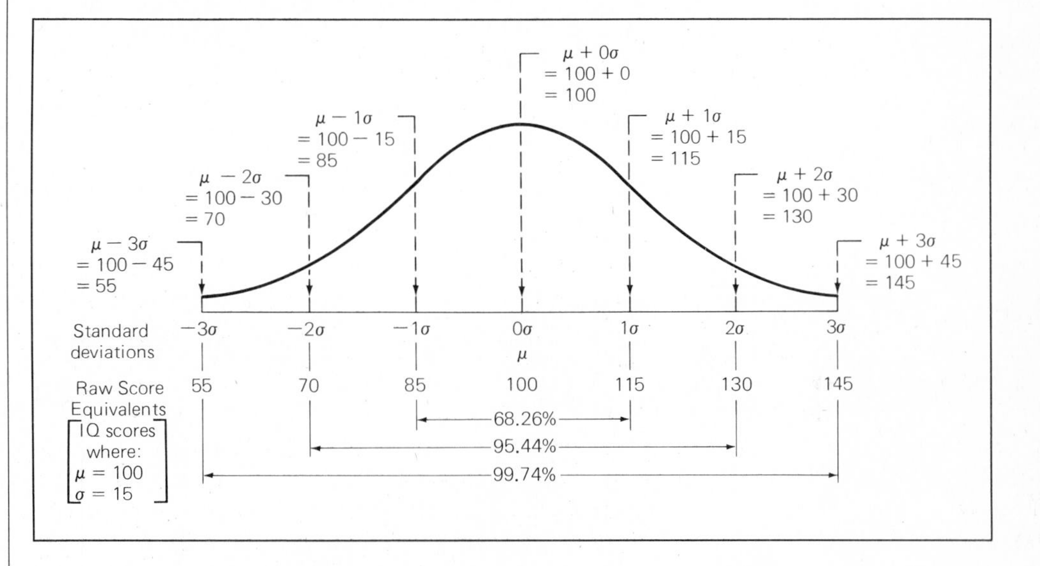

FIGURE 3-6

Properties of the standard deviation in a normal distribution. The standard deviation units are shown in relation to raw-score equivalents (IQ scores). The derivation of the raw-score equivalents is shown above the curve. The approximate percentage of raw scores lying above and below the mean are shown below.

Suppose you administer an IQ test to a population of 1000 people and find that the empirical distribution of scores is approximately normally distributed as in Figure 3-6, with a mean of 100 ($\mu = 100$) and standard deviation of 15 ($\sigma = 15$). Think of the standard deviation as one chunk of successive IQ scores lying on the abscissa, with a spread of 15 IQ points, such as the chunk from 100 to 115. Conversely, you can think of a range of 15 IQ points as *one* standard deviation unit ($1\sigma = 15$, or equivalently, $\sigma = 15$). Then a range of 30 IQ points is *two* standard deviation units (2 times 15 equals 30, or $2\sigma = 30$), and 45 IQ points is *three* standard deviation units (3 times 15 equals 45, or $3\sigma = 45$). Now you can think of being some number of standard deviation units away from the mean, such as one, two, or three standard deviation units *above* the mean (represented $+1\sigma$, $+2\sigma$, $+3\sigma$), or one, two, or three standard deviation units *below* the mean (represented -1σ, -2σ, -3σ). The IQ scores lying within one standard deviation *above and below* the mean (-1σ to $+1\sigma$) are the scores from 85 to 115. The score 85 is the raw-score equivalent of -1σ, obtained by subtracting one standard deviation unit from the mean ($\mu - 1\sigma$, or $100 - 15 = 85$). The score 115 is the raw-score equivalent of $+1\sigma$, obtained by adding one standard deviation unit to the mean ($\mu + 1\sigma$, or $100 + 15 = 115$). The derivation of more raw-score equivalents is shown in Figure 3-6.

Now the role of the standard deviation can be seen in light of the properties of the normal distribution. For example, one property of the normal distribution is that 68.26% of the scores lie within within one standard deviation above and below the mean. This indicates that in the approximately normal distribution of IQ scores in Figure 3-6, about 68.26% of the IQ scores must be within one standard deviation above and below the mean. Thus, if there are 1000 people represented in Figure 3-6, you will find about 683 of them (68.26%) with IQs between 85 and 115. Another way to see this is to remember that the approximately normal distribution shown in Figure 3-6 is a frequency polygon, in which the height of the curve above a particular IQ score corresponds to the number of people who have that IQ. You can think of the people with a given IQ as being "stacked up" over that IQ score in the graph, as in Figure 3-7. Thus you could

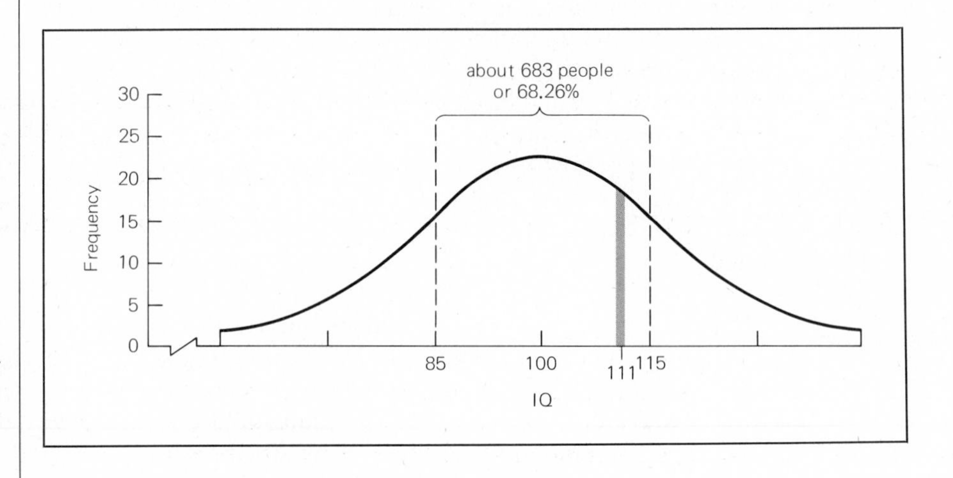

FIGURE 3-7
This approximately normal distribution of IQ scores is a frequency polygon in which the height of the curve above a particular IQ score corresponds to the number of people who have that IQ. For example, if there are 1000 scores, you will find about 21 people stacked up with IQs of 111, and about 683 people with IQs between 85 and 115.

directly count the number of people with IQs between 85 and 115. Further properties of the normal distribution are that 95.44% of all scores lie within two standard deviations above and below the mean, and 99.74% of all scores lie within three standard deviations above and below the mean. Translated into raw scores, 68.26% of the people have IQs between 85 and 115, 95.44% of the people have IQs between 70 and 130, and 99.74% have IQs between 55 and 145.

Even when a distribution is not very well approximated by a normal distribution, there is often close to 68% of the scores lying within $\pm 1\sigma$, close to 95% lying within $\pm 2\sigma$, and close to 99% lying within $\pm 3\sigma$. In general, regardless of the shape of the distribution, the standard deviation serves as a useful measure of variability that takes into account the magnitude of each score. Consider the following sets of raw scores along with their calculated standard deviations:

Set	*Raw Scores*	μ	σ
A	2, 3, 4, 4, 5, 8	4.33	1.89
B	2, 4, 6, 6, 8, 14	6.67	3.77
C	52, 53, 54, 54, 55, 58	54.33	1.89
D	52, 54, 56, 56, 58, 64	56.67	3.77
E	3, 4, 4, 4, 5, 6, 7, 7	5	1.41
F	1, 3, 3, 3, 5, 7, 9, 9	5	2.83
G	53, 54, 54, 54, 55, 56, 57, 57	55	1.41
H	52, 53, 53, 53, 55, 57, 59, 59	55	2.83

Notice that the broader the distribution the larger the standard deviation.

Computational formulas for the variance and standard deviation

Formulas 3.6 and 3.7 for the variance and standard deviation are technically referred to as *definitional* formulas, since they serve to precisely define the mathematical operations corresponding to the reasoning that went into developing them. In most applications, however, the *computational* formulas below provide an easier method of calculation. By applying a little mathematical finesse to the definitional formula for the variance, you can derive the following computational formula for the population variance[4]:

FORMULA 3.8

computational formula for population variance

$$\sigma^2 = \frac{\Sigma X^2 - \dfrac{(\Sigma X)^2}{N}}{N}$$

The computational formula for the population standard deviation is:

[4]For mathematical proof:

$$\sigma^2 = \frac{\Sigma(X-\mu)^2}{N} = \frac{\Sigma(X^2 - 2X\mu - \mu^2)}{N} = \frac{\Sigma X^2 - \Sigma 2X\mu - \Sigma\mu^2}{N}$$

$$= \frac{\Sigma X^2 - 2\mu\Sigma X - N\mu^2}{N}$$

$$= \frac{\Sigma X^2 - 2\left(\dfrac{\Sigma X}{N}\right)\Sigma X - N\left(\dfrac{\Sigma X}{N}\right)^2}{N}$$

$$= \frac{\Sigma X^2 - 2\dfrac{(\Sigma X)^2}{N} - \dfrac{N(\Sigma X)^2}{N^2}}{N} = \frac{\Sigma X^2 - \dfrac{(\Sigma X)^2}{N}}{N}$$

FORMULA 3.9

computational formula for population standard deviation

$$\sigma = \sqrt{\frac{\Sigma X^2 - \frac{(\Sigma X)^2}{N}}{N}}$$

The difference between the definitional and computational formulas lies in the numerator. The computational formulas require you to square each score to get X^2, sum the squared scores to get ΣX^2, and then subtract the square of the sum of scores, $(\Sigma X)^2$, divided by N.

A good way to keep track of each operation when using the computational formulas is to list each score and its square in two columns, labeled X and X^2, respectively. Next, sum each column and square the sum of the X column. Be sure to label all the numerical values with their symbolic terms, as with the freshman anxiety data below:

	X	X^2
	86	7396
	44	1936
	43	1849
	40	1600
	30	900
$N = 10$	86	7396
	69	4761
	45	2025
	37	1369
	10	100
	$\Sigma X = 490$	$\Sigma X^2 = 29{,}332$

$$(\Sigma X)^2 = (490)^2 = 240{,}100$$

Then, substitute into the computational formula and complete the calculations. Thus the population variance is 532.2:

$$\sigma^2 = \frac{\Sigma X^2 - \frac{(\Sigma X)^2}{N}}{N} = \frac{29{,}332 - \frac{240{,}100}{10}}{10}$$

$$= \frac{29{,}332 - 24{,}010}{10}$$

$$= \frac{5322}{10} = 532.2$$

To find the standard deviation, simply take the square root of the variance:

$$\sigma = \sqrt{\frac{\Sigma X^2 - \frac{(\Sigma X)^2}{N}}{N}} = \sqrt{532.2} = 23.07$$

The computational formulas are superior to the definitional formulas when N is fairly large because you do not have to perform many "error-inducing" subtractions when finding $X - \mu$. If an electronic calculator is used, the computational formulas are far superior.

When to use the range, variance, and standard deviation

The scale of measurement guides the selection of measures of variability. With nominal data there is no useful measure of variability since the categories of scores cannot be meaningfully ordered. With ordinal data, the range is the most useful measure of variability. Calculating the variance and standard deviation requires you to subtract the mean from each score and then add the deviations. This operation is useful only when you can assume that equivalent entities are being added. This is true only with interval/ratio data. With interval/ratio data, although the range may be used, the variance and standard deviation are preferred because they convey more information. To summarize these preferences:

Scale of measurement	*Preferred measure of variability*
nominal	none
ordinal	range
interval/ratio	variance and standard deviation

KEY CONCEPTS

1 Measures of central tendency are single scores that describe where the scores in a distribution stack up. Three measures of central tendency are the mode, median, and mean.

2 The mode is the most frequently occurring score. It is used more often with nominal data, although it may be used as a quick measure of central tendency with ordinal and interval/ratio data. The mode is unstable.

3 The median is the middle score in a distribution. The median is most appropriate with ordinal data and with interval/ratio data that are extremely skewed.

4 The mean is the arithmetic average. The mean of a population is abbreviated μ, and the mean of a sample is abbreviated $\bar{X}$. The mean is the balance point of a distribution; the sum of the deviations from the

mean is always zero. The mean is appropriate with interval/ratio data that are not extremely skewed.

5 Measures of variability are single numbers that describe the spread of a distribution. Three measures of variability are the range, variance, and standard deviation.

6 The range is the distance between the highest score and the lowest score and is obtained by subtracting the lowest score from the highest score and adding one. The range is most appropriate for ordinal data, although it may be used as a quick measure of variability for interval/ratio data. The range is unstable.

7 The population variance, σ^2, is the mean squared deviation from the mean. It is calculated with the definitional formula on the left or the computational formula on the right:

$$\sigma^2 = \frac{\Sigma(X - \mu)^2}{N} \quad \text{or} \quad \sigma^2 = \frac{\Sigma X^2 - \frac{(\Sigma X)^2}{N}}{N}$$

8 The population standard deviation, σ, is the square root of the variance:

$$\sigma = \sqrt{\sigma^2}$$

9 Taking the square root of the variance converts back to the original units of measurement.

10 An interpretation of the standard deviation is seen in the normal distribution: 68.26% of the scores will lie within one standard deviation above and below the mean; 95.44% will lie within two standard deviations above and below the mean; and 99.74% will lie within three standard deviations above and below the mean.

11 Raw scores may be interpreted as a number of standard deviation units above or below the mean, and vice versa.

EXERCISES

1 Calculate the mode, median, and mean for each of the following groups of scores:

a. 7, 4, 3, 2, 5, 8, 3, 1
b. 118, 4, 3, 2, 4, 8, 3, 1
c. 11, 6, 4, 0, 0, 12, 1, 2, 0
d. 11, 22, 22, 10, 18, 15, 13, 15, 15, 12, 10, 7, 12
e. 14, 21, 16, 3, 14, 22, 14, 9, 9, 18
f. 1.6, 2.5, 2.5, 3.8, 3.6, 5.0, 4.7, 3.2, 4.2, 4.2, 2.7

2 Calculate the mode, median, and mean of the data in the frequency distribution below:

X	f
60	1
61	1
62	0
63	1
64	2
65	3
66	2
67	1
68	2
69	4
70	5
71	1
72	1

3 For each of the following hypothetical research situations, identify the scale of measurement, and calculate the preferred measure of central tendency.

a. You measure the maximum speed (in miles per hour) of 15 automobiles traveling through a school zone where the maximum speed is posted at 20 miles per hour. The speeds are as follows:

21	55	19
17	20	15
15	22	20
16	21	18
18	17	47

b. There were 16 airline personnel who loaded cargo for one year without wearing protective devices over their ears. You classify them as having high, medium, low, or no hearing impairment. Your data are as follows:

high	high	medium	low
medium	medium	low	high
medium	no	high	medium
high	low	medium	medium

c. You examine the school psychologist's diagnoses of ten children classified during the first month of school as requiring special education. Each child's diagnosis is given below:

mentally retarded	conduct disorder
conduct disorder	conduct disorder
attention deficit	anxiety disorder
mentally retarded	conduct disorder
conduct disorder	mentally retarded

d. In a relaxation study, prior to initiating any specific relaxation training you record the systolic blood pressure of eight persons with hypertension. The blood pressures were:

162.5	178.7	159.0	152.1
187.8	164.1	190.7	174.3

e. You observe 20 children in several classroom leadership roles. After observing all the children in all the roles, you rate each child on a scale from 1 to 5, based on your judgment of their classroom leadership. You use ratings of 1, 2, 3, 4, or 5. Children you judge very high in classroom leadership are given a rating of 5, those judged very low are given a rating of 1, and those judged as having intermediate levels of leadership are assigned intermediate ratings. The ratings are as follows:

2	1	3	1	4
2	1	1	2	2
1	5	1	2	3
1	1	3	1	1

4 A recent advertisement for a new 6-hour speed-reading course stated that the "average" reading speed of the persons who completed the first 6-hour course was 1000 words per minute with at least 80 percent comprehension. Having been one of those who completed this first course, you have access to all of the reading speeds of the persons who completed the first course with you. The reading speeds for all those who completed the first course are as follows:

1000	900	800	1000	900	850
650	1000	1050	800	1000	850
700	750	800	850	900	950
600	1100	950	700	750	650

Calculate the mode, median, and mean. Which measure of central tendency was being reported in the advertisement? Which measure of central tendency do you believe would more accurately reflect the "average" reading speed?

5 Calculate the range for each of the following groups of scores:
a. 7, 4, 3, 2, 4, 6, 3, 1
b. 12, 10, 2, 4, 5, 3, 0, 12, 1, 7
c. 7, 4, 3, 2, 4, 6, 3, 42
d. 42, 26, 25, 25, 38, 32, 38, 44
e. 14, 0, −5, −2, 8, 7, 6, 6, 13, 16, 14

6 Calculate the mean and the variance and standard deviation using the computational formulas (Formulas 3.8 and 3.9) for each of the following groups of data:
a. 7, 4, 3, 2, 4, 6, 3, 1
b. 12, 10, 2, 4, 5, 3, 0, 12, 1, 7
c. 37, 34, 33, 32, 34, 36, 33, 31
d. 92, 90, 82, 84, 85, 83, 80, 92, 81, 87
e. 10, 5, 5, 8, 6, 8, 7, 9, 5, 8, 6
f. 14, 0, −5, −2, 8, 7, 6, 6, 13, 16, 14
g. 70, 65, 65, 68, 66, 68, 67, 69, 65, 68, 66
h. 54, 40, 35, 38, 48, 47, 46, 46, 53, 56, 54

7 Examine the mean and standard deviation for each group of data in Exercise 6. Why do some pairs of groups have the same mean but have different standard deviations? Why do some pairs of groups have different means but have the same standard deviations?

8 Calculate the variance and standard deviation using the definitional formulas (Formulas 3.6 and 3.7) for each of the following groups of data:

a. 7, 4, 3, 2, 4, 6, 3, 1
b. 37, 34, 33, 32, 34, 36, 33, 31
c. 14, 0, −5, −2, 8, 7, 6, 6, 13, 16, 14
d. 8, 6, 4, 4, 5, 5, 4.5, 4.5, 2, 3

9 Calculate the standard deviation for the data sets below using both the definitional and computational formulas (Formulas 3.7 and 3.9). Does one method of calculation appear easier than the other?

a. 4, 4, 4, 4, 4, 4, 4, 4
b. 42.7, 26.3, 25.3, 25.3, 38.4, 32.8, 38.7, 44.6
c. 356, 295, 324, 438, 444, 149
d. 2, 4, 6, 1, 4, 3, 8

10 Given a population of normally distributed scores whose mean is 50 and standard deviation is 7, determine the following:

a. The raw scores that cut off approximately 68% of the scores above and below the mean.
b. The raw scores that cut off approximately 95% of the scores above and below the mean.
c. The raw scores that cut off approximately 99% of the scores above and below the mean.

4

CHAPTER GOALS

After studying this chapter, you should be able to:

1 Transform raw scores into percentile ranks.

2 Transform raw scores into *z* scores.

3 Convert percentile ranks and *z* scores to raw scores.

4 Verbally define "*z* score" and describe its usefulness.

5 Define a normal distribution and describe when it arises.

6 Calculate the percentage or frequency of scores above or below a given *z* score in a normal distribution and between two *z* scores in a normal distribution.

7 Determine *z* scores that cut off a given percentage or frequency of scores in a normal distribution.

8 Determine raw scores that cut off a given percentage or frequency of scores in a normal distribution.

CHAPTER IN A NUTSHELL

Raw scores are more meaningful when they are compared with other scores in a distribution. Raw scores can be converted to transformed scores that indicate at a glance where a raw score stands in a distribution. Two types of transformed scores are: percentile ranks, or percentiles, which tell you the percentage of scores equal to or less than a given raw score; and *z* scores, which tell you the number of standard deviation units a given raw score lies above or below the mean.

The normal distribution is a theoretical, bell-shaped distribution, which can frequently be used to approximate empirical distributions arising in the behavioral sciences. The normal distribution is very useful because a specified percentage of scores lies between the mean and any given *z* score.

Introduction to transformed scores

Raw scores are meaningful only when compared with some preset standard or in relation to a distribution of scores. For example, suppose a friend comes up to you and proudly announces receiving a score of 50 on a statistics exam. Unless you have additional information, a score of 50 is meaningless. To provide a context, you need to know information such as the grade cutoff points, the modal/median/mean score, or how well the rest of the class did. If your friend said 50 is the second highest score, or that it is better than 94% of the others, then you can understand the pride.

transformed score

Context is efficiently conveyed by transforming the particular raw score into a new value, called a **transformed score.** You can look at a transformed score and immediately know how its raw-score equivalent stands in a distribution of raw scores. Two types of transformed scores are *percentile ranks* or simply *percentiles,* and *z scores.*

Percentile ranks (percentiles)

percentile rank

Transforming raw scores into percentile ranks gives the percentage of raw scores that are less than or equal to that particular raw score. For example, if a raw score of 50 can be transformed to a **percentile rank** of 94, this means 94% of the raw scores are less than or equal to 50. In other words, a raw score that is at the 94th percentile is one such that 94% of the raw scores are less than or equal to it. Many people confuse percentile ranks and percentiles with conventional "percentage" scores, such as the percentage of points obtained on an exam. For example, a "percentage" score of 60% on an exam merely tells you that you got 60% of the possible points correct. It tells nothing about how you scored compared with other people taking the exam.

To transform one or more raw scores into percentile ranks, begin by constructing a cumulative frequency distribution. For example, suppose you measure the heights (in inches) of 15 women in your statistics class and summarize the results as follows.

X	f	cf
60	1	1
61	0	1
62	1	2
63	1	3
64	3	6
65	4	10
66	3	13
67	1	14
68	1	15

Since a cf value represents the number of scores less than or equal to a certain raw score, percentile ranks may be calculated by converting the *cf* value to a percentage. This is actually a two-step process. The first step is to look at the cumulative frequency distribution and find the *cf* value corresponding to the raw score of interest. The second step is to divide this *cf* value by N and multiply by 100 to yield the percentile rank. For example, to calculate the percentile rank for a height of 65 inches, first find its corresponding *cf* of 10. Then divide 10 by $N = 15$ and multiply by 100. The percentile rank is 66.7:

$$\frac{cf}{N}(100) = \frac{10}{15}(100) = .667(100) = 66.7$$

The percentile rank of 66.7 means that 66.7% of the women are 65 inches tall or shorter. The percentile rank of 66.7 also tells you that 33.3% of the women are taller than 65 inches. In general, the formula for transforming a raw score into a percentile rank is:

FORMULA 4.1

$$\text{percentile rank} = \frac{cf}{N}(100)$$

Sometimes you may want to convert a percentile rank to a raw score. For example, suppose you want to determine what height corresponds to the 25th percentile. You need to invert the two steps used above. The first step is to rework Formula 4.1 and solve for *cf*:

FORMULA 4.2

$$cf = \frac{\text{percentile rank}}{100}(N)$$

Then search the cumulative frequency distribution for the raw score corresponding to the calculated *cf*. This raw score is the answer. Frequently, however, your calculated *cf* is not specifically listed in the *cf* column of the cumulative frequency distribution. In such cases, *locate the closest listed cumulative frequency value that is larger than the calculated cf.* The raw score corresponding to this listed *cf* value is the answer. For example, to find the raw score for the 25th percentile, use Formula 4.2. Divide the percentile rank of 25 by 100 and multiply by 15 to yield the *cf* of 3.75:

$$cf = \frac{\text{percentile rank}}{100}(N) = \frac{25}{100}(15) = .25(15) = 3.75$$

Since 3.75 is not listed in the *cf* column, the next larger *cf* value listed is 6. The raw score corresponding to the *cf* value of 6 is 64, and thus 64 is the raw-score equivalent for a percentile rank of 25.

The procedures given above for transforming raw scores to percentiles and percentiles to raw scores often involve considerable rounding off. You should therefore not expect to be able to take a raw score and convert it to a percentile and then convert that percentile back to exactly the original raw score. More precise methods involving interpolation are available but are not often used, so these methods are not discussed in this book.

z scores

z scores provide another way of determining where a particular raw score lies in a distribution. Transforming a raw score into a *z* score tells you how many standard deviation units that score lies above or below the mean of a distribution. You were doing this at the end of Chapter 3 when the standard deviation was discusssed in light of a normal distribution. In the example with IQ scores, $\mu = 100$ and $\sigma = 15$. Thus an IQ of 115 can be translated to $+1\sigma$ from the mean, an IQ of 85 can be translated to -1σ from the mean, an IQ of 130 is equivalent to $+2\sigma$, an IQ of 70 is equivalent to -2σ, and so on. A shortcut way to denote this equivalence is with a ***z* score.** A *z* score tells you how many standard deviation units a raw score lies from the mean. A positive *z* score indicates that the raw score is above the mean, and a negative *z* score indicates that the raw score is below the mean. For example, an IQ of 115 is equivalent to a *z* score of +1, an IQ of 85 is equivalent to a *z* score of −1, an IQ of 130 is equivalent to a *z* score of +2, and so on.

z score

It is easy to transform raw scores into *z* scores when the raw score is exactly 0, 1, 2, 3, and so on, standard deviation units above or below the mean, as are the IQs of 55, 70, 85, 100, 115, 130, and 145. But how can you determine the *z* score for IQs between these whole standard deviation units? For example, what are the *z* scores for IQs of 106, or 72? The general formula for transforming raw scores into *z* scores is:

FORMULA 4.3

$$z = \frac{X - \mu}{\sigma}$$

Although you already know an IQ of 130 is two standard deviations above the mean and is equivalent to a *z* score of 2, following Formula 4.3 gives the derivation:

$$z = \frac{X - \mu}{\sigma} = \frac{130 - 100}{15} = \frac{30}{15} = 2$$

Consider another example. Suppose you measure the weights of 83 women in your statistics class and find $\mu = 120.29$ pounds and $\sigma = 14.75$ pounds. A person with a body weight of 109 has what z score? You can use Formula 4.3 to find that a weight of 109 is .77 standard deviation units below the mean:

$$z = \frac{X - \mu}{\sigma} = \frac{109 - 120.29}{14.75} = \frac{-11.29}{14.75} = -.77$$

To convert a z score to its raw-score equivalent, simply solve Formula 4.3 for X:

FORMULA 4.4

$$X = \mu + z\sigma$$

For example, what weight corresponds to a z score of -1.5? Since a z score is the number of standard deviation units a raw score lies away from the mean, the raw-score equivalent for a z of -1.5 lies 1.5 standard deviation units below the mean. Thus the raw score is 1.5 times 14.75 pounds below 120.29, which is 98.16:

$$\begin{aligned} X = \mu + z\sigma &= 120.29 + (-1.5)(14.75) \\ &= 120.29 + (-22.13) = 98.16 \end{aligned}$$

Some people erroneously assume that you can only use z scores with normal or approximately normal distributions. In actuality, z scores may be calculated for distributions of any shape. However, you may not find 68.26% of the scores falling within one standard deviation unit above and below the mean, or 95.44% of the scores within two standard deviation units falling above and below the mean, or 99.74% of the scores falling within three standard deviation unit above and below the mean. z scores are very useful for comparing raw scores from different distributions, as well as for determining at a glance where a score lies in a distribution. For example, suppose you want to know if you are a better swimmer (measured in elapsed time) than a mathematician (measured in problems correct), or whether you are taller (measured in inches) than heavy (measured in pounds), or whether you do better on standardized achievement tests (SAT score) than on course work (GPA). Since all of these abilities/traits come from different distributions, it is impossible to compare raw scores. For example, if you swim 100 meters in 90 seconds and get 85 out of 100 math problems correct, are you a better swimmer or a better mathematician? Or how can you compare your SAT score with your GPA? To make such comparisons, consider first what it means to do well or score high. Scoring high is relative to the other scores in the distribution. To compare the scores in different distributions, first you need to determine how each score compares to the rest of the scores in its respective distribution. Since a z score is a common measure of where a score stands in comparison to the rest of the scores in a distribution, it serves your purpose.

Suppose you have a GPA of 3.3 and score a 735 on the SAT. To determine whether you did better on the standardized achievement test or in your course work, transform each score to a z score. The SAT has $\mu = 500$ and $\sigma = 100$. Suppose GPA at your school has $\mu = 2.6$ and $\sigma = .5$ grade points. The z score for your SAT score is 2.35:

$$z = \frac{X - \mu}{\sigma} = \frac{735 - 500}{100} = \frac{235}{100} = 2.35$$

The z score for your GPA is 1.4:

$$z = \frac{X - \mu}{\sigma} = \frac{3.3 - 2.6}{.5} = \frac{.7}{.5} = 1.4$$

Since your SAT score is more standard deviations above the mean than your GPA, you are better at taking achievement tests than at getting grades. This type of comparison can be used to compare scores in any two distributions of the same shape.

z scores and the normal distribution

z scores are especially useful in relation to normal and approximately normal distributions. Recall that the normal distribution is a theoretical distribution defined by a rather complex mathematical formula. The derivation of the formula assumes there are a large number of random and independent factors contributing to the ability/trait of interest. The normal distribution is very useful because it offers a close approximation to many empirical real-life ability/trait distributions. For example, due to the way IQ tests are constructed, for all practical purposes IQ is determined by a large number of random and independent factors, such as genetic contributions and contributions of the environment, and may be thought of as being approximately normally distributed. There are uncountable abilities/traits in the behavioral sciences which are approximately normally distributed.

The normal distribution is symmetrical, unimodal, and bell-shaped. It is important to remember that a normal distribution implies only a particular shape and does not imply a specific mean or standard deviation. For example, SAT scores form an approximately normal distribution with $\mu = 500$ and $\sigma = 100$, and IQ scores also form an approximately normal distribution, but with $\mu = 100$ and $\sigma = 15$. Therefore the "normal distribution" refers to a family of distributions with varying means and standard deviations, but all with the same shape.

One useful property of a normal distribution is that a specified percentage of scores falls between the mean and a given number of standard deviation units away from the mean. For example, in Figure 4-1, 34.13%

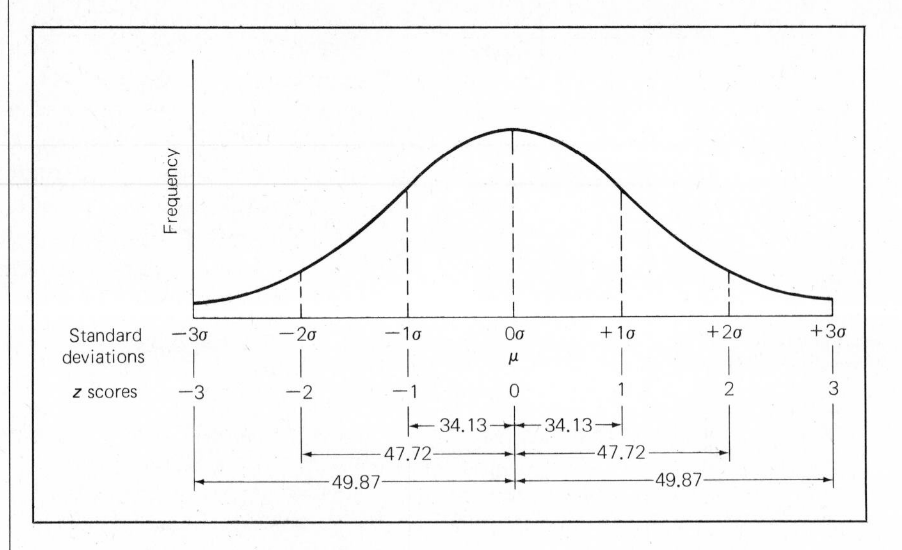

FIGURE 4-1
The symmetrical characteristic of the normal distribution shown in relation to the percentage of scores falling between the mean and the positive and negative z scores of 1, 2, and 3.

of the scores lie between the mean and $z = +1$, 47.72% of the scores lie between the mean and $z = +2$, and 49.87% of the scores lie between the mean and $z = +3$. Since the normal distribution is symmetrical, the same percentage of scores falls between the mean and a given number of standard deviations below the mean as between the mean and a given number of standard deviations above the mean. As seen in Figure 4-1, 34.13% of the scores lie between the mean and $z = -1$, 47.72% of the scores lie between the mean and $z = -2$, and 49.87% of the scores lie between the mean and $z = -3$.

In general, the percentage of scores between the mean and any given z score in a normal distribution may be determined from Table z in the Appendix. To use Table z, first round off to the hundredths place. If there are no digits at the tenths and/or the hundredths place, add zeros to the right of the decimal as placeholders. Then look down the column labeled z for the digits in the ones and tenths places of your z score. This leads you to the proper row. Then find the column labeled with the digit in your z score at the hundredths place. The four-digit value listed at the intersection of the row and column corresponding to your z score is the percentage of scores falling between the mean and the z score. For example, to find the percentage of scores between the mean and $z = 1.5$, place a zero in the hundredths place, making $z = 1.50$. Look down the leftmost column of Table z for 1.5. Since the digit at the hundredths place is 0, look at the very top row for the column having a 0 at the hundredths place. This is the column labeled .00. The number listed at the intersection of the row and column is 43.32%, which is the percentage of raw scores lying between the mean and $z = 1.5$. This is shown pictorially in Figure 4-2. When solving problems requiring use of Table z, we urge you to *draw a diagram* similar to that in Figure 4-2

percentage of scores between the mean and a positive z

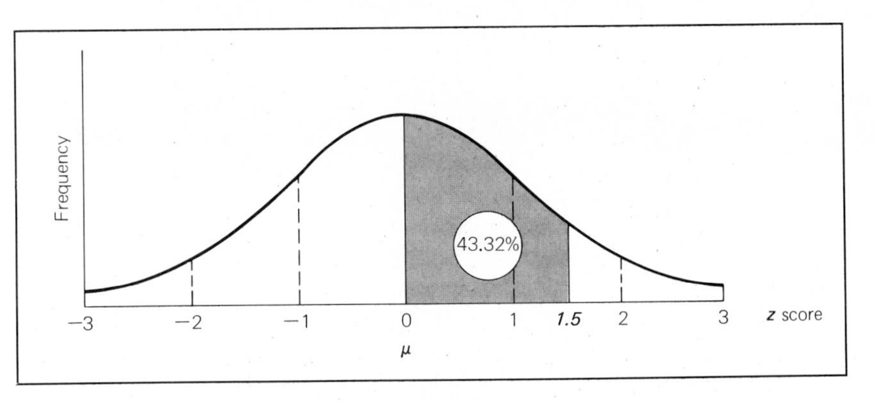

FIGURE 4-2
The shaded area is the percentage of scores derived from Table z for $z = 1.5$.

so that you can better conceptualize what you are looking for.

percentage of scores between the mean and a negative z

Notice that Table z lists only **positive z scores** because it represents the right half of the normal distribution. Since the normal distribution is symmetrical, it is not necessary to list the negative z scores. The percentage of scores between the mean and a **negative z score** is exactly the same as the percentage of scores between the mean and a positive z score of the same magnitude. For example, to find the percentage of scores between the mean and $z = -2.47$, look up a z value of 2.47 in Table z and find 49.32%. Remember that the percentage of scores between it and the mean is *not negative,* because all percentages are positive. Figure 4-3 depicts

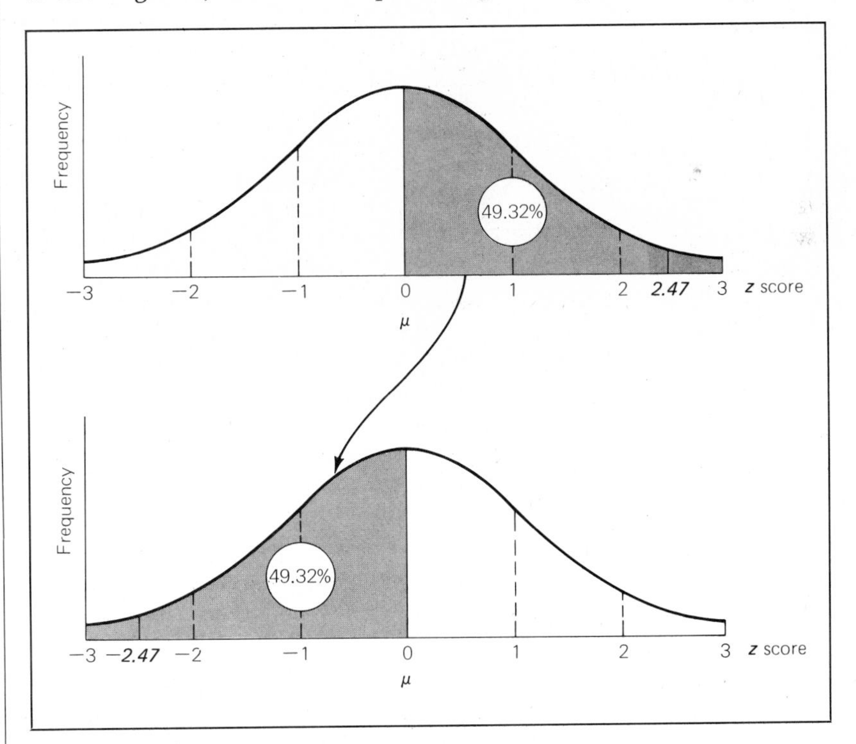

FIGURE 4-3
The shaded area is the percentage of scores that were derived from Table z for $z = 2.47$. The same percentage holds for $z = -2.47$, because the normal distribution is symmetrical.

the symmetry of the normal distribution and demonstrates why Table z may be used to obtain the percentage of scores for both positive and negative z scores.

To find the percentage of scores lying **above** a given z score, you can take further advantage of the symmetrical character of the normal distribution. Exactly 50% of the scores lie above the mean and 50% lie below the mean. The process of finding the percentage of scores above a given z score depends on whether you have a positive or negative z score.

percentage of scores above a positive z

For example, to find the percentage **above a positive z score** such as 2.53, Table z reveals that 49.43% of the scores fall between it and the mean. Since 50% of the scores fall above the mean, the percentage above $z = 2.53$ is found by *subtracting:* 50% minus 49.43% leaves .57% as the answer. This subtraction process is shown in Figure 4-4.

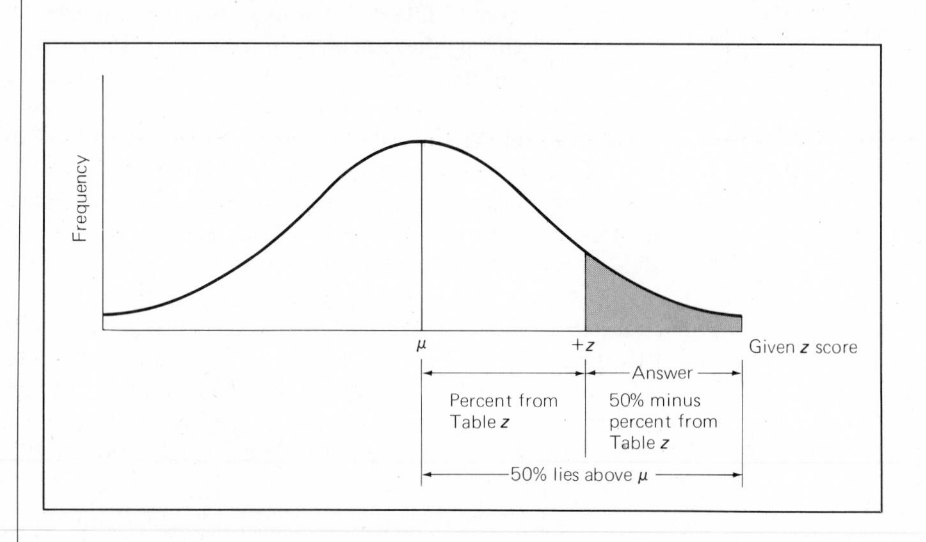

FIGURE 4-4

To determine the percentage of scores above a positive z score, subtract the percentage found in Table z from 50%. The shaded area under the curve is the desired percentage.

percentage of scores above a negative z

To find the percentage of scores **above a negative z score,** say $-.75$, look in Table z to determine that 27.34% of the scores lie between the mean and $z = -.75$. Since 50% of the scores lie above the mean, you can find the percentage above $z = -.75$ by *adding:* 27.34% plus 50% sums to the answer of 77.34%. This addition process is shown in Figure 4-5.

The same addition and subtraction processes are used to find the percentage of scores lying *below* a given raw score, but in reverse fashion. The addition process is used with positive z scores, whereas the subtraction process is used with negative z scores. For example, to find the percentage of scores below $z = 1.20$, Table z reveals that 38.49% of the scores lie between it and the mean. Since 50% of the scores fall below the mean,

percentage of scores below a positive z

the **percentage below a positive z** score is found through *adding:* 50% plus 38.49% gives 88.49% as the answer (see Figure 4-6). Conversely, to find

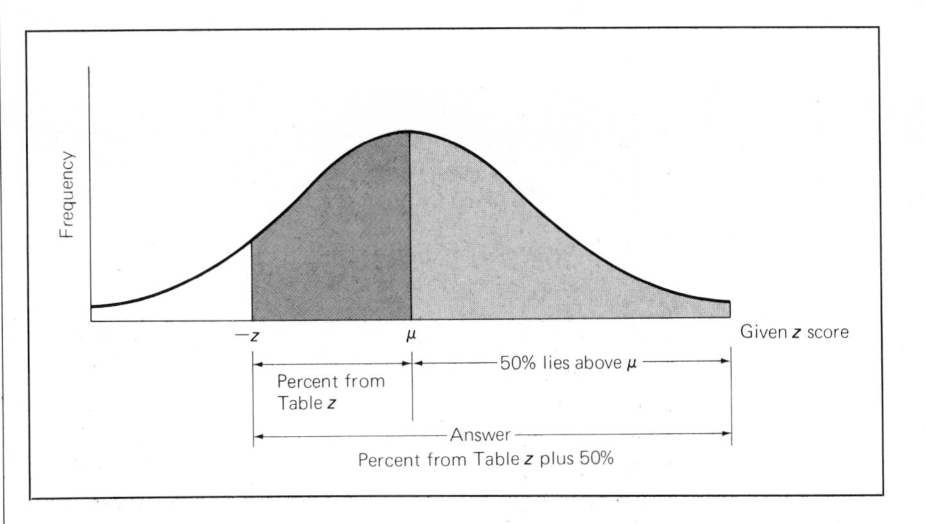

FIGURE 4-5
To determine the percentage of scores above a negative z score, add the percent found in Table z to 50%. The shaded area under the curve is the desired percent.

percentage of scores below a negative z

the **percentage below a negative** z score, say $z = -1.67$, Table z reveals that 45.25% of the scores lie between it and the mean. Since 50% of the scores lie below the mean, you can find the percentage below a negative z score by subtracting: 50% minus 45.25% leaves 4.75% as the answer. This process is shown in Figure 4-7.

percentage of scores between z's on opposite sides of the mean

The addition and subtraction processes used above are also used to find the percentage of scores **between two** z **scores.** Whether you add or subtract, however, depends on whether the two z scores are **on opposite sides of the mean,** or on the same side of the mean. When the two z scores are on opposite sides of the mean, one z will be negative and the other will be positive. For example, suppose you want to find the

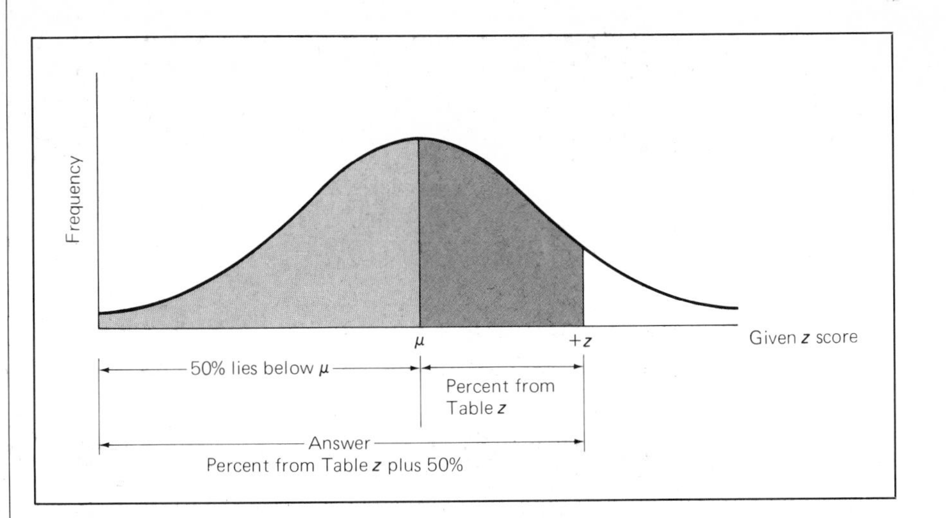

FIGURE 4-6
To determine the percentage of scores below a positive z score, add the percentage found in Table z to 50%. The shaded area under the curve is the desired percentage.

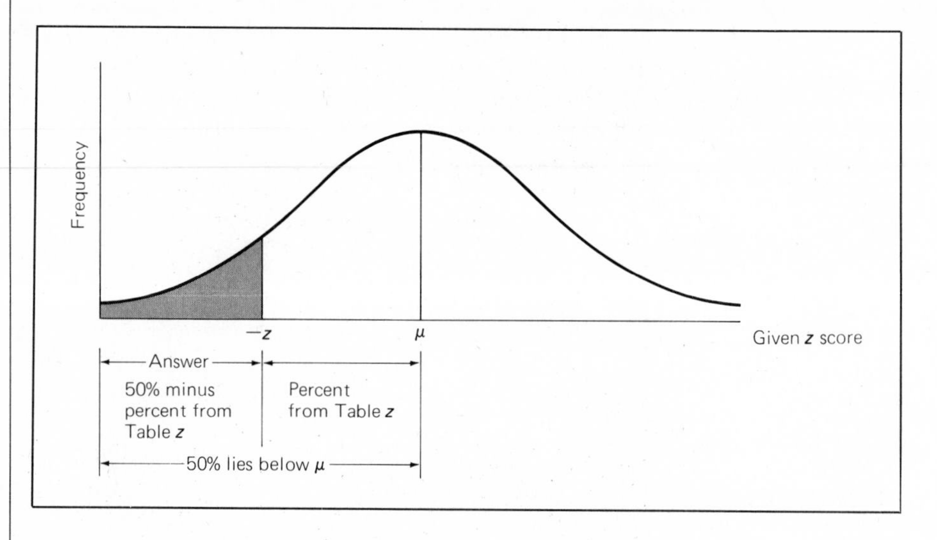

FIGURE 4-7

To determine the percentage of scores below a negative z score, subtract the percent found in Table z from 50%. The shaded area under the curve is the desired percent.

percentage of scores between the z scores of -1.5 and 2.5. First find the percentage of scores between the mean and $z = -1.5$ from Table z, which is 43.32%. Next, find the percent between the mean and $z =$ 2.5, which is 49.38%. Then the total percentage of scores between two z scores is found by *adding:* 43.32% plus 49.38% yields 92.70% as the answer. Figure 4-8 illustrates this process.

percentage of scores between z's on the same side of the mean

When the two z scores are **on the same side of the mean,** both will be positive or both will be negative. For example, to find the percentage of scores between z scores of 2.0 and 3.0, look in Table z to find that 47.72% falls between the mean and $z = 2.0$, and 49.87% falls between the mean and $z = 3.0$. The percentage of scores between the two z scores

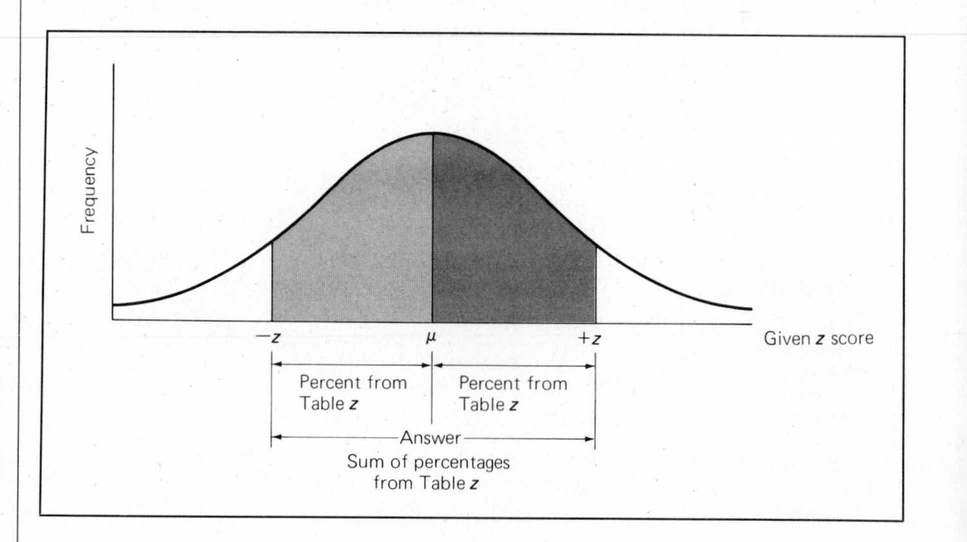

FIGURE 4-8

To determine the percentage of scores lying between z scores on opposite sides of the mean, add the percentages found in Table z. The shaded area is the desired percentage.

is the difference between the larger percentage and the smaller percentage, found by *subtracting* the smaller percentage from the larger: 49.87% minus 47.72% leaves 2.15% as the answer. Figure 4-9 represents this process. The same technique is used to determine the percentage of scores between two negative z scores.

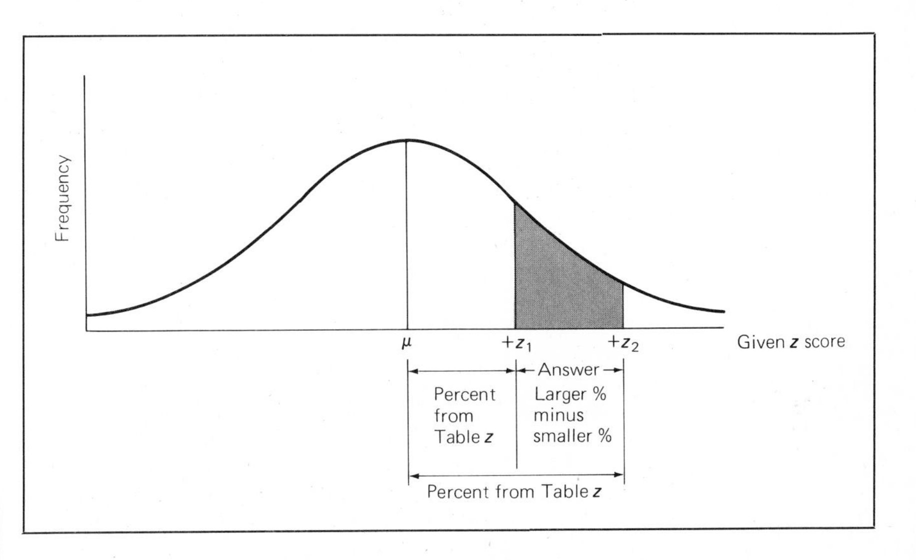

FIGURE 4-9 *To determine the percentage of scores between z scores lying on the same side of the mean, subtract the smaller percentage found in Table z from the larger percentage. The shaded area represents the desired percentage.*

Practical applications of z scores in approximately normal distributions

Transforming raw scores into z scores and using the information in Table z has many applications when you can assume the distribution to be approximately normally distributed. Several examples follow.

converting raw scores to percents in approximately normal distributions

Suppose you are interested in finding the percentage of college women who have body weights between 100 and 130 pounds. If you have an empirical frequency distribution, you could simply count the women with weights between 100 and 130 pounds to solve this problem. But even if you do not have an empirical frequency distribution of the weights, you can still solve this problem because you can assume the weights of college women are approximately normally distributed with $\mu = 120$ pounds and $\sigma = 15$ pounds. You already know how to find the percentage of scores between any two given z scores in an approximately normal distribution, but in this problem you are only given raw scores. The solution to problems like this involves two major steps. The first step is to transform the raw scores into z scores, so you can use Table z. The second step is to determine the relevant percentages from Table z that are appropriate to the problem. For this problem, the first step is to transform the scores of 100 and 130 pounds to z scores:

$$\text{For 100 pounds:} \quad z = \frac{X - \mu}{\sigma} = \frac{100 - 120}{15} = \frac{-20}{15} = -1.33$$

$$\text{For 130 pounds:} \quad z = \frac{X - \mu}{\sigma} = \frac{130 - 120}{15} = \frac{10}{15} = .67$$

In the second step look in Table z for $z = -1.33$ and find 40.82%, and find 24.86% for $z = .67$. Since you are finding the percentage of scores between z's lying on opposite sides of the mean, you must add the percentages to determine that 65.68% of college women have body weights between 100 and 130 pounds. Figure 4-10 illustrates this problem.

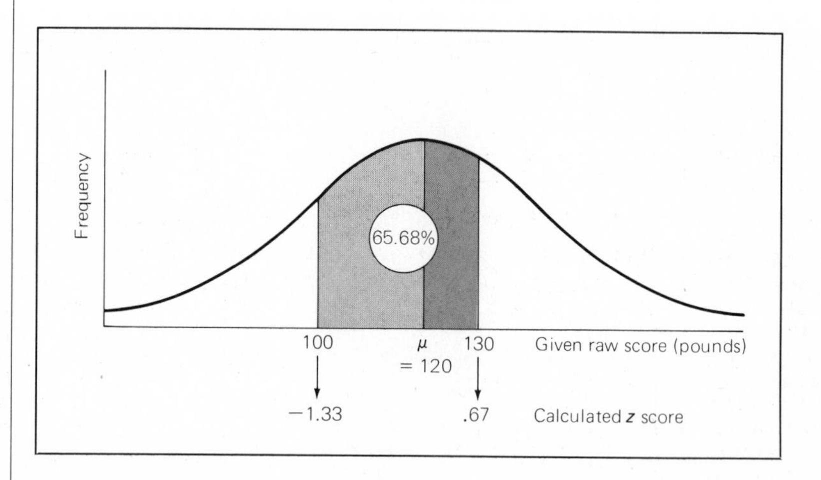

FIGURE 4-10

To determine the percentage of raw scores lying between two raw scores on opposite sides of the mean, transform the raw scores into z scores and add the percentages found in Table z. The shaded area under the curve is the percentage of college women weighing between 100 and 130 pounds.

Now suppose you are interested in finding the percentage of women having weights less than 90 pounds. First transform the raw score to a z score:

$$\text{For 90 pounds:} \quad z = \frac{X - \mu}{\sigma} = \frac{90 - 120}{15} = \frac{-30}{15} = -2$$

Second, from Table z find 47.72%, the percentage of scores between the mean and a $z = -2$. The percentage of weights below a $z = -2$ is 50% minus 47.72% or 2.28%. Figure 4-11 illustrates this problem.

Assume you are interested in finding the percentage of college women weighing between 95 and 105 pounds. First transform the weights to z scores:

$$\text{For 95 pounds:} \quad z = \frac{X - \mu}{\sigma} = \frac{95 - 120}{15} = \frac{-25}{15} = -1.67$$

$$\text{For 105 pounds:} \quad z = \frac{X - \mu}{\sigma} = \frac{105 - 120}{15} = \frac{-15}{15} = -1$$

Second, from Table z, 45.25% is the percentage of scores between the mean and $z = 1.67$, and 34.13% is the percentage between the mean and

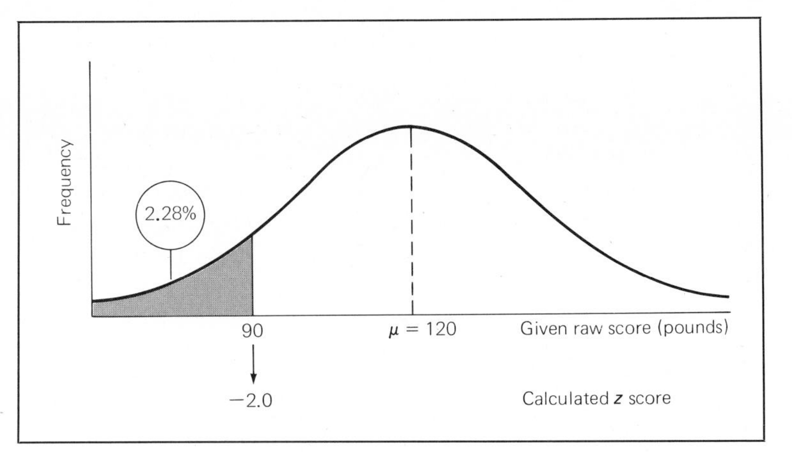

FIGURE 4-11
To determine the percentage of scores lying below a given raw score, transform the raw score into a z and find the percentage in Table z. The shaded area under the curve is the percentage of college women weighing less than 90 pounds.

$z = -1$. Since you want to find the percentage of scores between z's lying on the same side of the mean, subtract the smaller percentage from the larger, leaving 11.12%. Figure 4-12 illustrates this problem.

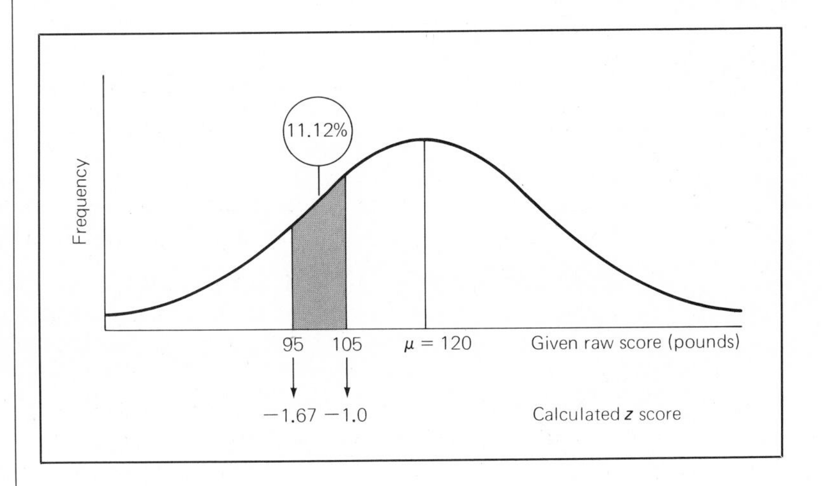

FIGURE 4-12
To determine the percentage of scores between two raw scores lying on the same side of the mean, transform the raw scores into z's and subtract the smaller percentage found in Table z from the larger percentage. The shaded area under the curve is the percentage of college women weighing between 95 and 105 pounds.

converting percents to raw scores in approximately normal distributions

Numerous problems arise where you must solve the inverse of the above sort of problem: you are given a percentage of raw scores and asked to solve for the raw score that cuts off that percentage. For example, suppose you are interested in starting an intellectual club that will only admit people with IQs in the top 10% of the population. What is the lowest IQ you will admit to the club? If you do not have an empirical frequency distribution, you can still proceed if you know that IQ is approximately normally distributed with $\mu = 100$ and $\sigma = 15$. You can reverse the process you used to solve for a percentage in an approximately normal distribution. First, determine the percentage that needs to be looked up in the body of Table *z*, and find the *z* score that corresponds to it. Second, convert the *z* score to its raw-score equivalent. In this problem, if you erroneously look up 10% in the body of Table *z*, you will find a *z* score such that 10% of the scores lie between it and the mean, and 40% lie above it. This is not what you want. Remember, Table *z* only gives the percentage of scores between the mean and a given *z* score. In this example, you want to find a *z* such that 10% of the scores remain *above z*. Draw a diagram such as Figure 4-13 to help you avoid any confusion

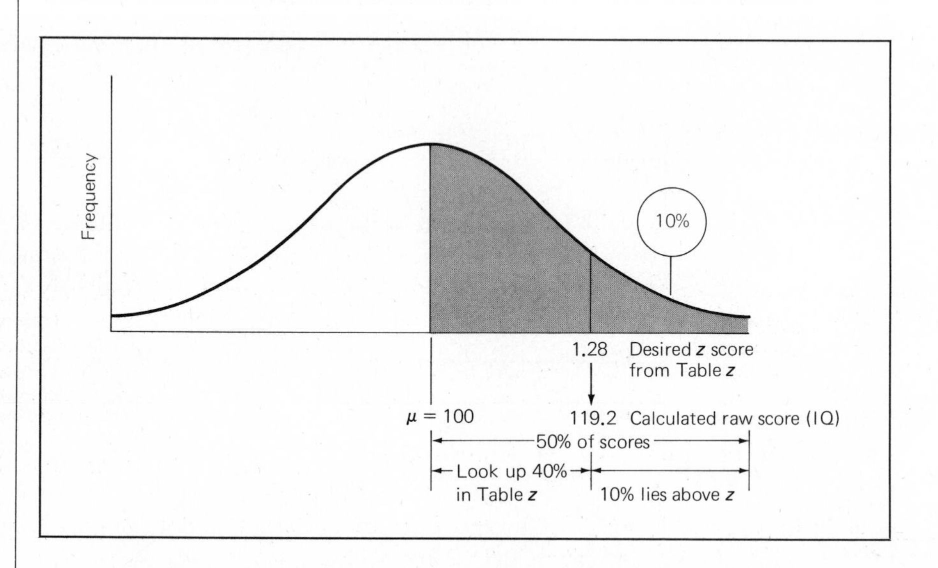

FIGURE 4-13
Always draw a diagram to determine the percentage to look up in Table z. In this example, although you want to find the IQ which cuts off the top 10% of the people, do not look up 10% in Table z. Table z gives only the percentage of scores between the mean and a specified z score. Thus, the shaded area under the curve is the percentage to look up to find its corresponding z score.

as to which percentage to look up in Table *z*. Since 50% of the scores lie above the mean in a normal distribution and you want to cut off the upper 10%, you can determine by subtraction that 40% lies between the mean and the desired *z* score. Thus 40% is the proper percentage to look up in the body of Table *z*. The closest value to 40% you can find in Table *z* is 39.97, which has a corresponding $z = 1.28$. The second step is to convert this *z* score to its raw-score equivalent using Formula 4.4:

$$X = \mu + z\sigma = 100 + (1.28)(15) = 119.2$$

Therefore, the cutoff score for your club is 119.2.

Similarly, you may wish to start an "Average Joe's Club," comprised only of people with IQs in the middle 50% of the population as measured with IQ. Only 25% above the mean and 25% below the mean will be admitted. The first step is to find the closest value corresponding to 25% in the body of Table *z*, which is 24.86%. The corresponding $z = .67$. Since about 25% of the IQs fall between the mean and $z = .67$, then, by symmetry, roughly 25% of the IQs will fall between the mean and

$z = -.67$. Thus, the qualifying IQs for your club are the raw-score equivalents of z's equal to $-.67$ and $.67$.

$$\text{For } z = -.67: \quad X = \mu + z\sigma = 100 + (-.67)(15) = 90$$
$$\text{For } z = .67: \quad X = \mu + z\sigma = 100 + (.67)(15) = 110$$

Therefore, anyone with an IQ from 90 to 110 will be eligible to be an "Average Joe." This is illustrated in Figure 4-14.

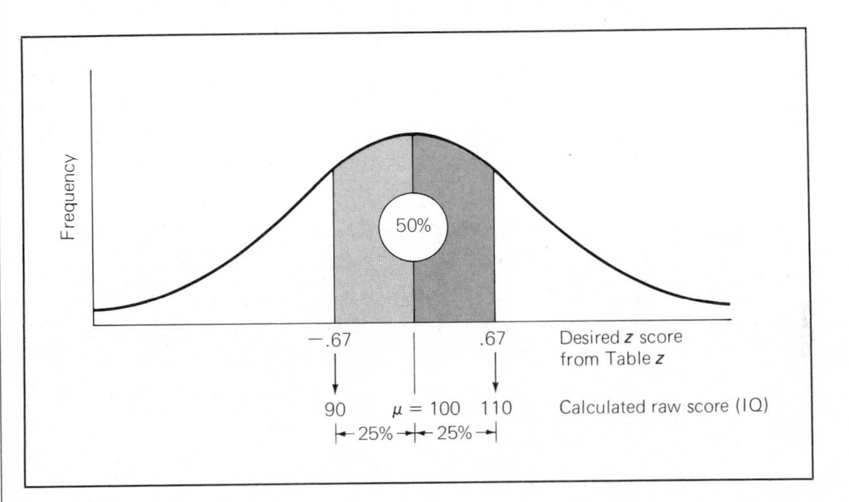

FIGURE 4-14
Finding the IQ qualifications of the "Average Joe's Club," in which only the middle 50% are eligible.

To summarize, you have learned how to solve two general types of practical problems in approximately normal distributions. In one you are given raw scores or z scores and asked to calculate a percentage. In the other you are given some percentage and asked to calculate raw-score cutoffs. Table z is used in solving both types of problems when you can assume an approximately normal distribution. The general strategy for solving each of these types of problems is summarized below:

general strategy for problems with approximately normal distributions

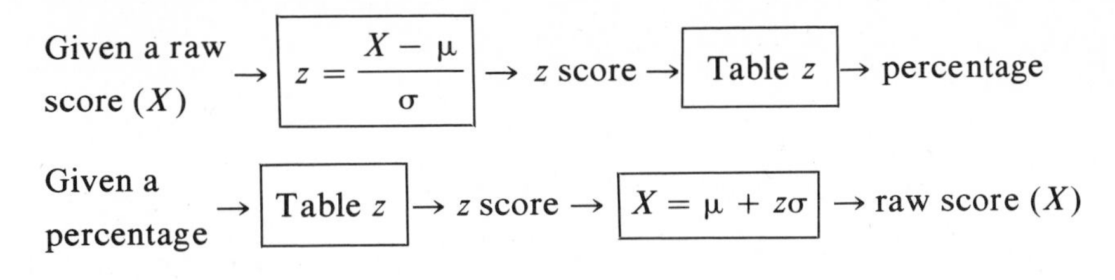

One final problem is a spinoff of the preceding types of problems. In the new type of problem, you want to find the frequency of scores in addition to the percentage. If a particular problem asks for the frequency of scores instead of the percentage, you can calculate the frequency provided that you know N. For example, you previously found that 11.12% of women weigh between 95 and 105 pounds. If you have to find the number of women in a class of 83 who have weights between 95 and 105 pounds, you can determine the answer by taking 11.12% of $N = 83$. This is equivalent to multiplying 83 by .1112, which gives the answer of 9.23 women. When you want to find a percentage of N, first convert the percentage to a proportion[1] by moving the decimal point two places to the left. For example, 11.12% means 11.12/100, which is .1112. Then multiply N by this decimal. Similarly, you can determine the raw score that cuts off a certain number of the population by first converting the frequency to a percentage. For example, if you want to find the weight that cuts off the five lightest women in your class, first convert the frequency of 5 to a proportion by dividing by N and then move the decimal point two places to the right to obtain a percentage:

$$\frac{5}{83} = .0602 = 6.02\%$$

Then look up the closest value to 43.98% (50% − 6.02%) in Table z to find $z = -1.55$. The raw-score equivalent for $z = -1.55$ is 76.75 pounds. Thus 76.75 pounds is the weight separating the five lightest women from the rest. This process is illustrated in Figure 4-15.

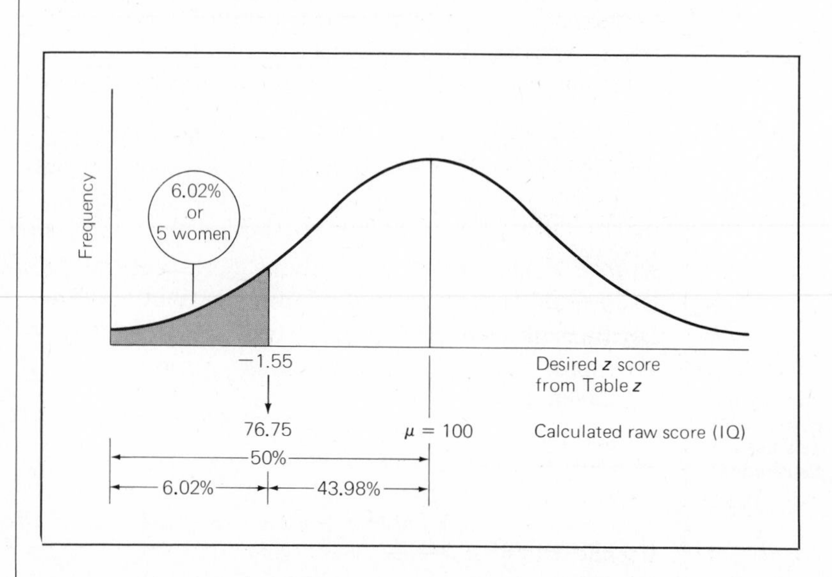

FIGURE 4-15

In problems dealing with score frequencies instead of percentages, first convert the frequency to a percentage. The shaded area under the curve represents the five college women weighing less than or equal to 76.75 pounds.

[1]A proportion is a fraction of 1. In contrast, a percent is a number out of 100. A percent can be converted to a proportion by dividing by 100 (moving the decimal point two places to the left). A proportion may be converted to a percent by multiplying by 100 (moving the decimal point two places to the right).

KEY CONCEPTS

1 A transformed score provides context, telling where its raw-score equivalent falls in a distribution of scores.

2 Percentile rank, or percentile, refers to the percentage of scores that are less than or equal to its raw-score equivalent. Raw scores are transformed to percentile ranks by making a cumulative frequency distribution and then calculating:

$$\text{percentile rank} = \frac{cf}{N}(100)$$

3 A z score refers to the number of standard deviation units away from the mean that its raw-score equivalent lies. z scores are useful in comparing raw scores in different distributions of the same shape. Raw scores are transformed to z scores by:

$$z = \frac{X - \mu}{\sigma}$$

4 The normal distribution is a theoretical distribution and is defined mathematically. It approximates many empirical distributions of abilities/traits. Approximately normal distributions arise when an ability/trait is determined by the sum of many random, independent factors.

5 There are an infinite number of normal distributions, each with a different mean and standard deviation. However, they are all the same shape and thus share the property that the same percentage of scores occurs between the mean and a given z score. These percentages are listed in Table z.

6 Table z can be used to calculate the approximate percentage of scores between the mean and a given z score in empirical distributions that can be approximated by the normal distribution.

EXERCISES

1 Use the data in the frequency distribution below to make the following calculations:

a. Calculate the percentile rank of a raw score of 4.
b. Calculate the percentile rank of a raw score of 9.
c. Calculate the raw score corresponding to the 25th percentile.
d. Calculate the raw score corresponding to the 50th percentile.
e. Calculate the raw score corresponding to the 75th percentile.

X	f
0	1
1	0
2	1
3	2
4	1
5	6
6	7
7	12
8	10
9	5
10	5

2 Buzz A. Field commutes to his statistics class by flying his helicopter to the campus landing pad. The numbers of minutes that it has taken Buzz to fly to campus each class session for the past six weeks are recorded below:

8	14	14	12	10	8
9	9	10	7	11	12
11	10	9	15	10	11

a. Calculate the percentile rank of a raw score of 8.
b. Calculate the percentile rank of a raw score of 13.
c. Calculate the raw score corresponding to the 50th percentile.
d. Calculate the raw score corresponding to the 90th percentile.

3 Given a normal distribution whose mean is 50 and standard deviation is 6.5, transform the following raw scores to their corresponding z scores:

57 43.5 75 64.1 57.5

4 Given a normal distribution whose mean is 18.5 and standard deviation is 3, calculate the raw scores corresponding to the following z scores:

1.5 3.25 −1.5 −.67 2.33

5 Given a normal distribution whose mean is 75 and standard deviation is 20, make the calculations below:

a. Calculate the percentage of scores between the mean and the following raw scores:

95 45 84 37 77.6 28.25

b. Calculate the percentage of scores above the following raw scores:

95 45 84 37 77.6 28.25

c. Calculate the percentage of scores below the following raw scores:

95 45 84 37 77.6 28.25

d. Calculate the percentage of scores between the following raw scores:

28.25 to 37 45 to 84 77.6 to 95

6 Given a normally distributed population of 400 scores whose mean is 500 and standard deviation is 50, make the calculations below:

a. Calculate the frequency of scores between the mean and the following raw scores:

380 590 415 515 646

b. Calculate the frequency of scores above the following raw scores:

380 590 415 515 646

c. Calculate the frequency of scores below the following raw scores:

380 590 415 515 646

d. Calculate the frequency of scores between the following raw scores:

380 to 415 415 to 590 515 to 646

7 Persons applying to the Ph.D. program in computer science are required to take the GRE (Graduate Record Examination) aptitude test as part of their admission requirements. Applicants are not considered for admission if their quantitative ability scores fall below the "cutoff" score. Given that the quantitative scores of all persons taking the test is normally distributed with a mean of 530 and a standard deviation of 128, make the calculations below:

a. Determine the quantitative ability score that cuts off the top 10 percent. This score is the Ph.D. program's cutoff score.

b. College Joe obtained a quantitative ability score of 720. Does he have a chance to be admitted to the Ph.D. program?

c. College Jane also took the GRE test and applied to the Ph.D. program. However, all she knows is that her z score on quantitative ability is 2.13. Does she have a chance to be admitted to the Ph.D. program?

d. Assuming that 7200 people took the GRE test, determine the quantitative ability score that cuts off the top 600 people.

e. When College Joe took the GRE test he obtained a verbal ability score of 740. Given that the verbal ability scores of all persons taking the test are normally distributed with a mean of 510 and a standard deviation of 157, determine whether College Joe was better in verbal ability or in quantitative ability.

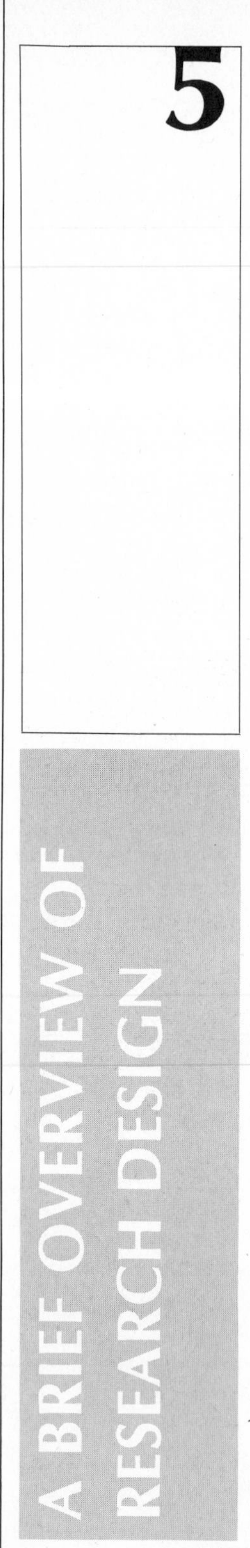

CHAPTER GOALS

After studying this chapter, you should be able to:

1 Define and give an example of each of the following terms: hypothesis; operational definition; independent variable; dependent variable; extraneous variable; confounding; and placebo.

2 Describe the naturalistic method and the experimental method and point out the relevant differences between them.

3 Describe the advantage of the experimental method over the naturalistic method when you are attempting to infer causality.

4 Describe, given a specific hypothesis, how you might set up an appropriate research study with each of the following methods (if appropriate): naturalistic method; two-sample independent design; two-sample dependent design; multiple-sample design; and one-sample design.

5 Describe the purpose of randomization in experimental design and apply it in a given situation.

6 Describe the role of inferential statistics in research.

CHAPTER IN A NUTSHELL

Researchers collect data to gather information and to test ideas or hypotheses about the relationships between objects, events, and processes. The changeable attributes of these objects, events, and processes are called variables. Subjective, ill-defined variables are quantified by operational definitions, which state what operations must be performed to measure the variables. To make inferences about relationships between variables in populations, random samples from the population usually are drawn, and then inferential statistics is applied to try to rule out the possibility that relationships observed in the sample are due to random chance factors, called sampling error. Two common research methods are the naturalistic method and the experimental method. In the naturalistic method the researcher merely observes, without intervention, the relationship between variables. In the experimental method the researcher actively manipulates variables and attempts to control unwanted sources of variation. The variable the researcher manipulates is called the independent variable, and the measured variable is called the dependent variable. Unwanted sources of variation are called extraneous variables. If the value of the dependent variable changes when the independent variable is manipulated, the researcher may infer that changes in the independent variable cause changes in the dependent variable. It is not so easy to infer causality in the naturalistic method because it is unclear which variable is the causal agent,

and it is also difficult to rule out the possibility that changes in some extraneous variable are causing the observed changes.

Four types of experimental design are discussed. In the two-sample independent design, subjects are randomly assigned to one of two groups so that on the average the groups are the same on all variables. The level of the independent variable is then manipulated in one of the groups (the experimental group), while it is left unchanged in the other group (the control group). A subsequent difference between groups in the dependent variable may be attributed to the change in the independent variable. In the two-sample dependent design the experimental and control groups are matched by pairs so that everyone in the control group is paired with someone in the experimental group who has a similar score on the dependent variable. This can be done by measuring the level of the dependent variable (or a related measure) before assigning subjects to the groups or by using only one group of subjects and measuring these subjects under both control and experimental conditions. In a multiple-sample design or "parametric design," one control group is used with several different experimental groups, with each experimental group receiving a different level of the independent variable. This design can be used to test the relative effects of several different levels of the independent variable. In a "one-sample design" a random sample is drawn from a population and the independent variable is manipulated in that sample. The resulting level of the dependent variable is compared with a known value of the dependent variable in the population.

Introduction to research in the behavioral sciences

theory

Empirical research involves the collection of data. Researchers in the behavioral sciences collect data by performing experiments and by conducting systematic field studies. One reason for collecting data is to answer practical questions, such as "Who will win the next presidential election?" or "What effect does Valium have on driving performance?" Data may also be collected to help develop and test theories. A **theory** is a conceptual framework that serves two main functions. One function is to *explain* how things work in the physical, social, cultural, behavioral, and mental realms of the world. The second function is to *predict* what will happen if certain changes occur. Theories are evaluated and improved by using them to make predictions and then seeing whether the predictions agree with the facts. The facts are the data collected in research projects—that is, the research "results." If the predictions and data agree, the theory is said to be "supported by the results" and needs no change. But if the data differ from the predictions, the theory is said to be "negated by the results" and must either be modified or abandoned.

Regardless of whether behavioral science researchers collect data to answer practical questions or to test theories, they usually have some expectation about the outcome of their research project. These expectations, or predictions, are called *hypotheses*. As you will see, it is often difficult to determine whether the results of a research project agree with the **hypothesis.** Since there are usually several alternative explanations for any observed result, it is usually not immediately clear whether the data collected support or negate a hypothesis. Researchers use techniques of inferential statistics called **hypothesis tests** to help them decide upon the implications of the results.

hypothesis

hypothesis tests

All researchers seek order in the realms they investigate. They try to find the relationships between variables. A **variable** refers to any attribute whose value can change from person to person, or from time to time within the same person, or both. For example, degree of learning, socioeconomic level, religious preference, amount of violence viewed on TV, or candidate voted for are all variables. An example of a practical question which a research project might help answer is how the variable of social aggressiveness in children changes according to the variable of amount of violence viewed on TV. "Are children who watch a lot of violence on TV more, equally, or less aggressive than children who watch little violence?" A more relevant but more difficult question to answer is, "Can aggressiveness in children be reduced by altering the amount of violence they watch on TV?" This is a more difficult question to answer because it involves the notion of **causality.** (The first question seeks to determine whether there is any kind of relationship, whereas the second question seeks to determine whether the relationship is a causal one.) If changes in one variable cause changes in the second variable, then you can alter the level of (value on) the second variable by directly manipulating (changing) the first variable.

variables

causality

Thus researchers investigate relationships among variables. You might ask, though, that since a variable is anything whose value can change (and this probably includes nearly everything you can think of), don't we need to be more precise about what behavioral science researchers actually do? You are correct; some further qualification is in order. First, the behavioral science researcher is interested in variables relating to the behavior of humans and other germane animals. Second, the behavioral science researcher is interested only in variables that can be reliably and precisely measured on some scale of measurement, either nominal, ordinal, or interval/ratio. Because of the greater amount of information conveyed in interval/ratio scales, researchers like to use interval/ratio scales whenever possible. However, it is relatively more difficult to measure some variables on an interval/ratio scale than it is to measure others. For example, suppose you are interested in the relationship between coffee consumption and anxiety. It is relatively easy to measure the variable of coffee consumption on an interval/ratio scale by simply counting the number of ounces of coffee (brewed in a specified manner) a person drinks in a given period of time, say, a day. But it is harder to measure anxiety. How can you quantify such a subjective state?

To solve the problem of measuring vague, ill-defined, subjective

operational definition

concepts such as anxiety, researchers have come to rely on operational definitions. An **operational definition** defines a concept in terms of the operations that must be performed to measure it. For example, you can measure anxiety by asking a subject to give you a number on a scale from 1 to 10, that describes the level of anxiety experienced at a given moment. The number 1 can stand for complete lack of anxiety, and 10 can stand for the maximum conceivable level of anxiety for that person. You define anxiety as "the number the subject gives you"—an operational definition.

There are many operational definitions of anxiety. The one to choose depends on the particular research situation in hand. One possibility is to operationally define anxiety in terms of one of its measurable physiological counterparts. For example, we know that as the subjective report of anxiety on a scale from 1 to 10 increases, heart rate increases, blood pressure rises, muscle activity increases, and several other physiological measures change. You might operationally define anxiety in terms of any of these physiological measures. You can measure heart rate in beats per minute, and as heart rate increases, you can infer that anxiety is increasing (as long as you assume nothing extraneous is occurring to increase heart rate, such as higher room temperature or exercise). The proper operational definition to choose is the one judged most valid under the circumstances, that is, the one that is a measure of the desired concept and not of something else. For example, you probably would not want to measure anxiety by asking your subject to state a number from 1 to 10 while in a group situation, because such a measurement will be affected by the impression the subject is trying to impart to the members of the group. The subject might be very anxious but might state a low number simply to display to the group members an aura of confidence. You would also not want to measure anxiety by determining heart rate if the research design requires the subject to be involved in vigorous physical exercise.

When researchers collect data regarding the relationships between two or more variables, two problems typically arise. The first problem concerns the fact that relationships between variables in entire populations are usually only statistical relationships. That is, people differ from each other, and a relationship that holds true for one person might not hold true for everyone else, or a relationship might manifest itself to a different degree for different people. For example, consuming 16 ounces of coffee might make you a little anxious, but it might make another person moderately anxious, and it might even have another effect on someone else. Thus, the conclusions you make are not necessarily true for any one individual, but are only true on the average, for the "average" individual. Therefore, you might conclude, "On the average, people who drink coffee are more anxious than people who do not drink coffee." The type of relationships you usually find are "average," "typical," or "normative" types of relationships. It is important to realize that an instance of the "average" situation might not exist in reality. For example, if the average number of offspring in United States families is 2.2, it is clear you will never find such an "average" family. The average is merely a statistical description.

The second problem that frequently arises when examining relationships is that researchers typically want to determine the relationship existing between variables *in the entire population.* The problem is that the researcher has to rely on data from samples. For example, you might be interested in the effect of coffee consumption on anxiety for everyone in, say, the United States. But it is impossible or impractical to go out and measure the variables of coffee consumption and anxiety for everyone in the United States. It would take too long and be too expensive. The practical recourse used by researchers is to infer the sort of relationship that exists in the entire population by measuring the relationship between the variables in a sample. If the sample is representative of the population, then the relationship observed in the sample can be considered representative of the relationship in the population. The simplest way to get a representative sample is to obtain a **random sample.** A random sample is a sample in which everyone in the population has an equal chance of being chosen for the sample. You could take a random sample of the population of the United States if you put each individual's name on a small piece of paper, put the pieces of paper in a vat, mixed them up, and then drew out as many as you wanted for your sample. Of course, it would be more practical to have a computer do an analogous operation so you would not have to bother with so many pieces of paper and a vat!

random sample

Given that you have a random sample and that you can measure a relationship between variables in that sample, you can then infer the relationship that exists in the population. Here lies the origin of the second problem: The relationship you observe in your sample might be representative of the relationship in the population; or it might be attributable to the vagaries of random sampling and be merely the result of chance. For example, you might observe in your sample that high coffee consumption relates to increased anxiety. Is this due to a relationship between coffee consumption and anxiety in the general population? Alternatively, the relationship could be due to a random selection of some people in your sample who drink a lot of coffee and who, just by chance, are also anxious, and to a random selection of other people who drink little coffee and who are not anxious. Is it possible that there is no relationship in the population between coffee consumption and anxiety, and that the results in your sample are due solely to chance? If you look at the entire population, might people who drink a lot of coffee have the same initial anxiety levels, on the average, as people who drink little coffee? If there truly is no relationship in the population between coffee consumption and anxiety, and you find a relationship in your sample, then your observed relationship results from chance fluctuations in random sampling.

When you make a chance observation in a sample that does not

sampling error

correspond to the true state of affairs in the population, you have what is called a **sampling error.** Sampling errors are very common in behavioral research. This is where hypothesis testing enters. Hypothesis tests provide a reasonable way of deciding whether your results are representative of a true relationship in the population or are due to chance sampling error. Hypothesis testing and sampling error will be thoroughly discussed in Chapter 6.

There are basically two different methods of investigation used in behavioral science research: the naturalistic method and the experimental method. Both are useful and each has its relative advantages and disadvantages.

The naturalistic method

naturalistic observation

Suppose you are interested in the effect of coffee consumption on anxiety. It might occur to you to go out and draw a random sample from the population and measure the variables of coffee consumption and anxiety as they occur naturally. If you do this, you are using the naturalistic method, or what is frequently referred to as **naturalistic observation.** The essence of this method is that you systematically measure the variables as they naturally occur in the population. You do not manipulate or control variables. You do nothing to change the world. You merely observe objects, events, or processes, and measure and record them.

Naturalistic observation is frequently used when it is difficult to manipulate variables to change the course of events. An excellent example of a science based solely on the naturalistic method is astronomy. Astronomers presently must be content to sit back and observe, without attempting to change the course of heavenly events in order to "see what would happen 'if' . . ."

scatter plot
correlation coefficient

The type of data obtained using the naturalistic method is a record of the relationship between changes in two or more variables. There are several ways in which these relationships can be summarized. Chapter 11 discusses how to use **scatter plots** and the **correlation coefficient** to describe the results of naturalistic observations. A scatter plot is a graphical method for describing the relationship between two variables. A correlation coefficient is a single number describing the relationship between two variables. Both the scatter plot and the correlation coefficient allow you to summarize the relationship you observe in your sample. There are many different possible types of relationships that might be observed between coffee consumption and anxiety level. High coffee consumption could be related to anxiety in a strong positive way, such that people who drink a lot of coffee tend to be very anxious on the average. Alternatively, the variables could be related in a weak positive way, such that people who drink a lot of coffee would have a mild tendency to be anxious. Another possibility is that the variables might be related in a weak or strong negative fashion, such that people who drink a lot of coffee tend to be less anxious than people who drink little coffee. Still another possibility is that there may be no relationship between coffee consumption and anxiety. Besides describing relationships between two variables, the correlation coefficient can be used to make predictions from a known value of a variable to an unknown value of a related variable. For example, if you find a relationship between the amount of coffee consumed and anxiety level, you can use this information to predict a person's anxiety if you know the amount of coffee that person consumes. The statistical tool which enables you to make such predictions is called **regression**, which is discussed in Chapter 12.

regression

In the naturalistic method, as in other methods, the hypothesis testing procedures of inferential statistics enable you to decide rationally whether the relationship you observe in your sample is indeed representative of the relationship that exists in the population or is due to sampling error. Since behavioral science researchers generally deal with random samples, it is always possible that any relationship observed in a sample cannot be extended, or generalized, to the population. There is always the possibility that the relationship observed in the sample could have arisen from sampling error. For example, psychologists have found that there is no relationship in the population between red hair and a hot temper. Yet it is *reasonably likely* for you to draw a small sample of people and find that those with red hair also by chance happen to be hot-tempered. Deciding whether or not sample results are due to chance is a recurrent theme in inferential statistics. The purpose of hypothesis tests is to help you decide. In the following chapters you will become familiar with many hypothesis tests.

The naturalistic method is useful because you observe variables in a natural setting. No artificiality is introduced by removing the situation to a laboratory and isolating the variables. It can sometimes be relatively quick and cheap to use. The major drawback of the naturalistic method is that it allows you only to observe relationships that *naturally exist* between variables. Simple observation such as this *does not allow you to directly infer causality*. You might observe that high coffee consumption relates to high anxiety in your sample, but you cannot infer it is the coffee that causes anxiety. The reason you cannot infer causality is that there are several possible interpretations for why coffee consumption and anxiety appear to be related in your sample. For example, coffee consumption may indeed cause anxiety, but it could also be true that people who are anxious may be drawn to drinking coffee, perhaps because it makes them feel better. There is still another possibility—namely, that there may be some third factor that causes both anxiety and increased coffee consumption. This third factor could conceivably be the type of social setting in which a person lives. For example, people with high-pressure jobs may experience anxiety because of their jobs and they may drink coffee because their co-workers also drink coffee to maintain alertness and resist fatigue. Since with the naturalistic method you can only observe relationships between naturally existing levels of variables, it is difficult to differentiate among the various possible causal factors. If you want to be able to say that changes in one variable *cause* changes in another, a different method of investigation is useful. This method is the experimental method.

The experimental method

The experimental method differs from the naturalistic method because the essence of the experimental method is the use of *control* and *manipulation* of variables. Suppose you want to establish that coffee consumption causes people to be more anxious. Essentially, you need to show that you can change anxiety level by changing coffee consumption. You must simultaneously rule out the role of other *confounding* factors, such as job stress, that might cause the change in anxiety. You can eliminate the influence of these confounding factors by making sure they remain constant. To do this, you need two samples of subjects that are about equal on all relevant abilities/traits. One way to arrange this is to randomly assign a large number of subjects to each of the two samples. You can do this by taking a large number of non-coffee drinking subjects from the population and by flipping a coin for each subject. If the coin lands "heads," the subject is assigned to one sample, and if the coin lands "tails," the subject is assigned to the other sample. Thus each subject is equally likely to be assigned to either group. Random assignment assures that the two samples will on the average be about equal for any ability/trait you care to name (for example, age, sex, height, weight, IQ, anxiety level, job stress, and so on). This idea of **randomization** is frequently used to equate two or more samples. In effect, you control the differences between the two groups by the use of randomization. Call one sample the **control group** and the other sample the **experimental group.** The control group serves as a standard for judging the effect of a manipulation you will introduce in the experimental group. Notice how the term "group" is substituted for "sample." In this text we use the words "sample," "group," and "condition" interchangeably. Thus the control "group" is really a "sample" which will be subjected to certain "conditions." Another common term used in this context is "treatment group," referring to a group in an experiment. This term is appropriate because the groups in an experiment will be treated differently by arranging for a change in one particular variable.

randomization

control group
experimental group

Suppose that for the experimental group you now introduce a change in one of the variables. Let us say that you manipulate coffee consumption by requiring subjects in the experimental group to drink a predetermined large amount of coffee every day. Since you are only interested in the effect of drinking coffee and not in the "forced consumption of liquid," you require the control group to consume an equal amount of some psychotropically[1] inactive coffee substitute, but no coffee. Thus the only difference between the experimental group and the control group is coffee. Since you initially had non-coffee drinkers in both groups, and you assume the control and experimental groups started out the same on all other abilities/traits, you expect them to remain the same except for the effect of the variable you manipulate—coffee consumption. After you manipulate coffee consumption, you measure anxiety levels in the two groups. If you observe that anxiety is the same, then you can conclude that coffee consumption has no effect on anxiety. But if you observe that the experimental group shows higher anxiety than the control group, you might conclude that coffee consumption increases anxiety, since coffee consumption is the only difference between the two groups.

An interesting factor in the coffee experiment is the use of a coffee substitute to make sure both groups are equated in terms of being required to consume liquids. The coffee substitute in this study is a psychotropically inactive substance and is called a **placebo.** Placebos are often used in drug research to control for "subject expectation," an extraneous or confounding variable that may make interpretation of the results difficult. The subjects in each group do not know whether they are getting an active drug or a placebo, so the expectation of drug effects, or the mere presence of a pill, liquid, or other substance, cannot affect one group any differently than another. The groups are treated exactly the same except for changes in the one variable whose effect you wish to examine.

placebo

To summarize, this type of experimental method proceeds by first forming two groups of subjects that are about equal on every ability/trait. One way to do this is by random assignment of subjects. Since the groups are assumed to be equal on all variables, you are said to be "controlling for **extraneous variables.**" Extraneous variables are unwanted sources of variation that might influence and confound the interpretation of a research project. You then manipulate the **independent variable.** The independent variable is the variable that you want to show will cause changes in the **dependent variable.** The independent variable is what you manipulate *before* you collect the data, and the dependent variable is what you measure and record as your data. Thus coffee consumption is what you change before you collect data (the experimental group will have coffee and the control group will not), and the dependent variable is the subjective perception of anxiety, rated on a scale of 1 to 10. If the dependent variable changes, you can say that it "depends" on the independent variable, and so you can infer that changes in the independent variable cause changes in the dependent variable. For example, you can say that changes in coffee consumption cause changes in anxiety. You can make such an inference if, by the use of inferential statistics, you can rule out the possibility

extraneous variable

independent variable

dependent variable

[1]Psychotropic means "having some psychological effect due to direct neurophysiological activity."

that your results are due to sampling error. Incidentally, in a naturalistic observation you have *two* dependent variables because you are measuring both variables and you are deliberately manipulating neither. Naturalistic observation is thus concerned with observing relationships between two (or more) dependent variables. This nomenclature is also appropriate because of the difficulty of inferring causality in naturalistic observation. In an experiment you can more easily infer whether the level of the dependent variable "depends" on the level of the independent variable.[2] However, in a naturalistic observation you are not sure what is the causal relation, if any, between variables. Thus both variables are "dependent," and the question of exactly what they depend on is left open.

two-sample independent design

Two-sample independent designs. The type of experimental design we have just described is called a **two-sample independent design.** The two groups are initially equated by random assignment of subjects. In this type of design you generally do not know the population parameters of the dependent variable. For example, in the coffee study you do not know the average anxiety level of the general population. This is another reason why you need to use a control group. If you just select a single group of non-coffee drinking subjects and require them to consume a large amount of coffee, you do not have anything with which to compare the resultant anxiety level. You do not know whether it is higher than, lower than, or the same as that of the general population of non-coffee drinkers. Thus, in a two-sample design the control group serves to provide a reference point, or baseline, for the dependent variable. If you observe a difference in anxiety between the control and experimental groups, you may interpret this observed result in one of two ways. One interpretation is that the difference is a direct effect of your manipulation of the independent variable. The other interpretation is that the difference is due to sampling error. Although your two groups probably have the same level of the dependent variable to start with (because of randomization), it is always possible that the groups initially differ by some amount due to chance random sampling error. There are hypothesis-testing procedures in the two-sample independent design to help you decide whether your results are due to chance. If certain assumptions about the population are met, you can use the two-sample independent t test described in Chapter 8. If these assumptions cannot be met, then you will probably be able to use the less powerful Mann-Whitney U test described in Chapter 9.

[2]We do not want to imply that you can readily infer that changes in the independent variable cause changes in the dependent variable whenever you use the experimental method. Much experimental research is very complex and there always lurks the possibility of subtle, uncontrolled extraneous variables, and problems with logical inference. We simply mean to say that we are more certain as to cause and effect with the experimental method than with the naturalistic method.

A germane criticism of the two-sample independent design is that the researcher is not sure that the two groups are really initially equal before the independent variable is manipulated. Randomization is generally very effective, but in some cases you might start with a large chance discrepancy between the two groups on some important variable, particularly on the dependent variable. You are not really certain the two groups are equal in anxiety before you manipulate the independent variable, so you have to rely on the techniques of inferential statistics to decide whether to attribute observed differences between the two randomly selected groups to the effect of manipulating the independent variable or to the effect of sampling error. You might ask why not simply measure the dependent variable beforehand to ensure that the two groups are equal, or at least to see how much they initially differ by chance. This is a useful suggestion, but such pretesting does not eliminate the problem, and it also has its own disadvantages. First, for practical reasons, it is frequently impossible to make "before" measurements of the dependent variable prior to manipulating the independent variable. For example, the subjects may not be available, or you may not have the time or money necessary. A second disadvantage is that pretesting does not rule out the possibility that chance factors will confound the results. For example, it is possible to measure anxiety in a group of subjects, and then some time later, without intentionally changing anything, to measure anxiety again and get a different measure simply because of moment-to-moment chance fluctuations in anxiety level. Anxiety level changes over time in response to random extraneous events, some of which are internal, some external, and many of which are unknown. This is also a type of sampling error. What we are really trying to do is to assess a subject's overall, or average, anxiety level. Since anxiety level changes from time to time, when you measure anxiety at any one given moment you are actually taking a sample of the subject's anxiety. This measurement may accurately reflect the subject's general anxiety level or it may be distorted by the influence of chance factors in sampling—a sampling error. Thus even if you do give an anxiety pretest to ensure that your control and experimental groups are the same on anxiety before you manipulate coffee consumption, it is still possible that any difference in anxiety observed after the administration of coffee is due to chance factors, or sampling error. However, you should not let this criticism alone discourage you from ever giving a pretest, since sampling error is a problem in all research.

sensitization

A third potential disadvantage is that making a "before" measurement may itself change the dependent variable, or it may **sensitize** the subjects to the manipulation you plan. For example, suppose you obtain a "before" measure of anxiety by asking subjects to give you a number from 1 to 10 describing their subjective perception of anxiety. Such a measurement might cause subjects to start thinking about their anxiety level and make them more sensitive to anything that might even remotely alter that level.

Introducing a "before" measurement of anxiety might cause coffee to have a greater effect on anxiety than it would otherwise have. Making a "before" measurement might even have the effect of itself increasing anxiety, regardless of coffee consumption, because it might focus attention on anxiety. When you start worrying about worrying, you are in trouble!

two-sample dependent or matched design

Two-sample dependent designs. Often used in behavioral science research is a variation of the two-sample independent design which incorporates the idea of giving a pretest. This is the **two-sample dependent design,** which may also be called the **two-sample matched design.** This design can be used when it is feasible to make "before" measurements, and sensitization is not seen as a problem.

matching

Since it is very helpful if the control and experimental groups are equal on the dependent variable before you manipulate the independent variable, you might want to equate the two groups by matching pairs of subjects, either on the dependent variable or on some related variable. For example, suppose you can obtain only a very few subjects for your research project. You do not want to rely on randomization to equate your experimental and control groups because randomization works best with larger numbers of subjects. You can equate the two groups by **matching** them on the dependent variable. First give all the subjects a pretest on anxiety and then list the subjects in order of increasing anxiety. Find pairs of subjects who are very close to each other in anxiety ratings. Then randomly assign one of the members of each pair to the control group and the other member to the experimental group, say by flipping a coin. Thus, for each subject in the control group there is a subject in the experimental group who is "matched," or very nearly equal, on the dependent variable of anxiety. Therefore the initial mean anxiety ratings for the two groups will be about equal before you manipulate the independent variable.

If you cannot make a beforehand measurement on the dependent variable itself, you can often measure some related ability/trait and match the groups on this variable. For example, if you have a measure related to anxiety, say "neurosis," you might pair up subjects on that. The more the ability/trait you use to pair up subjects is related to the dependent variable, the more likely it is that the groups will be equal on the dependent variable.

With the two-sample dependent design as well as the two-sample independent design, after you manipulate the independent variable and collect your data by measuring the dependent variable, you may interpret any observed difference between groups in one of two ways. One interpretation is that the difference is a result of your manipulation. The other interpretation is that the difference is the result of sampling error; that is, the observed difference between groups is due to random fluctuation over time and has nothing to do with a change in the independent variable.

You want to rule out this latter interpretation by using an appropriate hypothesis test. If certain assumptions about the population parameters are met, you can use the two-sample dependent t test described in Chapter 8 to help you decide whether your observed results are due to chance. If these assumptions about population parameters cannot be satisfied, then you can probably use the Wilcoxon test described in Chapter 9.

Another application of the two-sample dependent design is when instead of pairing up two different groups of subjects, you use only one group of subjects and test them under two different conditions. For example, you could measure the subjects' initial anxiety level in a "before" condition (or control group, if you wish). Then, in an "after" condition (or experimental group), you could have all subjects consume a large amount of coffee and then remeasure their anxiety level. The anxiety level in the "before" condition serves as the standard by which to judge the effect of coffee on anxiety in the "after" condition.[3] In this application of the two-sample dependent design, in which the same subjects participate in both the "before" and "after" conditions, the subjects "serve as their own control."

Multiple-sample design

Multiple-sample designs. An extension of the two-sample independent and dependent designs discussed above is a **multiple-sample design.** For example, you might be interested in finding out if various doses of coffee have different effects on anxiety. Perhaps you think that only large doses have the effect of increasing anxiety level, whereas small doses have very little, if any, effect. You do not have to carry out a large number of two-sample designs, each with its own control group, to assess the effect of each different dose of coffee. A more efficient design is to have one control group and several different experimental groups. Each experimental group can be administered a different dose of coffee. You might randomly assign non-coffee drinking subjects to, say, five different groups. One group serves as a control and consumes placebo only, 36 ounces. Another group gets a low dose of coffee (say 8 ounces of coffee per day and 28 ounces of placebo); another group gets a moderate dose (16 ounces coffee and 20 ounces placebo); another group gets a higher dose (24 ounces coffee and 12 ounces placebo); and the final group gets a high dose of coffee (36 ounces coffee and no placebo). The placebo is used to equate

[3]In this example there is a confounding "order effect." That is, subjects are *first* tested without coffee and *later* retested with coffee. Any differences in anxiety between the two groups could be due to the order of testing rather than the effect of coffee. For example, as mentioned before, simply asking subjects to report anxiety level may have an effect on subsequent anxiety levels. To help control for this "order effect," a more sophisticated experimental design would use "counterbalancing." For example, you could take half the subjects and test them first with coffee and later with no coffee, and you could take the remaining half of the subjects and test them first with no coffee and later with coffee. You could then compare the average anxiety level of both coffee groups with the average anxiety level of both no-coffee groups.

liquid consumption in all the groups, so as to eliminate the effect of this confounding variable. You can now compare anxiety ratings to see the effects of varying doses of coffee. Such a multiple-sample design, or **parametric design,** is very efficient because several different levels of the same independent variable are compared simultaneously. There is an efficient hypothesis test for these multiple sample designs which is known as the one-way analysis of variance (or ANOVA), discussed in Chapter 13. The one-way ANOVA is an extension of the two-sample *t* test to accommodate more than two groups. Of course, there are different ways to design an experiment to investigate the effect of varying doses of coffee. For example, instead of randomly assigning different subjects to each of the groups, you could match subjects, or you could use the same subjects and test them under each different level of the independent variable. Different forms of the one-way analysis of variance are available to handle these other related designs, which are also referred to as multiple-sample or parametric designs.

parametric design

One-sample designs. There are other experimental designs which can be used to determine the effects of manipulating independent variables. An alternative design which is appropriate in some situations is called a **one-sample design.** Suppose you know the population parameters, specifically the mean and standard deviation, of the dependent variable. For example, in the coffee-anxiety study, assume that you know the mean and standard deviation of anxiety ratings for the population. You might have actually measured the entire population or you might have had enough experience with large samples to be able to infer these population parameters reliably. Situations like this can occur when you are dealing with IQ, for example, which has a known population mean and standard deviation. In the one-sample design, you take a random sample from the population and manipulate the independent variable in your sample. For a coffee-anxiety study, you might take a random sample and then administer a high dose of coffee (more than the average amount consumed in the population) to all the subjects in the sample. Afterwards you measure anxiety and compare it to the anxiety level of the entire population. If your sample's anxiety level is higher than that of the population, one interpretation is that increased coffee consumption caused it. You can see that this design has several limitations. First, since you do not have a control group, it is impossible to have a placebo to control for the effect of requiring subjects to consume large amounts of liquid. Subjects might become more anxious simply because they consume a large amount of liquid, regardless of whether or not it is coffee. (This might particularly be true if there are no restrooms nearby!) Or they might become anxious because they *think* coffee will make them anxious and not because of the actual consumption of coffee. It is necessary to have an experimental design with a control group to rule out these alternative explanations.

one-sample design

With the one-group design, sampling error is a second possible interpretation for why the sample results might differ from the population. Your sample might by chance have an anxiety level that is above that of the population because of fluctuations in random sampling. To decide whether your result is due to the manipulation you introduced, or to sampling error, you can use one of the one-sample hypothesis tests. If you happen to know the population standard deviation, you can use the z test described in Chapter 7. If you do not know the population standard deviation, as is often the case, then you can estimate it from your sample by using the one-sample t test also described in Chapter 7.

As with the two-sample design, you cannot eliminate the possibility of sampling error by giving your sample a pretest on the dependent variable to determine whether it initially has the same mean as the population. There is always the possibility of sampling error *over time,* as well as possible effects of time-related extraneous variables. If you do determine that your sample has the same mean as the population at the outset of your experiment, it is still possible that the sample mean might change thereafter due to sampling error or due to uncontrolled extraneous factors. For example, suppose you give a pretest on the dependent variable and assure yourself that your sample's anxiety level is the same as that of the population. Suppose that immediately afterward there is an accident at a local nuclear power facility, which increases the anxiety level of your sample, independent of the effect of the coffee. How can you distinguish the effect of the coffee from the effect of the accident? This is admittedly an extreme example, but because of the effect of changes in uncontrolled extraneous variables, most of which are unknown, there will always be some chance fluctuation between tests administered at different times. There will always be chance day-to-day fluctuations in the variables measured, and you can never be sure when or why these fluctuations occur. Because the one-sample design lacks a control group, the problems of interpreting the results are even harder than with the two-sample design. For example, it is possible for some extraneous factor to cause a change in the population parameters themselves, thus changing the standard with which you will compare your sample. When you have no control group, you are implicitly assuming that the population parameters remain constant, which may not be true. A good example of this problem might be a study in which the dependent variable is the Scholastic Aptitude Test (SAT) score. If you determined that the population mean was 500 last year, and you use this value as a standard to assess a special enriched curriculum in a sample of twelfth-grade students this year, the population mean might have shifted in the meantime—to, say, 450. If your sample has a mean of 500, then you might erroneously conclude that the curriculum has no effect on SAT scores, when it actually might have produced a 50-point increase.

We must emphasize that there is not always *one best* research design for a given research question. The research design you choose, and hence the type of statistical analysis you use to examine your results, depends on a large number of considerations. These considerations include such things as time, money, other resources available, knowledge of population parameters, knowledge of possible extraneous sources of influence, the particular research question you are asking, and so on. Research design is itself a very complex field about which many textbooks have been written, and a complete discussion is beyond the scope of this book. Our aim is to present a brief overview to introduce you to the subsequent topics in this text. Interested students should consult one of the many available texts on research methodology for a more complete discussion.

chi-square test

Other statistical tools and applications

Chi-square. Another useful hypothesis test is the **chi-square test** discussed in Chapter 10. (Chi is pronounced "kī" as in *pi* = 3.1416.) This test can be used with either the naturalistic or experimental methods. There are two applications of chi-square. One is to see whether a sample with a given frequency distribution (in discrete categories) on one variable is a random sample from a population with a known or inferred frequency distribution. For example, if you know the population frequency distribution for the number of cups of coffee consumed per day, you can decide whether a sample is representative of that population. A useful application might be to take a sample of subjects with high anxiety levels and see whether their coffee consumption differs from that of the general population.

The second application of chi-square is to determine from a sample whether the frequency distribution of scores on one variable follows the same pattern for several different levels of another variable. You can thus see whether changes in the level of one variable affect the values of the other variable. For example, you can determine whether subjects with high coffee consumption have different patterns of anxiety ratings (falling into categories of, say, high anxiety, medium anxiety, and low anxiety) than do low coffee consumers.

Chi-square differs from the other hypothesis tests in this book because it deals with frequency data only. That is, the data must be the number of subjects or scores falling into a particular category. The data cannot be ordinal or interval/ratio data.

Other applications of statistical techniques. Just as you can apply a chi-square test to data from either naturalistic observations or experiments, all the other tools mentioned in this chapter (z test, t tests, Mann-Whitney U test, Wilcoxon test, chi-square, correlation, regression, ANOVA) can be used with data from either naturalistic observations or experiments. You can use the hypothesis testing procedures (z test, t tests, Mann-Whitney U, Wilcoxon test, chi-square, ANOVA) to decide whether your observed results are due to sampling error. Regardless of whether your data are collected using the naturalistic method or the experimental method, groups can be compared with hypothesis-testing procedures. For example, using the naturalistic method, you might observe that a sample of "delinquent" adolescents differs from a sample of "normal" adolescents on a measure of attitude toward authority figures. You might apply the two-sample independent t test to decide whether this observed difference reflects a true difference between all delinquents and normals, or if it is due to sampling error.

KEY CONCEPTS

1. Researchers collect data to help test hypotheses. A hypothesis is a prediction of a specific outcome or a proposed explanation for an outcome.

2. Researchers are concerned with discovering the relationships between variables. A variable refers to any attribute whose value can change.

3. An operational definition defines a concept in terms of what must be done to measure that concept.

4. Random samples are useful because they are in a sense representative of the populations from which they are drawn. Using the hypothesis-testing tools of inferential statistics, you can draw inferences about populations from random samples.

5. An ever-present problem in drawing inferences about populations from random samples is sampling error. A sampling error occurs when sample characteristics do not accurately represent population characteristics because of chance fluctuations in random sampling.

6. There are two different methods of investigation in behavioral science research. These methods are the naturalistic method and the experimental method. In the naturalistic method the researcher merely observes and records the relationship between two dependent variables.

7. In the experimental method the researcher controls extraneous variables, manipulates the independent variable, and measures the dependent variable. The independent variable is the variable the researcher wants to show has a causal influence on the dependent variable. The researcher first manipulates the independent variable and then collects the data. The data recorded is a measure of the dependent variable. Extraneous variables are unwanted sources of variation that might affect the dependent variable and confound the interpretation of the experiment.

8. In the two-sample independent experimental design, the researcher takes two groups that are equal on all variables (this is usually arranged by

randomization) and changes the level of the independent variable in the experimental group. The level of the independent variable is left unchanged in the control group. Thus the control group serves as a reference for judging changes in the experimental group. Any subsequent difference between groups in the dependent variable is either the result of the manipulation of the independent variable or the result of sampling error.

9 The two-sample dependent experimental design is the same as the two-group independent design except that the control and experimental groups are matched pairwise, subject by subject. This ensures that the groups are about equal on the dependent variable before the independent variable is manipulated. Matching can be achieved by using the same subjects and measuring the dependent variable "before" and "after" manipulation of the independent variable, or by pairing up different subjects in the two groups on the basis of their scores on the pretest of the dependent variable or of some related measure.

10 In a multiple-sample design there are usually one control group and several experimental groups, each with different levels of the independent variable. This is also called a parametric design because it is possible to investigate the effect of a range of different levels of the independent variable.

11 In the one-sample design a random sample is drawn from the population and the independent variable is manipulated in the sample. The resulting level of the dependent variable is compared with the known level of the dependent variable in the population.

12 A placebo is a psychotropically inactive substance that is administered to the control group in drug experiments in order to control extraneous variables and help make the control and experimental groups equal on all variables except the independent variable.

13 The role of inferential statistics is to allow researchers to make inferences from samples to populations. Inferential statistics offer tools to help decide whether results observed in samples are representative of the population, or whether they are due to sampling error. These techniques are called hypothesis tests.

EXERCISES

For each of the hypothetical research situations described in Exercises 1 to 4, indicate whether the research employs the naturalistic method or the experimental method.

1 You are interested in whether early childhood obesity is related to the early introduction of solid food to the diet during infancy. You examine the early histories of a group of 50 obese and normal-weight children and determine the age (in months) at which solid foods were introduced.

2 You are interested in whether there is some genetic predisposition for the later development of a specific type of schizophrenia. In a long-term study of many identical twins (same genes) and fraternal twins (different genetic makeup) reared apart with normal foster parents, you observe the incidence of schizophrenia in adulthood. You observe that for 17 identical twins, among which one twin developed schizophrenia, the other twin also developed schizophrenia in 13 of the cases. But of the 21 fraternal twins, where one twin developed schizophrenia, only 5 of the other twins developed schizophrenia.

3 You are a physician examining the effect that a new drug for morning sickness during pregnancy has on the newborn infant. For the next 30 women who are three months pregnant and who complain of morning sickness, you prescribe the new drug to 15, and provide no medication for the other 15. You administer a test of infant development to the newborn children at seven days of age.

4 As an attorney for several clients who are suing for physical injuries sustained in automobile accidents, you are very interested in the use of particular words and phrases that might bias witnesses' estimates of auto speeds as they try to reconstruct the accident scene. You take 30 new law students and show them a film of two cars colliding. One hour after viewing the film, you ask 10 of the students, "How fast were the cars going when they *bumped* into each other?". You ask another 10 students, "How fast were the cars going when they *crashed* into each other?". The remaining 10 students you ask, "How fast were the cars *racing* when they *crashed* into each other?"

For each of the hypothetical research situations employing the experimental method described in Exercises 5 to 8: (a) identify the independent and dependent variables; (b) identify the hypothesis being examined; (c) identify at least two extraneous variables that could influence the results; and (d) name the specific type of experimental design employed.

5 Memory loss (amnesia) is one of the prominent side effects of electroconvulsive shock therapy (ECT). Hoping to reduce such amnesia, you decide to examine whether the general placement of the electrodes on the head may affect the amount of memory loss. You select 30 female patients suffering from a specific type of depression to participate in your study. None of these patients has ever received ECT. The patients are then randomly divided into three groups of 10. All patients are given a general anesthetic (put to sleep) prior to ECT. In one group the electrodes are placed on both the right and left sides of the head; in the second group the electrodes are placed only on the right side; and in the third group the electrodes are placed only on the left side. The same intensity and duration of the ECT is administered to the patients while they are under anesthesia. After the anesthesia wears off, you measure the amount of time (in minutes) it takes for the patient to remember who she is.

6 A marriage and family counselor suspects that television viewing inhibits communication and thereby increases the number of arguments between husbands and wives. The counselor designs an experiment to examine this speculation. Ten childfree couples married less than 18 months volunteered to participate in the study. All couples had been referred to the counselor. As part of the study, the counselor asked the couples to keep a log for one week of the number of hours they daily spent watching TV together, and to record the number of daily arguments they had. It was determined that all the couples had large color TV sets, and watched TV together each evening from 6:00 P.M. to at least 11:00 P.M. Then the counselor asked the couples to restrict their TV viewing to no more than 30 minutes each evening. All of the couples complied for a 28-day period and recorded the number of daily arguments they had during the last week of this period. The researcher then compared the average number of arguments the couples had the week prior to the reduction in TV viewing time with the average number of arguments they had the last week of the TV reduction period.

7 You are interested in determining the effect, if any, that consumption of one 12-ounce can of beer has on the braking time (or reaction time, RT) of motorcycle drivers. You advertise to pay volunteers $10.00 for participating in your experiment, which amounts to a half-hour simulated motorcycle ride. Thirty persons who responded to your ad show up. All had at least three years of experience driving motorcycles and drinking beer, and all were between the ages of 25 and 29. You randomly divide the subjects into two groups of 15. The subjects in one group consume 12 ounces of a popular beer on an empty stomach, and 15 minutes later they perform in the simulated motorcycle ride. The subjects in the other group consume 12 ounces of a placebo beer on an empty stomach. The placebo liquid tastes and looks like the popular brand, but the ethyl alcohol has been removed. These subjects also perform in the simulated motorcycle ride 15 minutes later. You measure each subject's braking time in one particular situation in the simulated ride, which involves peripheral vision, and compare the RTs of subjects in the "real beer" group with the RTs of subjects in the placebo group.

8 A person hears a randomly generated sequence of 12 digits, such as 3, 5, 9, 8, 8, 1, 0, 3, 4, 2, 8, 7, and is asked to report them, in order, immediately after the last digit is uttered. Thousands of previous trials with college students reveals that the average number of digits recalled in their proper order is seven. To test a new mnemonic aid you have devised which involves the representation in an exaggerated image of each digit interacting with the other numbers in the sequence, you select a sample of 30 college students to participate in an experiment. You give all the subjects instructions and practice in your mnemonic imagery and then present them with a 12-digit test sequence to recall. You then compare the number of digits recalled in proper order with the previously established norm of seven.

9 You hypothesize that watching violence on TV increases aggressiveness in pre-school-aged children (3–4 years old).

a. Identify the independent variable and the dependent variable you are interested in.

b. Operationally define the independent variable and the dependent variable.
c. Design a study employing the naturalistic method to test your hypothesis. Identify at least two extraneous variables that might influence your results.
d. Design an experiment employing a one-sample design to test your hypothesis. Identify at least two extraneous variables that could influence your results.
e. Design an experiment employing a two-sample independent design to test your hypothesis.
f. Design an experiment employing a two-sample dependent design to test your hypothesis.
g. Design an experiment employing a multiple-group design to test your hypothesis.
h. Regardless of which research method you use, there is always one prominent opposing explanation for any result you might obtain. Identify this important possible explanation.

6

INTRODUCTION TO HYPOTHESIS TESTING

CHAPTER GOALS

After studying this chapter, you should be able to:

1. Use the hypothesis-testing procedure, describe what is meant by arguing by contradiction, and list the four steps involved in the procedure.
2. Verbally and symbolically state the null and alternative hypotheses of a research problem.
3. Define alpha and describe an alpha, or Type I, error.
4. Describe when to retain and when to reject the null hypothesis.
5. Explain why you must retain the null hypothesis, and not accept it.
6. Define probability, and calculate the probability that an event will occur.
7. Define independent events.
8. State the multiplicative rule, and calculate the probability that two or more independent events will occur in a simple situation.
9. Define beta and describe a beta, or Type II, error.
10. Define power.
11. Describe the relationship between alpha, beta, and power.
12. Determine the appropriate level of alpha in different research situations.

CHAPTER IN A NUTSHELL

It is usually impossible to measure an entire population of subjects to see whether it differs from some other population on a given characteristic, trait, or ability. When you cannot measure an entire population, the next best procedure is to make inferences about the population characteristic on the basis of a sample from that population. Unfortunately, the sample results reflect the effects of chance errors in random sampling (sampling error) in addition to the true population characteristics. A procedure called hypothesis testing is used to decide whether the results seen in a sample are representative of the true population characteristic or are due mainly to the chance effects of sampling error. In hypothesis testing you make an initial assumption that the sample results are due to sampling error. Then, you use the laws of probability to calculate the likelihood of getting the sample results. If the likelihood of getting the sample results by chance seems so slim as to be implausible, you may conclude that the initial assumption is incorrect and that the sample results are probably not due to sampling error.

Introduction to hypothesis testing

Suppose you are employed by a large pharmaceutical company and you want to know if one of the company's newer drugs has the side effect of influencing IQ. You begin by drawing a random sample of 16 people and administering the drug. Drug administration is the independent variable since that is what you are manipulating. After an appropriate period of time, you measure the dependent variable (IQ) and obtain the following IQ scores:

108, 120, 97, 97, 122, 111, 124, 94
116, 123, 94, 98, 102, 120, 102, 96

The sample mean IQ is 107.75. Since you know that the general population from which your sample is chosen has a mean IQ of 100, your sample after taking the drug clearly scored higher than the population. May you conclude that the drug influenced IQ?

Before making any conclusions, you must realize there are two possible explanations for why the sample mean IQ is higher than the population mean IQ. One explanation is that the drug increases the IQ. The other explanation is that by some chance factor, before the drug is administered, the sample mean IQ just happens to be 7.75 IQ points higher than the population mean and the drug has nothing to do with the sample's higher mean. For example, perhaps the subjects you randomly selected had a higher mean IQ than the general population to begin with. Whenever the sample and population means differ by chance to begin with, it is one example of a **sampling error**. Sampling errors are a major cause of difficulty in interpreting experiments.

sampling error

Having a sampling error does not imply you made a mistake in choosing your sample. Although it is ideal to begin with equal sample and population means, in practice, no matter how carefully you draw your sample, the sample mean will usually be different. You may verify this for yourself by collecting the heights of everyone in your statistics class. Consider the class as your population, and compute the mean height. Then randomly select several small samples and compute the mean height of each. It will be very unlikely to find even one of the sample means equal to the population mean.

Another demonstration of the prevalence of sampling error is to flip a fair coin a number of times, scoring a head as 1 and a tail as 0. If you were to continue flipping the coin forever, you would get half heads, and thus the mean number of heads would be .5. After all, if you do not get half heads, your coin is not "fair." For the entire population of flips (that is, all of the flips that the coin could ever make), the mean number of heads should be .5. However, in your sample of just a few coin flips, it is unlikely that you will get heads exactly half the time. Thus your mean will usually be different from the population mean of .5. Experiment by flipping a coin ten times, repeating each series of ten flips several times. Consider each series of ten flips as a sample, and calculate the mean number of heads for each sample. You can observe the prevalence of sampling errors by comparing the sample means to the population mean.

Although it is possible to reduce the influence of sampling error by proper research designs, the researcher is always faced with the possibility that any difference between two means may be due to chance factors rather than to the treatment of concern. For example, a reasonable way to reduce the influence of sampling error in testing for drug effects is to give an IQ test to all the subjects before giving the subjects the drug, and then administer the drug and retest the subjects' IQs after an appropriate period of time has passed. Then you can see if there is a difference between the mean IQ of the first test and the mean IQ of the second. This research design will tell you whether or not the subjects were near the population mean IQ to begin with, and you can get some idea of the effect of the drug by looking at the *difference* between the means of the first and second tests. Unfortunately, this experiment is by no means foolproof. First, it is possible that simply taking the test the first time influences performance on the retest. For example, some subjects may learn how to do better on the IQ test by the second time around, and may even remember some test items. Thus, the pretest might introduce a systematic, confounding source of extraneous variation. Even if you could eliminate this systematic error of experimental design, there is a second problem. Due to *chance factors,* it is possible that the subjects may have different scores on the retest even if they do not take the drug. It is unlikely that a person will score exactly the same on a test administered at two different times because of moment to moment chance fluctuations in processes that influence test scores. Also, during that time between tests the subject may have certain undetermined experiences that affect performance on the retest, such as learning experiences, fatiguing experiences, and so on. Thus you cannot rule out the possibility of chance factors affecting the results of an experiment no matter how sophisticated and well-conceived the study is. Then how can you decide whether the difference is due to the drug, or to sampling error? Inferential statistics can help guide you to the proper conclusion.

Returning to the drug study, you can see that the sample mean IQ after the drug is administered, $\bar{X} = 107.75$, is not equal to the general population mean, $\mu = 100$. Now ask yourself this question, "Pretending no drug was administered, is it possible to randomly select a sample of 16 subjects from a population having $\mu = 100$ and by chance alone get a sample in which $\bar{X} = 107.75$?" If it is possible, then you must think twice before concluding that the difference between μ and $\bar{X}$ is due to the drug. If it is not possible, or at least *not likely,* that the difference is due to sampling error, then you can place more confidence in the conclusion that the drug caused the difference.

Inferential statistics enters into the problem here because the crux of the matter is to find out just *how likely* it is that the obtained difference is due to chance. To do this, you have to make inferences from samples to populations. In this example you are not merely interested in the fact that the sample mean is higher than the mean of the general population, because the difference can be due to sampling error. You are really interested in the effect of the drug on everyone in the general population. What you are asking is, "*If everyone in the population takes the drug, will it have an effect on the mean IQ*?" You could answer this question if you could administer the drug to everyone in the population and then measure the individual IQs. However, you cannot administer the drug to everyone. You must rely on your sample to decide whether the drug will have an effect on the mean of the population. If the drug *has no effect* on IQ, then the population of people administered the drug will have the same IQs as the general population. In effect, you have merely taken a sample from the general population. In this case, the fact that your sample mean IQ is higher than the general population mean is a chance result of sampling error. On the other hand, if the drug *has an effect* on IQ, then the population of people administered the drug will have a different mean IQ from the general population. In this case, the fact that your sample mean is 7.75 IQ points higher than the mean of the general population might be due to drug-induced changes in the population from which the sample is drawn.

Trying to decide whether the drug has an effect on the population mean is really the same as trying to decide whether the sample arose from the general population or from a population with a mean which is different from the general population. You are therefore trying to decide between the following two competing hypotheses:

1. The drug has no effect on the dependent variable of IQ. This is the same as saying the sample arose by chance from the general population having $\mu = 100$, and the difference between the sample mean of 107.75 and the population mean of 100 is due to sampling error.
2. The drug has some effect on the dependent variable of IQ. This is the same as saying the sample arose from a different (hypothetical) population having a mean IQ which is not 100, and the reason why the sample mean is different from the general population mean of 100 is that the drug changes the mean IQ of those who take it.

These alternative hypotheses are depicted in Figure 6-1.

If everyone in the population takes the drug, is the population mean IQ equal to 100, or to some other value? You have gone beyond a simple description of a distribution of scores. Your conclusions are really inferences about population parameters. These inferences are based on sample statistics. The field of inferential statistics guides you to the proper conclusion by a process known as **hypothesis testing.** For each and every result obtained through naturalistic observation or the experimental method, there are two competing hypotheses, and you must decide which one is more likely correct. By convention, one of these hypotheses is referred to as the **null hypothesis,**

hypothesis testing

null hypothesis, H_0

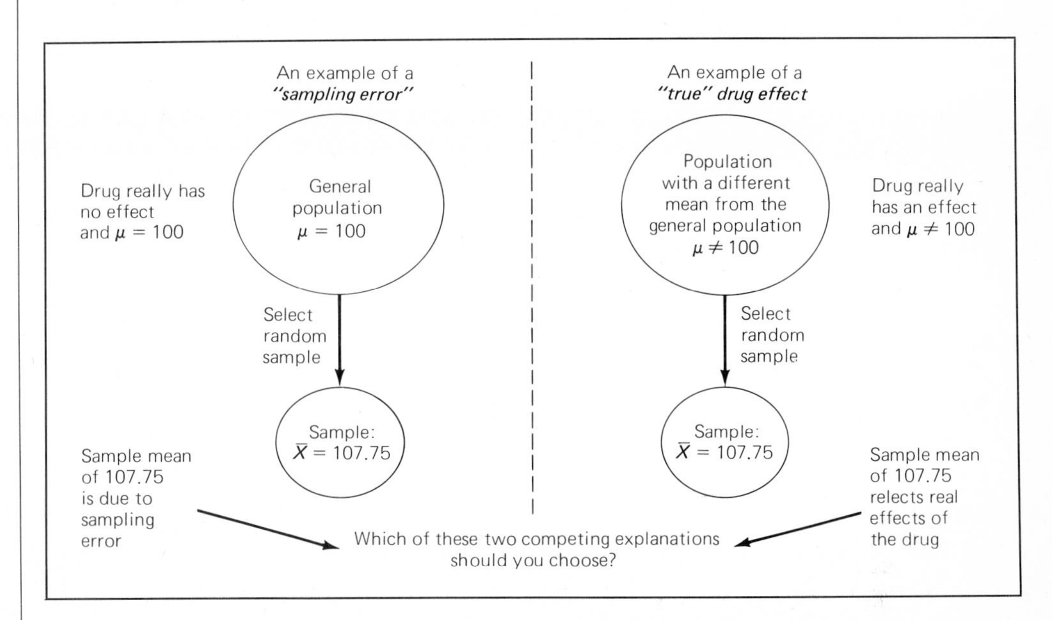

FIGURE 6-1
The two competing explanations for the results of the drug experiment.

alternate hypothesis, H_A

and is abbreviated H_0. The other is the **alternate hypothesis,** and is abbreviated H_A (some texts use H_1 instead of H_A). The null hypothesis always assumes the independent variable really has no effect. (Hence the word *null* as in null and void; the treatment has no effect, there is no difference between treatment and nontreatment and so on.) The alternate hypothesis assumes the opposite conclusion: the independent variable really has an effect. For example, in the drug study you are trying to decide whether the null hypothesis or the alternate hypothesis is most likely correct. These hypotheses are:

H_0: The drug has no effect. The sample comes from the general population having $\mu = 100$.

H_A: The drug has an effect. The sample comes from a different population having $\mu \neq 100$.

The formal process for deciding between two competing hypotheses may be illustrated with a simple experiment about two bags of poker chips. Suppose you are being held captive by a psychotic Las Vegas tycoon who requires you to perform correctly in a guessing game to get your daily meal. His whimsical tasks vary from day to day. On one particular day you are given two bags of poker chips and told one bag contains 100 red chips and 100 white chips. The other bag also contains 200 poker chips, but has an unequal number of red and white chips. The tycoon, in a rare exhibition of generosity, allows you to take a random sample. The bags are not labeled, so you do not know which bag is which. With your eyes closed, you must reach into one bag and select one chip at random. You look at the chip, record the color and then put it back into the same bag. You shake the bag and select another chip, record its color, and replace it. You do this five times using the same bag and note that the five chips you select are all red ones. Your task is to decide whether the bag you have selected is the one with equal numbers of red and white chips, or the other bag. These two possibilities are shown in Figure 6-2. If you make an incorrect decision, you do not get fed for the day. What reasoning would you follow in making your decision?

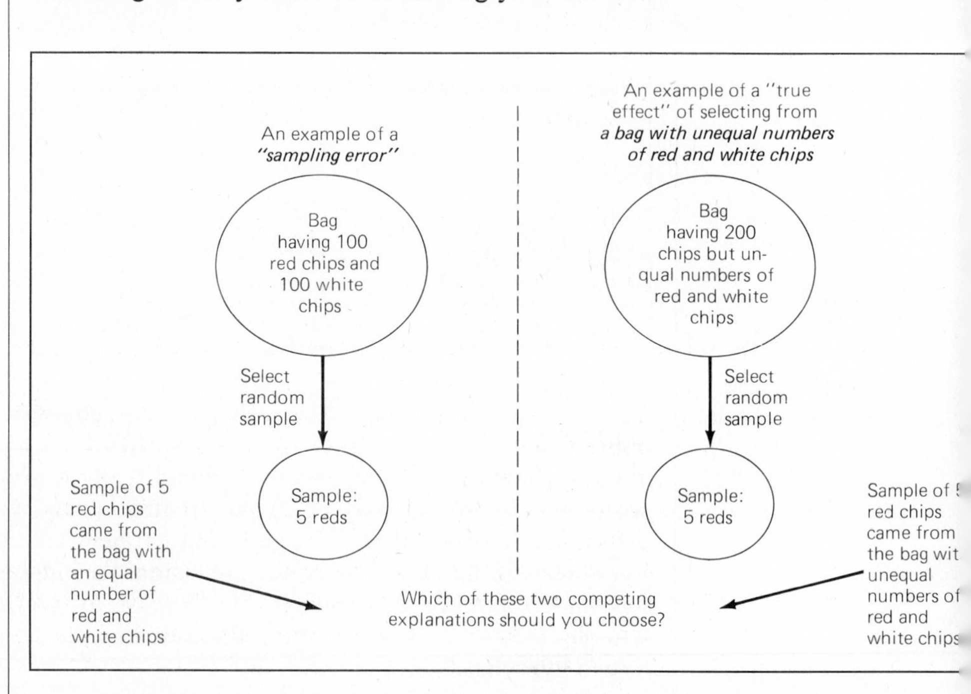

FIGURE 6-2
The two competing explanations for the results of the experiment with two bags of poker chips.

It seems unlikely that five red chips in a row are randomly selected from a bag containing equal numbers of red and white chips. It seems more likely that the bag you selected from actually contains more red than white chips. But can you conclude that your chips came from the bag with unequal numbers? Since you do not know the relative proportions of red and white chips in this bag, it might even contain more white chips than red! You must rely on the information that one bag has *equal* numbers of chips, since you do not know the relative proportions of chips in the other bag. Your task is to decide between the null and alternate hypotheses:

H_0: The five chips come from the bag having equal numbers of red and white chips.

H_A: The five chips come from the bag with an unequal number of red and white chips.

The null hypothesis is that the chips come from the bag having equal numbers of red and white chips because this is the hypothesis that states there is *no difference* in the numbers of red and white chips. Since you do not know whether the unequal bag has more red chips or more white chips, you must rely on your sample of five red chips and *determine how likely* it is that they could be selected at random from the bag with equal numbers. If it is unlikely enough to be implausible, then you can reject the null hypothesis, retain the alternate hypothesis by default, and conclude that you selected from the unequal bag. To determine the likelihood of such an event, you need to become acquainted with a tool used in inferential statistics to measure likelihood, called **probability**. Afterward, we will discuss the hypothesis-testing process in more detail and return to this experiment with two bags of chips.

probability

Probability

p

Although there are many definitions of probability, they all contain the notion that probability, abbreviated ***p***, is a measure of the likelihood of events. The definition we will use is:

FORMULA 6.1

$$\text{probability of a particular event occurring} = \frac{\text{number of possible outcomes}^{1} \text{ in which the particular event occurs}}{\text{total number of possible outcomes}}$$

For example, consider the event of "heads" occurring with a flip of a fair coin. On a single flip there are two possible outcomes (heads or tails), and there is only one possible outcome in which a head occurs. The probability of a head is .5; that is,

$$\text{probability of a head} = \frac{1}{2} = .5$$

Now consider the probability of drawing an ace at random from a standard deck of shuffled cards. There are 52 possible outcomes (one for each different card in the deck), and four outcomes in which an ace occurs (ace of spades, ace of hearts, ace of diamonds, and ace of clubs). The probability of an ace is .077:

$$\text{probability of an ace} = \frac{4}{52} = \frac{1}{13} = .077$$

Notice that probability is always a value between 0 and 1. If there is no way a particular event can occur, then the probability is 0:

[1] An outcome is a single result of some simple situation or research problem, such as flipping a coin.

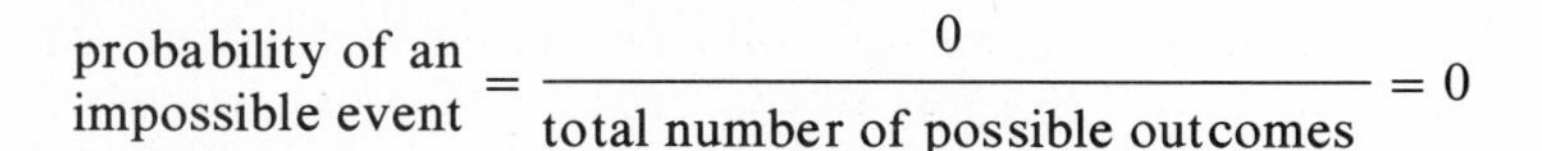

$$\text{probability of an impossible event} = \frac{0}{\text{total number of possible outcomes}} = 0$$

For example, the probability of obtaining a "7" on a roll of a single die is 0 since there is no way such an event can occur. On the other hand, if a particular event always occurs, then each possible outcome will be an instance of that event, and the probability is 1:

$$\text{probability of a sure event} = \frac{\text{number of possible ways in which a sure event may occur}}{\text{total number of possible outcomes (this will be the same value as in the numerator)}} = 1$$

For example, the probability of obtaining a "head" on a toss of a two-headed coin is 1 since no other outcome is possible. Thus, a probability of 0 means an event will never occur, a probability of 1 means the event will always occur, and probabilities between 0 and 1 represent intermediate degrees of likelihood. Likelier events will have probabilities closer to 1. For example, it is much more likely to flip a head with a fair coin ($p = .5$) than to draw an ace from a shuffled deck of cards ($p = .077$).

There is a very useful manipulation of two or more separate probabilities called the **multiplicative rule:**

FORMULA 6.2
multiplicative rule

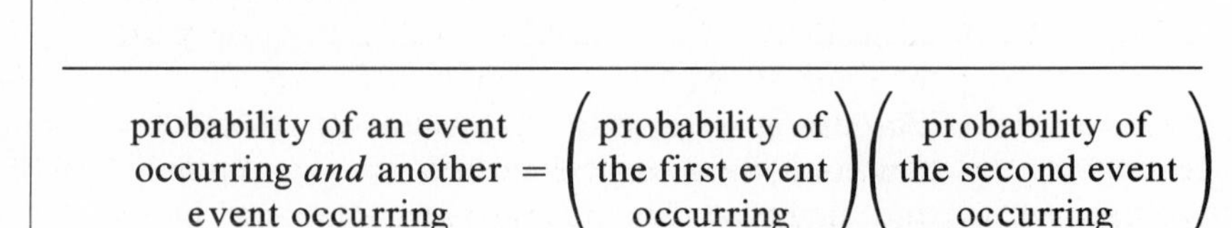

$$\text{probability of an event occurring \textit{and} another event occurring} = \left(\text{probability of the first event occurring}\right)\left(\text{probability of the second event occurring}\right)$$

For example, suppose you want to find the probability of flipping a fair coin and getting a head, and then flipping it again and getting another head. Following the multiplicative rule, the probability of flipping two heads in a row is the probability of flipping the first head (.5) times the probability of flipping the second head (.5). Thus, the probability of flipping two heads in a row is .25:

$$\text{probability of a head and another head} = \left(\text{probability of a head}\right)\left(\text{probability of a head}\right) = (.5)(.5) = .25$$

The multiplicative rule may be extended to determine the probability of flipping three heads in a row, because this is the same as flipping two heads in a row, and then flipping another head. Multiply the probability of getting two heads in a row by the probability of flipping the third head. The probability turns out to be .125:

$$\text{probability of three heads in a row} = \left(\text{probability of a head and another head}\right)\left(\text{probability of a head}\right) = (.25)(.5) = .125$$

Alternatively, you can multiply the separate probabilities by each other:

$$\text{probability of three heads in a row} = \left(\text{probability of a head}\right)\left(\text{probability of a head}\right)\left(\text{probability of a head}\right)$$

$$= (.5)(.5)(.5) = .125$$

You can find the probability of flipping as many heads in a row as you want by extending the multiplicative rule in this manner. You may also use the multiplicative rule to calculate other probabilities, such as the probability of drawing an ace from a deck of cards, replacing it and shuffling, and then drawing a king. The probability of drawing an ace and then a king is .006; that is,

$$\text{probability of an ace and a king} = \left(\text{probability of an ace}\right)\left(\text{probability of a king}\right) = (.077)(.077) = .006$$

independent events

Notice that you may use the multiplicative rule only when the particular events are **independent** of each other. Independent events do not influence the probability of the occurrence of one another. For example, flipping a head the first time does not affect the probability of flipping a head another time. The probability of flipping a head on either toss is still .5. The two events are independent. On the other hand, with the card drawing experiment, if you get an ace on the first draw and do not replace it, you affect the probability of getting a king on the second draw because there are only 51 cards left. The probability of getting a king is then 4/51, or .078, and not .077. These two events are thus not independent and hence are called **dependent.** When the occurrence of an event affects the probability of a subsequent event, the events are not independent and the multiplicative rule as formulated in Formula 6.2 cannot be used.

dependent events

The hypothesis-testing procedure

argument by contradiction

The hypothesis-testing procedure uses what is commonly called an **argument by contradiction.** To argue by contradiction, you make an initial assumption and then search for a contradiction. If you observe a contradiction, you can conclude that the initial assumption is false. You frequently argue by contradiction without realizing it. For example, suppose you live in an apartment with only one window having an awning virtually surrounding it. If you look out the window you see only the ground, but none of the surrounding scenery or sky. If you want to determine whether or not it rained by looking out the window, you first assume that it really has rained. Then you check the ground to see if it is wet. If the ground is not wet, or if it appears *very unlikely* that it is wet, then you have a contradiction, and can reject your initial assumption. By default, you can conclude that it has not rained.

Now that you are familiar with measuring likelihood and arguing by contradiction, return to the two bags of chips. The null and alternate hypotheses are:

H_0: The five chips come from the bag having equal numbers of red and white chips.
H_A: The five chips come from the bag with an unequal number of red and white chips.

To apply the argument by contradiction to this situation, you first make an initial assumption. For your initial assumption, assume the null hypothesis is true. Then search for contradictions. You selected five red chips out of one bag. Since you are assuming the five chips come from the bag having equal numbers of red and white chips, you can calculate the likelihood of this event by using the multiplicative rule. What is the probability of selecting five red chips in a row by chance alone? First, the probability of one red chip is .5; that is,

$$\text{probability of a red chip} = \frac{100}{200} = .5$$

From the multiplicative rule, the probability of selecting five red chips in a row is .03125; that is,

$$\text{probability of five red chips in a row} = (.5)(.5)(.5)(.5)(.5) = .03125$$

Five red chips in a row is not a very likely event. Multiply the probability of .03125 by 100% to find the likelihood in percent. You can see that this result would occur by chance only about 3% of the time, or three times out of every 100 times you sample. The next question is, "Is this result unlikely enough to be considered a contradiction?" If it is contradicting, then you can reject your initial assumption.

So far, you have succeeded in transforming your problem into one of determining whether .03125 is such an unlikely event that it can be considered contradicting. How unlikely should a result be to be considered contradicting? Researchers have resolved this problem by convention. If

$p < .05$

the probability of a result is **less than .05** (abbreviated $\mathbf{p < .05}$), then it is considered a very unlikely occurrence and therefore contradicting, and you can reject the initial assumption. Notice that this is an arbitrary decision criterion, but nevertheless it is commonly accepted. There are circumstances in which you might want to use a smaller value, such as .01. Convention permits this. Then if the probability of an event is less

$p < .01$

alpha, α

than .01 (abbreviated $\mathbf{p < .01}$) you can reject the initial assumption. The decision criterion is called **alpha** (abbreviated α), or the **alpha level.** Conventionally, the value of alpha cannot be greater than .05. The most common levels of alpha are .05 and .01. Later on we will discuss when it is appropriate to use .01 versus .05 levels of alpha.

You can now compare the probability of getting five red chips in a row ($p = .03125$), with an alpha set at .05. Since the probability of your result is less than alpha, you decide the event is so unlikely that it contradicts the initial assumption. You therefore *reject* the null hypothesis. Then, by default, you decide that the alternate hypothesis is the better of the two hypotheses: the five chips come from the bag with an unequal number of red and white chips.

An interesting problem occurs when the probability of your observed results is *equal to or greater* than the alpha level. For example, suppose that you randomly select only four red chips in a row from one bag. If you replace each chip after every selection, the probability of four red chips in a row is .0625:

$$\text{probability of four red chips in a row} = (.5)(.5)(.5)(.5) = .0625$$

Since the probability of this result is greater than $\alpha = .05$, by convention this does not contradict your initial assumption that the null hypothesis is true. You should retain the null hypothesis because there is insufficient contradictory evidence to reject it. Notice that you simply have not gathered sufficient contradictory evidence against the null hypothesis. You have not proved that it is really true, or even that it is the one and only possible hypothesis consistent with your results. An argument by contradiction can only show that the initial assumption is false. It can never show that it is true because you are looking only for contradictory evidence and not supporting evidence. Therefore, you never really *accept* the null hypothesis as being true, you merely **retain** it as the better of the current two hypotheses. If you accept the null hypothesis as proved, then you are overlooking the fact that there is usually a tremendous number of other hypotheses that are also consistent with your results. However, like the null hypothesis, you cannot conclude that any one of the other hypotheses is true. Selecting four red chips in a row is consistent with the hypothesis that they came from a bag with equal numbers of red and white chips (H_0), and is also consistent with the H_A that they came from a bag with unequal numbers of red and white chips. Just because you cannot reject H_0, you cannot say it is true! All you can say is that your results are not inconsistent with H_0. To make this point clearer, take the example of looking at the ground from your window and deciding whether or not is has rained. Suppose you see the ground is wet. Does this tell you for sure that it has rained? Of course not! Someone could have wet the ground with a hose, a fire hydrant could have been opened, or whatever. Any one of these alternatives could be true, and you cannot decide between them on the basis of your limited observation. If the ground is dry, you can conclude that it has not rained. But if it is wet, you cannot use this as "proof" that it has rained because there are many possible reasons

retaining the null hypothesis

why the ground is wet. The argument by contradiction only allows you to retain or *reject* the null hypothesis. It does not allow you to prove that the null hypothesis is true.

The hypothesis-testing process may be summarized:

1. Determine H_0 and H_A. Assume H_0 is true.
2. Set a value for alpha. Usually this is .05 or .01.
3. Calculate the probability of getting your results assuming H_0 is true.
4. Either reject or retain H_0. If the probability of your results is less than alpha, reject H_0; if the probability of your results is greater than or equal to alpha, retain H_0.

People typically raise important questions about the hypothesis-testing process. A common reaction is, "Why do you assume the null hypothesis is true? Why not assume the alternative hypothesis is true?" All right, suppose you start off and assume H_A is true. With the two bags of chips, assume that the five red chips come from the bag with an unequal number of red and white chips. Now see if this result contradicts H_A. How can you observe a contradiction and thereby reject H_A? You cannot calculate the probability of drawing five red chips in a row because you do not know the proportion of red chips in this bag. You cannot even calculate the probability of drawing one red chip from this bag. On the other hand, if you assume H_0 is true, you *can* calculate the probability of getting your results, and thereby reject H_0 if the probability is less than alpha. Bear this point in mind whenever you are trying to determine which of two competing hypotheses should be H_0. Use as H_0 the hypothesis that allows you to calculate the probability of getting the results you obtained by chance.

Errors in decision making

The hypothesis-testing procedure is by no means foolproof. It is possible that H_0 is either really true or really false. Not knowing its truth or falsity, you have to make a decision to either reject or retain H_0 on the basis of sample data. There are four possible outcomes when you make a decision. Two of them are correct decisions, and two of them are erroneous. In the poker chip example, the four possible outcomes are as follows. *First,* suppose H_0 is really true and you have the bag with equal numbers of red and white chips. Imagine you select four red chips in a row. If you set $\alpha = .05$, since the probability of the result by chance ($p = .0625$) is greater than alpha ($p > \alpha$), you decide to retain the true H_0. This decision is correct since H_0 is true. *Second,* suppose H_0 is really false and you have the bag with unequal numbers of chips. Imagine you select five red chips in a row. Since the probability of this result occurring by chance ($p = .03125$) is less than alpha ($p < \alpha$), you decide to reject the false H_0, and conclude that the five red chips come from the bag with equal numbers. This is a correct decision. *Third,* in contrast, suppose

H_0 is really true and you have the bag with equal numbers of chips. Imagine you select five red chips in a row. Since the probability of getting this result by chance (p = .03125) is smaller than alpha ($p < \alpha$), you reject the true H_0 and conclude you have the bag with unequal numbers. This is an erroneous decision. This type of error is called an **alpha error** or Type I error. *Fourth,* suppose H_0 is really false and you have the bag with unequal numbers of chips. Imagine you select four red chips in a row. Since the probability of getting this result by chance (p = .0625) is greater than alpha ($p > \alpha$), you decide to retain the false H_0 and conclude that you have the bag with equal numbers. This is an erroneous decision called a **beta error** or Type II error. These four possible outcomes of the decision-making process are summarized below:

alpha error (Type I error)

beta error (Type II error)

	If H_0 is really true	*If H_0 is really false*
The probability of your result is greater than or equal to alpha, so you decide to **retain H_0**:	Outcome: correct decision	Outcome: beta error (or Type II error)
The probability of your result is less than alpha, so you decide to **reject H_0**:	Outcome: alpha error (or Type I error)	Outcome: correct decision

You can see that an alpha error is an incorrect rejection of H_0, and a beta error is an incorrect retention of H_0. Both alpha and beta errors are bad because you make a mistake. With an alpha error, you actually determine the probability of making such a mistake when you set alpha. For example, suppose H_0 is really true and you have the bag with equal numbers of chips. In most samples, you will get about equal numbers of red and white chips when you draw five chips in a row from this bag. The probability of getting about equal numbers of each is relatively large, and is greater than alpha. But occasionally you can expect to select an unlikely string of the same color chips. You might get three, four, five, or more red chips in a row. In fact, since the probability of selecting five red chips in a row is .03125, you can expect to select five red chips in a row about three times out of every 100 times you draw samples of five chips in a row. Now, given that H_0 is true, anytime an unlikely event occurs that has a probability less than alpha, you will incorrectly reject H_0. Thus, if α = .05, then nearly five times out of 100 samples you will get very unlikely results by chance and wind up erroneously rejecting a true H_0. Consequently the probability of an alpha error *is* the level of alpha you set.

An important conclusion can be drawn from the fact that the alpha level you set is actually the probability of making an alpha error. Namely, you can reduce the probability of an alpha error by decreasing the alpha level. For example, remember that you get fed if you make the correct decision about the bag of poker chips from which you selected the five red chips. Now suppose the tycoon imposes an additional condition. If you make an alpha error and decide that you have the bag with unequal numbers when it really is not, you will have to sleep on a bed of nails that night. This payoff-cost contingency is:

	If H_0 is really true	*If H_0 is really false*
You decide to **retain H_0**:	correct decision (get fed)	beta error (no food, no bed of nails)
The only way you will have to sleep on a bed of nails is if you decide to **reject H_0** (by decreasing alpha you reduce this risk):	alpha error (bed of nails)	correct decision (get fed)

So naturally you wish to avoid rejecting a true H_0 because making an alpha error will hurt you. To reduce the risk of making an alpha error, you can decrease alpha to .01. You now have to get an extremely unlikely result by chance (one that has a probability of occurring which is less than .01) before you erroneously reject a true H_0. You must therefore weigh the payoff-cost conditions of making an alpha error in a given research situation before setting alpha.

There is a side effect, however, of decreasing alpha from .05 to .01. Although decreasing alpha has the advantage of reducing alpha errors, it makes it harder to reject H_0, regardless of its truth or falsity. To reject H_0 when alpha is low, you must observe a result that has a very low probability of occurring assuming H_0 is true. For example, you will have to observe an even larger preponderance of red chips in your sample. But an extremely large preponderance of red chips will still be unlikely to occur even if H_A is true and the bag actually contains, say, 60% red chips and 40% white chips. You will thus have a tendency to retain H_0 more often. On some occasions you will retain H_0 when it is really false and should be rejected. Therefore, decreasing alpha increases the probability of a beta error. Conversely, increasing alpha makes it easier to reject H_0, regardless of its truth or falsity. Although increasing alpha means you will tend to make more alpha errors, you will also make fewer beta errors. You can see that the probabilities of making alpha and beta errors vary *inversely*[2]: increasing alpha decreases the probability of a beta error;

[2]Inversely means in opposite directions. Two things vary inversely if one increases as the other decreases, and vice versa.

decreasing alpha increases the probability of a beta error.

A good question is whether or not you can determine the probability of making a beta error. The answer is no, for the same reason you never assume H_A is true. With an alpha error, H_0 is really true. The reason you can determine the probability of an alpha error is because you set the level of alpha. Assuming H_0 is true, this allows you to precisely specify how likely or unlikely an alpha error is. If H_0 is true and the probability of a result is less than alpha, you will make an alpha error. On the other hand, with a beta error, H_A is really true. In the poker chips example, a beta error occurs when you have the bag with unequal numbers of chips and you conclude it is the bag with equal numbers. You cannot determine the probability of a beta error because you do not know the proportion of red chips in the bag with unequal chips. Without knowing the specific situation under H_A you cannot determine beta.

Since the level of alpha you set may either increase or decrease the probability of making a beta error, you must also consider the payoff-cost conditions of making a beta error in a study before you set alpha. For example, if you make a beta error and conclude you have the bag of unequal numbers when it really is not, suppose your captor makes you stack bricks for nine hours. To reduce the risk of making a beta error such as this, you can increase alpha and set it at .05. This payoff-cost contingency is:

	If H_0 is really true	*If H_0 is really false*
The only way you will have to stack bricks nine hours is if you decide to **retain H_0** (by decreasing alpha you reduce this risk):	correct decision (get fed)	beta error (stack bricks)
You decide to **reject H_0**:	alpha error (no food, no stacking bricks)	correct decision (get fed)

power

Another important concept in hypothesis testing is **power.** Power is the probability of rejecting a false H_0. It is the probability of making the correct decision when H_0 is really false:

	If H_0 is really true	*If H_0 is really false*
You decide to **retain H_0**:	correct decision	beta error (or Type II error)
You decide to **reject H_0**:	alpha error (or Type I error)	correct decision (power)

In contrast, if you make an incorrect decision when H_0 is really false, you have a beta error. Therefore, as the probability of making an incorrect decision increases, the probability of making a correct decision decreases. Thus, increasing beta decreases power, and decreasing beta increases power. Furthermore, since alpha and beta are also inversely related, decreasing alpha increases beta, increasing beta decreases power, and so decreasing alpha decreases power. Similarly, since increasing alpha decreases beta, increasing alpha will increase power. These relationships are shown below for increases and decreases in alpha:

relationships between alpha, beta, and power

	Alpha	*Beta*	*Power*
If you increase alpha:	.05	decreases	increases
If you decrease alpha:	.01	increases	decreases

Thus you can make your hypothesis test more powerful if you increase alpha, but not without some side effects. In setting the level of alpha, you must consider the cost of an alpha error, the cost of a beta error, and the payoff of correctly rejecting H_0 (power).

Now consider the hypothesis-testing procedure in the drug study. The first step is to determine H_0 and H_A. The two competing hypotheses are:

H_0: The drug has no effect. The sample comes from the general population having $\mu = 100$.
H_A: The drug has some effect. The sample comes from a different population having $\mu \neq 100$.

The hypothesis that the sample comes from the general population with $\mu = 100$ is chosen as H_0 not only because it states that the drug has *no effect,* but because it is the only hypothesis of the two that allows you to calculate the probability of getting your results by chance. If the sample comes from a population having a mean that is not 100, then you have no idea what the population mean really is. You need the value of the population mean to calculate the probability of obtaining your sample results.

There is an abbreviated way of expressing H_0 and H_A. If the drug has no effect on IQ, then if you administer the drug to the entire population, it will have no effect on the population mean IQ. Thus if you administer the drug to everyone in the general population and it has no effect on IQ, then you are actually drawing your sample of IQ scores from the general population which has $\mu = 100$. So your H_0 is that your sample comes from the general population of IQ scores with $\mu = 100$. Your H_0 can be symbolically stated:

$$H_0: \quad \mu = 100$$

This reads, "the mean of the population from which your sample comes is 100." On the other hand, if the drug really has some effect on IQ, then the mean IQ of the population of everyone who takes the drug must be different from that of the general population. So you can say that your sample comes from a different population consisting of people who take a drug known to affect IQ on the average. Their mean is then different from 100. Your H_A can be symbolically stated:

$$H_A: \quad \mu \neq 100$$

This reads: "The mean of the population from which your sample comes is not equal to 100."

The second step in the hypothesis-testing process is to set alpha, and the third step is to calculate the probability of getting your results by chance. However, completion of this third step must be delayed until Chapter 7 because it requires some additional knowledge. This is what the *z* test, *t* tests, Mann-Whitney *U* test, Wilcoxon test, chi-square, and ANOVA are all about.

The fourth step in the hypothesis-testing procedure is to make a decision to reject or retain H_0. The value selected for alpha reflects the researcher's desire to avoid alpha and beta errors and to correctly reject a false H_0. A consideration of the payoff-cost conditions in a specific research situation will guide you in selecting the appropriate level of alpha. For example, in the drug study, since there have been no other studies investigating the effects of the drug on IQ, you are doing investigative research and are piloting the way for future research. If the drug really has an effect, then you want to discover it. At least you want to open up the area to further research so that a sufficient number of studies can be done to determine the drug's effect on IQ as well as other potential ramifications. Therefore you want maximum power. If you decide the drug has an effect, you can go ahead and publish your results. Publication will stimulate other researchers to try to replicate your results. If you have made an alpha error it will be discovered when other researchers ultimately fail to replicate your findings. However, if you decide that the drug does not have an effect, then you probably will not be able to publish your results because journals normally do not publish results showing no effects. Consequently, no other researchers will be moved to replicate your results and the matter will probably be dropped. Thus a beta error will likely lead to termination of research on the topic. Since you do not want to overlook a potentially important discovery, you do not want to risk a beta error by saying there is no effect if there actually is. In a situation like this you should set alpha high, to .05, to decrease beta and increase power. Here, making a beta error is worse than making an alpha error.

There are research situations in which making an alpha error is worse than making a beta error. For example, in a situation in which there has already been considerable exploratory research, you may want to confirm that some alleged effect really exists. If the area is controversial, or if you believe the health of others or your professional reputation is at stake, then you want to be very confident that some effect really exists before you publicly claim it does. Under these circumstances, set alpha low, to .01 or less. This will minimize the risk of an alpha error, but at the expense of increasing beta and decreasing power.

KEY CONCEPTS

1 Hypothesis testing is a process used to decide whether or not a sample result is due to sampling error. A simple application is to decide whether a sample with a certain mean occurs by chance from a specified population. You begin by assuming your sample result did occur by chance from the general population, and then you calculate the likelihood of getting the sample mean you obtained. If this turns out to be so unlikely as to contradict the initial assumption, you can conclude that your initial assumption is wrong and that the sample comes from a population with a different mean than that of the general population.

2 The steps of the formal hypothesis-testing process are:

a. Determine H_0 and H_A. H_0 generally implies your results are due to sampling error.

b. Set a value for alpha. Usually this is .05 or .01.

c. Assuming H_0 is true, calculate the probability of getting your results by chance.

d. Either reject or retain H_0. If the probability of your results is less than alpha, reject H_0 and conclude your results are not due to sampling error. If the probability of your results is greater than or equal to alpha, you have not gathered sufficient contradictory evidence against H_0, so you should retain H_0.

3 Probability is a value between 0 and 1 that gives a measure of likelihood. The probability of a particular event is defined as the number of outcomes in which the particular event occurs divided by the total number of possible outcomes.

4 Independent events are events that do not influence the probability of occurrence of one another.

5 The multiplicative rule is used to calculate the probability of two or more independent events occurring together. The probability of two or more independent events occurring together is the product of their individual probabilities.

6 An alpha error occurs when you erroneously reject a true H_0. The probability of making an alpha error is the level of alpha you set.

7 A beta error occurs when you erroneously retain a false H_0. Beta is the probability of making a beta error. The exact value of beta usually cannot be determined.

8 Power is the probability of correctly rejecting a false H_0. In other words, power is the probability of detecting when a real effect exists.

9 As alpha increases (.05), beta decreases and power increases. As alpha decreases (.01), beta increases and power decreases.

EXERCISES

1 Given an unbiased die, answer the following:

a. What is the probability of tossing a one?
b. What is the probability of tossing three ones in a row?
c. What is the probability of tossing a one on the first roll, a two on the second roll, and a three on the third roll?

2 Given a regular deck of 52 cards, answer the following:

a. What is the probability of drawing a heart?
b. If you draw two cards, replacing the first card and shuffling before drawing the second card, what is the probability of drawing two hearts in a row?
c. If you draw two cards, replacing the first card and shuffling before drawing the second card, what is the probability of first drawing a heart and then drawing a spade?
d. If you draw two cards, replacing the first card and shuffling before drawing the second card, what is the probability of first drawing a heart and then drawing an ace?

For each of the hypothetical research situations described in Exercises 3 to 6, determine the appropriate null and alternate hypothesis.

3 You believe that stress affects the ability to recognize previously learned information, as in a multiple-choice test. You know that the mean of the population of sociology majors taking a standard multiple-choice test just before graduation is 104 correct (out of 135 items). You place 36 graduating seniors in sociology in a situation designed to induce stress and then test their recognition memory with the standard multiple-choice test. These seniors had not taken the standard test previously.

4 You want to test the hypothesis that passively observing an instructional film in shooting baskets from the free throw line influences free throw shooting in sixth-grade girls. You know that the population of sixth-grade girls makes a mean of 1.6 shots out of 10 their first time shooting from the free throw line. You show the instructional film to a group of 25 sixth-grade girls and then allow them to shoot 10 shots for the first time from the free throw line.

5 You are trying to determine whether hot black coffee can sober up people more effectively than no coffee. You know that an adult's liver will metabolize (break down) ethyl alcohol at the mean rate of 1 ounce per hour, or $\mu = 1$. You give a

sample of 40 adults an alcoholic beverage containing 1 ounce of ethyl alcohol, immediately followed by three 8-ounce cups of black coffee. You then examine the time (in minutes) it takes the liver to completely metabolize the alcohol.

6 You have a hunch that infants born in a darkened, quiet, and very warm (85°) labor room will evidence different developmental behavior patterns and strength from the population of infants born into the typical bright, noisy, and 72° labor room. Knowing that the population mean is a score of 20 on a particular infant development rating scale given at 7 days of age, you rate the seventh-day development of ten infants born under your prescribed conditions.

7 Test the hypothesis that a deck of 52 cards is fair (regular), given that four clubs were drawn in a row (with replacement and shuffling). That is, test the hypothesis that the probability of drawing a club is .25. Show the four steps of the hypothesis-testing procedure. Set alpha equal to .05.

8 A friend of yours is trying to determine whether or not a coin is fair. He tells you that his null hypothesis is "the coin is biased." What has your friend done wrong? Which step in the hypothesis-testing process would your friend be unable to perform?

9 You are interested in studying ESP. To do this, you use a special deck of 25 cards with five different symbols: five cards have a star; five cards have a circle; five cards have a triangle; five cards have a square; and five cards have a heart. A friend who claims to have ESP serves as your subject and correctly guesses the first three cards (you replace each card and shuffle before guessing another). You stop the test at this point. Do you have enough evidence to conclude that your friend is out of the ordinary (that is, he has ESP)? Use a stringent value for alpha and follow the hypothesis-testing procedure to make your decision.

10 After completing a study concerning suicide prevention showing no reduction in suicides after implementing a community telephone crisis intervention service, you conclude, "I accept the hypothesis that no benefit of the hot-line service exists in the area of suicide prevention." Criticize this conclusion by describing why you "retain" the null hypothesis rather than "accept" it.

11 Differentiate between "alpha" and an "alpha error," and define beta and power.

7

ONE-SAMPLE HYPOTHESIS TESTING

CHAPTER GOALS

After studying this chapter, you should be able to:

1. Define the sampling distribution of the mean.
2. State the central limit theorem and apply it to find the mean and standard deviation of a sampling distribution.
3. Describe the difference between a one-tailed test and a two-tailed test, and formulate the appropriate H_0 and H_A.
4. Perform a z test.
5. Estimate σ from a sample using both the definitional and computational formulas for s.
6. Perform a one-sample t test using both the definitional and computational formulas for t.
7. Describe when to perform a one-sample t test as opposed to a z test.
8. Describe the effect of varying the sample size on the sampling distribution of the mean.
9. Define critical value, region of rejection, and significant.
10. State the assumptions for the one-sample t test.

CHAPTER IN A NUTSHELL

This chapter addresses the problem of whether or not it is reasonable to conclude that a sample with a mean of $\bar{X}$ occurred by chance from a population having a mean of μ. The general hypothesis-testing procedure is employed. First, you assume that the sample comes from the given population with a mean of μ. Then set alpha and calculate the probability of randomly getting a sample mean as extreme as the one obtained. If this probability is less than alpha, reject H_0 and conclude that the sample mean did not come from the given population. You can calculate the probability of getting a sample mean by chance with the aid of the central limit theorem. The central limit theorem states that if you make a distribution of a very large number of sufficiently large random sample means from a population, then the distribution of sample means will be approximately normal, the distribution will have a mean equal to the population mean, and the distribution will have a standard deviation equal to $\sigma/\sqrt{N}$. The distribution of sample means is called the sampling distribution of the mean, or simply the sampling distribution. A sample mean can be transformed to a z score in the sampling distribution. The probability of getting the sample mean by chance can be calculated from the z score. A hypothesis test conducted in this way is called a z test. If the population standard deviation, σ, is not known, you cannot conduct a z test. You can, however, estimate σ from your sample and use a one-sample t test to determine the probability of getting the sample mean by chance. Since estimating σ introduces additional variability, the one-sample t test compensates by using a more stringent decision criterion.

Introduction to one-sample hypothesis testing

Let us return to the original drug study discussed in Chapter 6. In your sample of 16 subjects administered the new drug, you find $\bar{X} = 107.75$, whereas for the general population $\mu = 100$. The question is whether this difference is due to the drug or to sampling error. H_0 is that $\mu = 100$. H_A is that the drug has some effect, or that $\mu \neq 100$. The third step of the hypothesis-testing process is to calculate the probability of obtaining your result if H_0 is true. This is the step that could not be completed in Chapter 6. The present chapter will provide you with the knowledge necessary to complete Step 3 of the hypothesis-testing process in many research situations.

You need to calculate the probability of randomly getting a sample mean as extreme as $\bar{X} = 107.75$ from a population with $\mu = 100$. How can you determine the probability of getting this sample mean by chance? You need to know what sort of sample means you would get if you randomly selected many, many samples of 16 subjects each from a population having $\mu = 100$. In other words, given that H_0 is true and the drug has no effect, how often would you expect to find a sample mean as extreme as 107.75 in a sample of 16 subjects? If the probability of this event is less than alpha, you may reject H_0 and conclude that the drug has an effect. If it turns out that a mean as extreme as $\bar{X} = 107.75$ is greater than alpha, then you may retain H_0 and conclude that the drug has no effect.

One way to calculate the probability of observing by chance a sample mean as extreme as 107.75 is to draw many samples of 16 subjects each from the general population, calculate the mean IQ of each sample, construct a frequency distribution of how often each sample mean occurs, and then depict this with a frequency polygon. The resulting frequency distribution is called the **sampling distribution of the mean** or *sampling distribution* for short. It is important to distinguish between a *sampling distribution,* a **sample distribution,** and a **population distribution.** A sampling distribution is a frequency distribution of the means of many equal-sized samples randomly drawn from a population having a given mean. A sample distribution is a frequency distribution of the raw scores in one sample randomly drawn from a population. A population distribution is a frequency distribution of the raw scores of everyone in the population. These three kinds of distributions are depicted with frequency polygons in Figure 7-1. Notice that raw scores are graphed in both sample and population distributions, whereas means are graphed in a sampling distribution.

sampling distribution of the mean

sample distribution
population distribution

The concept of a sampling distribution plays a crucial role in the hypothesis-testing process because it enables you to calculate the probability of getting your results by chance when you assume H_0 is true. There are two methods for constructing a sampling distribution. One method is to actually go out and randomly select many equal-sized samples from the population, calculate each sample mean and tally the frequency with which each $\bar{X}$ occurs. If you really go out and calculate sample means,

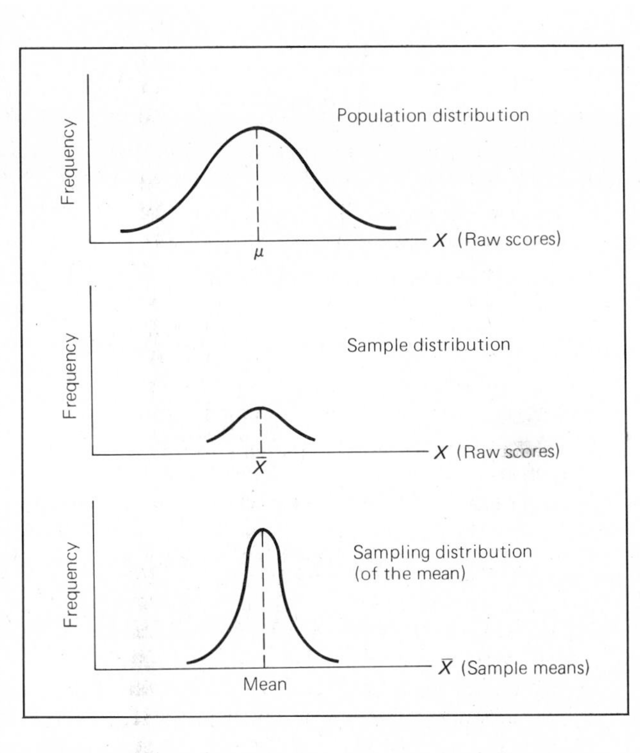

FIGURE 7-1
Frequency polygons depicting the population, sample, and sampling distributions.

empirical sampling distribution

the resulting sampling distribution is called an **empirical sampling distribution.** You could look at the empirical sampling distribution to determine the probability of selecting a sample having a mean as extreme as $\bar{X} = 107.75$. However, constructing an empirical sampling distribution is a very tedious and time-consuming task. Fortunately, the second method of constructing a sampling distribution is much easier. It is possible to mathematically infer several characteristics of a typical empirical sampling distribution. If you make a few assumptions, you can infer the shape of the typical empirical sampling distribution, its mean, and its standard deviation. In taking advantage of this second method you are inferring the characteristics of a **theoretical sampling distribution.** It is not an empirical sampling distribution because you do not go out and collect the data. However, such a theoretical sampling distribution nicely approximates the typical empirical sampling distribution. This theoretical sampling distribution is analogous to the normal distribution, another entity that is theoretical because it never occurs exactly in nature but also is a very useful approximation to many empirical distributions.

theoretical sampling distribution

The central limit theorem

The central limit theorem is mathematically derived and enables you to infer several characteristics of the theoretical sampling distribution. Since we are going to use the theoretical sampling distribution to approximate the typical empirical sampling distribution, the central limit theorem thus also approximates the characteristics of the typical empirical sampling distribution. Because the characteristics of the theoretical and typical empirical sampling distributions will be approximately the same, it is not necessary to distinguish between them for the purpose of calculating the probability of getting sample results by chance in Step 3 of the hypothesis-testing procedure. Therefore we shall follow convention and simply refer to the "sampling distribution" without qualifying whether it is empirical or theoretical. Simply stated, the **central limit theorem** is:

central limit theorem

With a sufficiently large sample size ($N \geq 25$), the sampling distribution of the mean is an approximately normal distribution with a mean of μ and a standard deviation (symbolized $\sigma_{\bar{x}}$) equal to $\sigma/\sqrt{N}$.

In this statement, μ is the mean of the population from which the samples are assumed to be drawn, σ is the standard deviation of this population, and N is the sample size. The standard deviation of the sampling distribution, $\sigma_{\bar{x}}$, is also called the **standard error of the mean.** Figure 7-2 illustrates the characteristics of the sampling distribution as inferred by the central limit theorem.

standard error of the mean
$\sigma_{\bar{x}}$

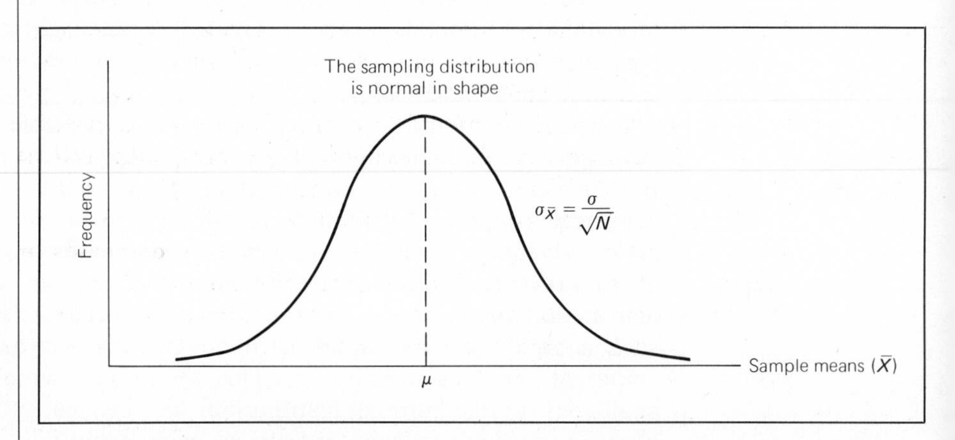

FIGURE 7-2
Characteristics of the sampling distribution (of the mean) as inferred from the central limit theorem.

There are three important things to note about the central limit theorem. The first is that it states that the sampling distribution is approximately normal. It is not so surprising that the sampling distribution is normal if the population distribution is normal. What is surprising, however, is that the central limit theorem assures that if you have a sufficiently large sample size ($N \geq 25$), the sampling distribution will be normal *regardless of the shape of the population distribution.*[1] For example, reaction time (RT) forms a very positively skewed distribution. If you measure the RTs of many samples of 30 subjects each, calculate the mean RT of each sample, and depict the frequency of the means with a frequency polygon, the shape of the polygon will be approximately normally distributed. This makes sense when you consider the nature of a skewed population distribution. The scores occurring in the elongation are rare, and therefore have a low probability of occurring. When you randomly select a sample from a skewed population, you expect to get very few of these rare scores in the sample. With a sufficiently large sample size, the rare scores have little effect on the sample mean because they are averaged in with and counterbalanced by the other more likely scores. For rare scores to have a sizable effect on the sample mean, several would have to occur in the same sample. This is extremely unlikely. Consequently, the weight that an occasional rare score has on the sample mean is washed out by the rest of the scores in the sample, and the sampling distribution approximates a normal distribution.

The second thing the central limit theorem says is that the mean of the sampling distribution is equal to the mean of the population, μ. This is clearer if you consider that the mean of the sampling distribution is calculated by summing all the means of samples drawn randomly from the population and dividing by the number of means. Since each sample mean is itself the sum of many raw scores (divided by the number of scores in the sample), when you sum all the sample means you are summing the raw scores as well. Thus you are adding up all the raw scores in the population but in a different way. Consequently, the mean of the sampling distribution is actually the same as the mean of the population, but it is calculated differently.[2]

The third thing to note about the central limit theorem is that it states that the standard deviation of the sampling distribution, $\sigma_{\bar{x}}$, is equal to $\sigma/\sqrt{N}$. This will not be mathematically proved here, but a little discussion will help give you an intuitive grasp for why $\sigma_{\bar{x}} = \sigma/\sqrt{N}$. Notice that as N increases, $\sigma_{\bar{x}}$ decreases—and as N decreases, $\sigma_{\bar{x}}$ increases. In other words, the smaller the size of the sample you select, the more variability you expect among sample means. This is because the fewer the raw scores in a sample, the more weight each raw score exerts on the sample mean.

[1] $N \geq 25$ is a convenient rule of thumb. If the population distribution is approximately normal, the sampling distribution will be approximately normal even with small sample sizes. As the population distribution departs from a normal distribution, a larger sample size is required to obtain an approximately normal sampling distribution. Even with extremely distorted population distributions, however, the sampling distribution will be approximately normal if $N \geq 25$.

[2] This clarification is intended to provide an informal idea about why the mean of the sampling distribution is equal to the population mean. A formal mathematical proof involves calculus.

As an extreme example, consider a sample size of one. The mean of such a sample is equal to the value of the one raw score. If you take many samples of one subject each, the sampling distribution will be the same as the population distribution. The central limit theorem accurately asserts that the mean of this sampling distribution is μ, and the standard deviation is $\sigma/\sqrt{N} = \sigma/\sqrt{1}$, or simply σ. Now consider a larger sample. Each raw score has less effect on the sample mean because the rest of the scores in the sample tend to wash out any extreme scores. Therefore the sample means will be less variable, tending to cluster more tightly around the mean of the sampling distribution. Another way to think of this is that the means of smaller samples are more subject to distortion by extreme scores than the means of larger samples, and hence there is more sampling error with smaller samples. You expect the mean of a larger sample to be close to the population mean because there is a better representation of the population in the sample and hence less sampling error. This conclusion is stated in the expression $\sigma_{\bar{x}} = \sigma/\sqrt{N}$. As N increases, the value of the denominator increases and forces $\sigma_{\bar{x}}$ to decrease. In other words, as N increases, the sample means will be less variable, and hence the sampling distribution will be tighter. See the example in Figure 7-3.

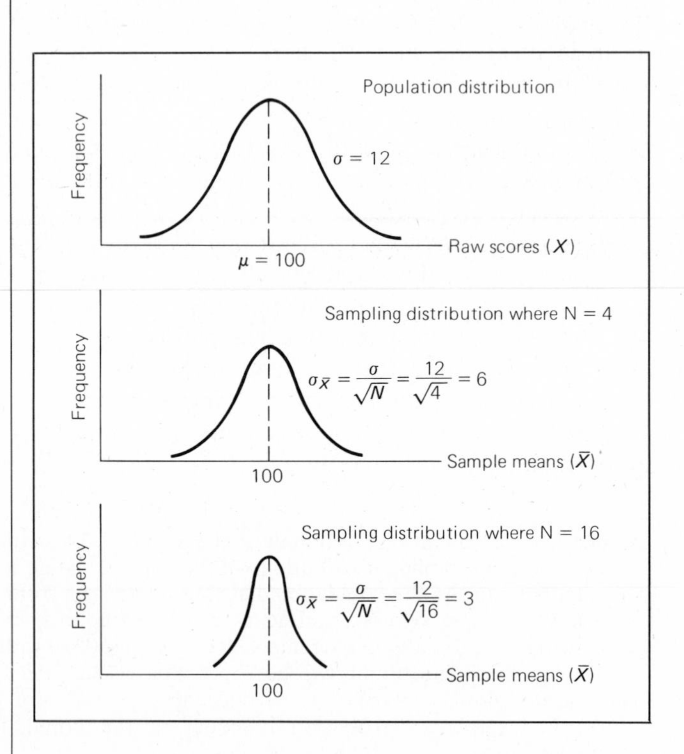

FIGURE 7-3

As the sample size increases, the standard error decreases.

Determining the probability of the results with the central limit theorem and the *z* test

Equipped with the central limit theorem, you are now in a position to determine the probability of randomly selecting from the general population a sample with a mean as extreme as $\bar{X} = 107.75$. If you assume H_0 is true and you select many samples of size 16 from a population having $\mu = 100$, the central limit theorem tells you the sampling distribution is normal and has a mean of 100 and a standard deviation of $\sigma/\sqrt{N}$. Since the standard deviation of the population of IQ scores is $\sigma = 15$ and the sample size is $N = 16$, the standard deviation of the sampling distribution is $\sigma_{\bar{x}} = 3.75$:

$$\sigma_{\bar{x}} = \frac{\sigma}{\sqrt{N}} = \frac{15}{\sqrt{16}} = \frac{15}{4} = 3.75$$

Even though N is less than 25, the sampling distribution is normal because the population distribution is normal. You have to be concerned about N being greater than or equal to 25 only when you know or believe the population distribution is skewed. Since the sampling distribution is normal in this example, you can use Table *z* to calculate the percentage of scores occurring between any two scores in the sampling distribution. Remember that the scores in the sampling distribution are really means. Since you already know how to calculate the percentage of scores in a normal distribution, you can calculate the probability of getting a sample mean between any two values.

Now let us determine the probability of selecting a sample whose mean is 7.75 IQ units or more away from the mean of the sampling distribution. First, remember to draw a diagram of what you want to find. Figure 7-4 shows that you need to determine the probability that a sample mean lies in the shaded area of the sampling distribution. Notice that in addition to the shaded area in the upper tail there is a shaded area on the lower tail. Sample means of 92.25 and below are included in the

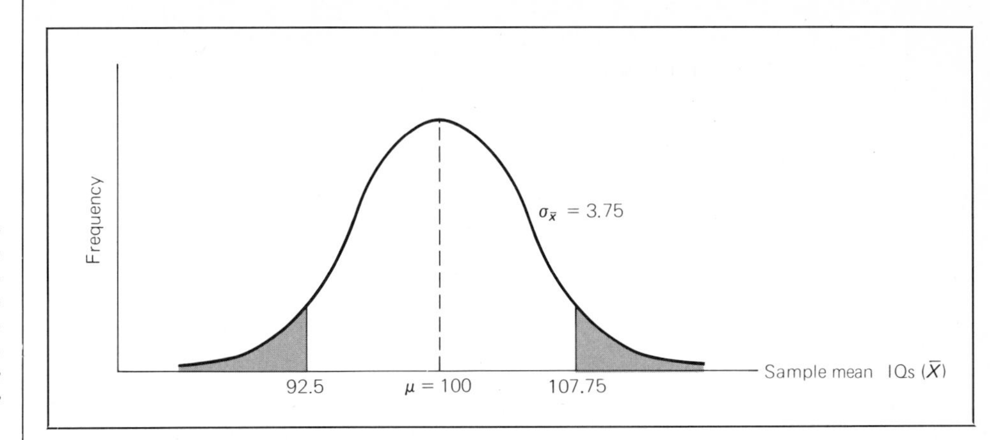

FIGURE 7-4

The sampling distribution for the IQ drug study. The sample means that are as extreme as the observed sample mean of 107.75 lie in the shaded areas under the curve.

shaded area because of the way the problem is stated above: "Determine the probability of selecting a sample whose mean is 7.75 units away from the mean of the sampling distribution." Sample means of 92.25 and below are also 7.75 units or more away from the mean of the sampling distribution. You can calculate the probability by using the techniques you learned for calculating percentage of scores in a normal distribution. First transform to z scores the sample means that cut off the shaded region, and then refer to Table z. Since you are dealing with a sampling distribution of the mean and not a distribution of raw scores as in Chapter 4, remember that the standard deviation of the sampling distribution is equal to $\sigma/\sqrt{N}$. Therefore, the formula to transform sample means to z scores is slightly different from Formula 4.3 for transforming raw scores into z scores. The formula to use with a sample mean is:

FORMULA 7.1

z test

$$z = \frac{\bar{X} - \mu}{\sigma_{\bar{x}}} = \frac{\bar{X} - \mu}{\dfrac{\sigma}{\sqrt{N}}}$$

Formula 7.1 tells you how many standard deviation units your sample mean lies away from the mean of the sampling distribution.

Substituting known values into Formula 7.1 tells you that the z score corresponding to your $\bar{X} = 107.75$ is 2.07. Thus your $\bar{X}$ is 2.07 standard deviation units away from the mean of the sampling distribution:

$$z = \frac{\bar{X} - \mu}{\dfrac{\sigma}{\sqrt{N}}} = \frac{107.75 - 100}{\dfrac{15}{\sqrt{16}}} = \frac{7.75}{3.75} = 2.07$$

Finding $z = 2.07$ in Table z reveals that 48.08% of the sample means lie between the mean of the sampling distribution and $\bar{X} = 107.75$. Since you are interested in the percentage of sample means above 107.75, subtract 48.08% from 50%, leaving 1.92% of the sample means in the shaded region of the upper tail. By symmetry, 1.92% of the sample means fall below 92.25 in the lower tail. Therefore, the total percentage of sample means that are as extreme as 107.75 is 1.92% + 1.92% = 3.84%. Converting 3.84% to a probability gives $p = 3.84/100 = .0384$.

Now you can finish the hypothesis-testing process for the drug experiment. Because this is exploratory research you set $\alpha = .05$. Since the probability of getting your results is less than alpha ($.0384 < .05$), reject H_0 and conclude that the drug has some effect on IQ. Since Formula 7.1 for calculating z is applied in Step 3 of the hypothesis-testing procedure

z test

to determine the probability of getting your results, this test is called a ***z* test.**

Using critical values to facilitate hypothesis testing

There is a shortcut method for conducting a z test. This procedure rests on an insight that can be gained from Figure 7-5. When the probability of getting a particular sample mean is less than alpha, you reject H_0. The shortcut stems from the fact that you can readily identify which sample means have a probability of occurring that is less than alpha. Any sample mean lying in the upper or lower tail-like elongations of the sampling distribution has a probability of less than alpha. Suppose that you shade the two tails so that only 5% of the sample means lie in the shaded regions of both tails combined: 2.5% lie in the lower tail and 2.5% lie in the upper tail. If you have a sample mean that falls into either shaded region, you immediately know the probability of getting a sample mean as extreme as yours is less than .05. You therefore reject H_0 whenever you get a sample mean that falls into the shaded regions. For this reason, these shaded regions are called the **regions of rejection.**

regions of rejection

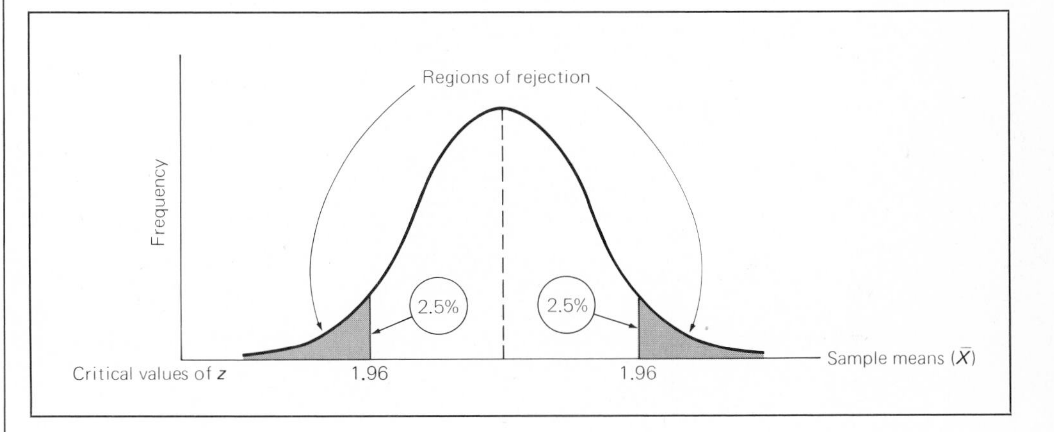

FIGURE 7-5
The shaded areas under the curve are the regions of rejection in the sampling distribution. The critical values of z are the absolute values[3] that cut off the regions of rejection.

After setting alpha, you can find the z scores that cut off the regions of rejection. If $\alpha = .05$, these z values cut off 2.5% of the sample means in each tail of the sampling distribution, so the total region of rejection is 5%. Looking up 47.5% (found by subtracting 2.5% from 50%) in Table z gives $z = 1.96$. The number 1.96 is called a **critical value of *z*,** and is symbolized z^*. This is the z score which cuts off 5% of the sample means in both tails combined. The critical value of 1.96 has the distinction that any sample mean with a calculated z score greater in absolute value[3]

critical value

z^*

[3]Absolute value refers to the magnitude of a number, regardless of sign. For example, the absolute value of -2 is 2, and the absolute value of $+3$ is 3.

than 1.96 has a probability of occurrence less than .05. Anytime the absolute value of your calculated z score is greater than 1.96, you know that the sample mean falls into the region of rejection and you reject H_0. Whenever a sample mean falls into the region of rejection, it is said to be **significant.** The word *significant* is merely an abbreviated way of saying that the probability of the sample mean occurring by chance is less than alpha. The alpha level you set is called the **level of significance.**

significant

level of significance

There is a modified version of the hypothesis-testing procedure which incorporates the shortened method for conducting a z test:

1. Determine the appropriate H_0 and H_A. (H_0 generally states that the sample mean occurred by chance from a population with a specified mean.) Assume H_0 is true.
2. Set alpha. Determine a critical value of z such that the proportion of sample means lying in the combined tails of the sampling distribution equals alpha.
3. Transform your sample mean to a z score in the sampling distribution specified in H_0. The formula is:

$$z = \frac{\bar{X} - \mu}{\frac{\sigma}{\sqrt{N}}}$$

Refer to the z score obtained from this formula as the *calculated* z score. Its absolute value is $|z|$.
4. If $|z| > z^*$, reject H_0 and conclude that the mean of the population from which your sample was drawn is not the value specified in H_0. This says that the treatment has an effect.

If $|z| \leq z^*$, retain H_0 and conclude that there is not sufficient evidence to say that the population from which your sample was drawn is different from that specified in H_0. This says that there is no reason to believe the treatment has an effect.

For an example of using this modified hypothesis-testing process, consider the original drug experiment in which you want to decide whether the drug affects IQ. This time, just for illustration, suppose you set $\alpha = .01$. The steps are:

1. H_0: $\mu = 100$ (The drug has no effect on IQ)
   H_A: $\mu \neq 100$ (The drug has some effect on IQ)
2. $\alpha = .01$. Determine the critical value of z so that 1% of the sample means lie in both tails combined. Thus 0.5% must be in each tail. Look up 49.5% (found by subtracting .5% from 50%) in Table z to get $z^* = 2.58$.
3. The calculated z score for your sample mean is:

$$z = \frac{\bar{X} - \mu}{\frac{\sigma}{\sqrt{N}}} = \frac{107.75 - 100}{\frac{15}{\sqrt{16}}} = 2.07$$

4. Since $|2.07| < 2.58$, retain H_0 and conclude that there is not sufficient evidence to say that the drug affects IQ.

One-tailed tests versus two-tailed tests

In the preceding examples you were trying to determine whether a drug has *some* effect, regardless of direction, on IQ. You will decide it has an effect if it either increases the mean IQ of the sample to a very high level or if it decreases the mean IQ to a very low level. Either result is unlikely enough to contradict the H_0 that the sample mean occurs by chance from a population with a $\mu = 100$. Since you are trying to determine whether there is *some* effect, you are looking for *either* very high or very low sample means to reject H_0. Therefore the region of rejection is split to include both the upper and lower tails of the sampling distribution. You may reject H_0 if the sample mean lies in either of the two tails. This type of test is called a **two-tailed test.** A two-tailed test is appropriate whenever you want to determine whether a treatment has *some* effect—that is, either to significantly raise or lower the sample mean. It does not matter which direction (up or down) the sample mean changes.

two-tailed test

There are certain research situations in which you want to determine whether a treatment has a specific effect in only one direction. This might be the case if you know beforehand that the treatment would probably change the sample mean in only one direction, or if you are interested in the result only if you get a change in one specified direction. For example,

Many previous experiments have been run on the same drug, and all showed an increase in IQ. You are trying to replicate these earlier results.

You are actively searching for a drug that increases IQ and have no interest in drugs that decrease it.

In these examples you are interested in showing that the drug has the effect of increasing IQ. On the other hand, consider the following example:

You are actively searching for a drug that decreases IQ and have no interest in drugs that increase it.

In this example you are interested in the result only if the drug decreases IQ.

In research situations in which you are looking for an effect in one specified direction, you should use a **one-tailed test.** A one-tailed test places the region of rejection in only one tail of the sampling distribution. For example, if you are searching for a drug that increases IQ, you are not interested in any sample that has a lower mean IQ than the general population mean, since this is not good evidence that the drug increases IQ. You are only interested in samples with high means. When you reject H_0, you want to be able to conclude that the drug *increases* IQ. Consequently, H_A must state that the drug increases IQ, and H_0 must state, in contrast, that the drug does not increase IQ.

one-tailed test

Whenever you want to conclude that the treatment *increases* scores, or that the treatment *decreases* scores, H_0 and H_A must be stated in specific directional terms. If H_A states that the treatment increases (decreases) the population mean, H_0 must therefore be that the treatment does not increase (does not decrease) the mean. When H_0 states that a treatment will not increase scores, this is the same as saying that the treatment either leaves the scores the same or decreases them. Thus there are two specific "null" statements made in H_0: either the population mean is equal to some specified value—for example, $\mu = 100$—or the population mean is less than this specified value—for example, $\mu < 100$. To take into consideration both null statements, H_0 may be written

$$H_0: \quad \mu \leq 100$$

You might ask, however, how to calculate the probability of getting a sample mean as extreme as yours when H_0 consists of two different statements such as the above. To answer, consider Figure 7-6. The H_0

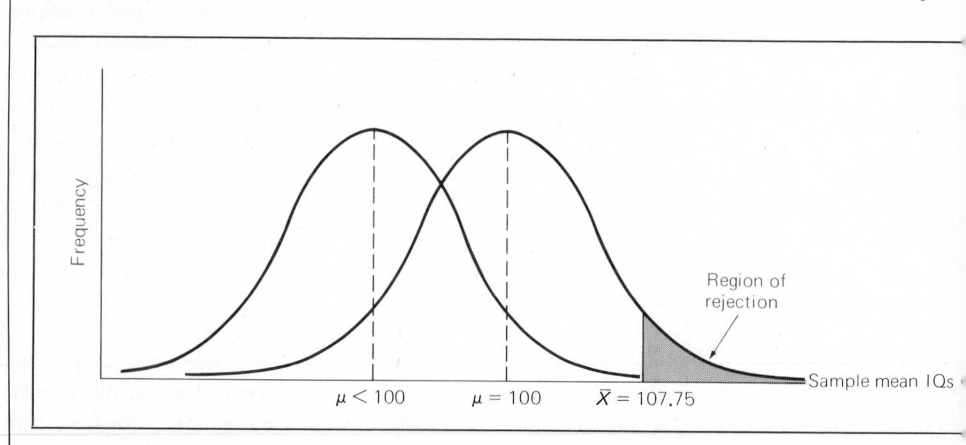

FIGURE 7-6

There are two null statements in a one-tailed test. In the example in which one null statement is $\mu < 100$ and the other is $\mu = 100$, the probability of a sample mean as extreme as 107.75 occurring from a sampling distribution with $\mu < 100$ is less than the probability of it occurring from a sampling distribution with $\mu = 100$.

of $\mu \leq 100$ consists of two null statements, $\mu = 100$ and $\mu < 100$. The sampling distribution on the right is for the null statement $\mu = 100$. You can transform your sample mean to a calculated z score using Formula 7.1, and if it falls into the region of rejection, you can reject this null statement. For the null statement $\mu < 100$, the population mean is not exactly specified. You do not know exactly where the sampling distribution referred to in this statement lies on the IQ scale. However, since you know its mean is less than 100, it must lie to the left of the sampling distribution with the mean of 100. Fortunately, when you can reject the null statement $\mu = 100$, you have sufficient information to reject the null statement $\mu < 100$. Notice in Figure 7-6, no matter how far away the left sampling distribution is on the IQ scale, whenever you are able to reject the null statement $\mu = 100$, you will also be able to reject the null statement $\mu < 100$. A sample mean that is high enough to be unlikely to occur in the right-hand distribution is even more unlikely to occur in the left-hand distribution. So with an H_0 of $\mu \leq 100$, it is sufficient to calculate the probability of your results using $\mu = 100$. If you can reject this null statement, you can reject the H_0 of $\mu \leq 100$.

To illustrate the formulation of a one-tailed test, suppose you are only interested in whether or not a drug *increases* IQ. H_A should state that the drug increases IQ, and H_0 should state that the drug does not increase IQ. Before symbolically stating H_0 and H_A, it is helpful to consider the conclusion you want to make. The only time you may conclude that a treatment has an effect is when you can reject H_0 and conclude that H_A is true. Therefore, you should symbolize H_A so that it specifically states the conclusion you want to reach. H_0 will then simply be the logical complement to H_A.

Keep in mind that regardless of whether you are conducting a one- or two-tailed test, H_0 *must contain a statement of equality*. For a two-tailed test, H_0 simply states that the population mean will be *equal to* an exactly specified value. For one type of one-tailed test, H_0 states that the population mean is less than or *equal to* an exactly specified value. For the other type of one-tailed test, H_0 states that the population mean is greater than or *equal to* an exactly specified value.

proving the null hypothesis

To digress briefly, H_A generally states that your treatment has the effect you are looking for, and H_0 states your treatment does not have such an effect. The above three general forms of H_0 suggest that you should never attempt to show that a treatment has *no* effect. Such an attempt is called trying to **prove the null hypothesis** and represents a fundamental error in logic. You can demonstrate that a statement about the real world is false by producing just one result that contradicts it, but you cannot demonstrate such a statement is true because this would require an examination of every situation in which the statement applies. For example, you can prove that the statement "All crows are black" is false just by finding one crow that is not black. But you would have to examine all crows, past, present, and future, and show that all of them are black in order to prove the statement true. Practically speaking, you cannot examine all present situations, much less future ones. Thus, although you can demonstrate that H_0 is false, you cannot demonstrate that it is true. This is another reason why you must *retain* H_0 rather than accept it.

The hypothesis-testing procedure is the same for one-tailed and two-tailed tests except in the statement of H_0 and H_A in Step 1, and in the placement of the region of rejection in Step 2. To illustrate the procedure for a one-tailed test, assume that you want to show that the drug increases IQ.

1. H_0: $\mu \leq 100$ (The drug does not increase IQ)
H_A: $\mu > 100$ (The drug increases IQ)

2. Let $\alpha = .05$. Since you are looking for a very high sample mean to reject H_0 and conclude that the drug increases IQ, you want to place all the region of rejection in the upper tail of the sampling distribution. Therefore, the critical value of z will be such that 5% of the sample means lie above it in the upper tail. Look up 45% (found by subtracting 5% from 50%) in Table z to get $z^* = 1.65$. Draw a diagram such as that in Figure 7-7 to verify that the sample mean lies in the direction specified by H_A.

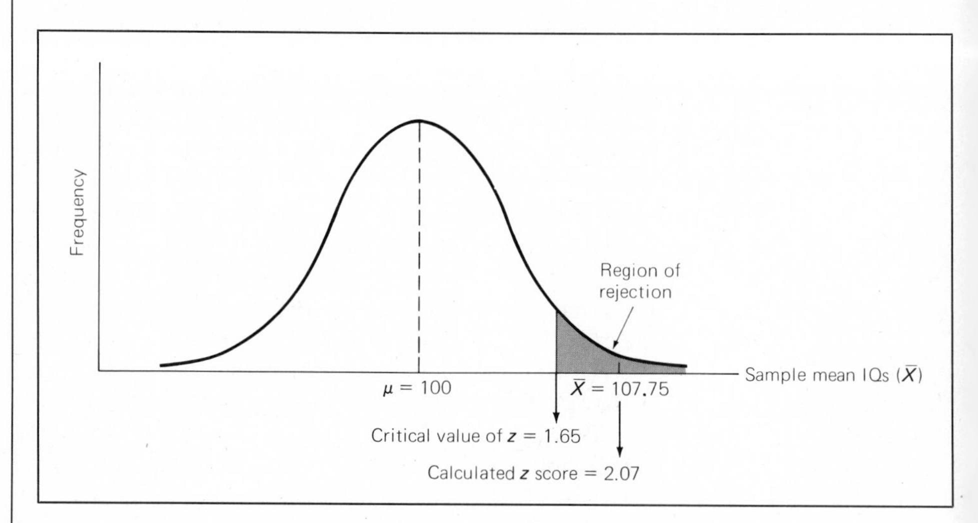

FIGURE 7-7

The shaded area under the curve is the region of rejection for a one-tailed test with $\alpha = .05$. Since the calculated z score falls into the region of rejection, reject H_0 and conclude that the drug increases IQ.

3. The calculated z score for your sample mean is:

$$z = \frac{\bar{X} - \mu}{\frac{\sigma}{\sqrt{N}}} = \frac{107.75 - 100}{\frac{15}{\sqrt{16}}} = 2.07$$

4. Since the sample mean lies in the direction specified by H_A, and $|2.07| > 1.65$, reject H_0 and conclude that the drug increases IQ.

For a one-tailed test, not only does $|z|$ have to be greater than z^* (i.e., $|z| > z^*$), but the sample mean must also lie in the appropriate direction to fall into the region of rejection. With a two-tailed test, $|z| > z^*$ is a sufficient condition to reject H_0. If $|z| > z^*$, then the sample mean will fall into the region of rejection. With a one-tailed test, however, $|z| > z^*$ is not a sufficient condition. The sample mean must also be in the appropriate direction before H_0 can be rejected. For example, to be able to conclude that a drug increases IQ, the sample must have a *higher* mean than that of the general population. You must be careful in carrying out Step 4 with a one-tailed test because you can only reject H_0 when the sample mean is in the one specified direction. Suppose you want to conclude that your treatment increases IQ. If you set $\alpha = .05$, then $z^* = 1.65$ for a one-tailed test. You select 25 people, administer the drug, and find a sample mean of 91. The calculated z score is -3.0:

$$z = \frac{\bar{X} - \mu}{\frac{\sigma}{\sqrt{N}}} = \frac{91 - 100}{\frac{15}{\sqrt{25}}} = -3.0$$

A careless individual might see that $|-3| > 1.65$ and reject H_0. This is a mistake because the sample mean did not increase, but rather it decreased, which is consistent with H_0. The way to avoid making such a mistake is to draw a diagram of the sampling distribution and mark the region of rejection, as in Figure 7-8. You can draw this diagram after completing

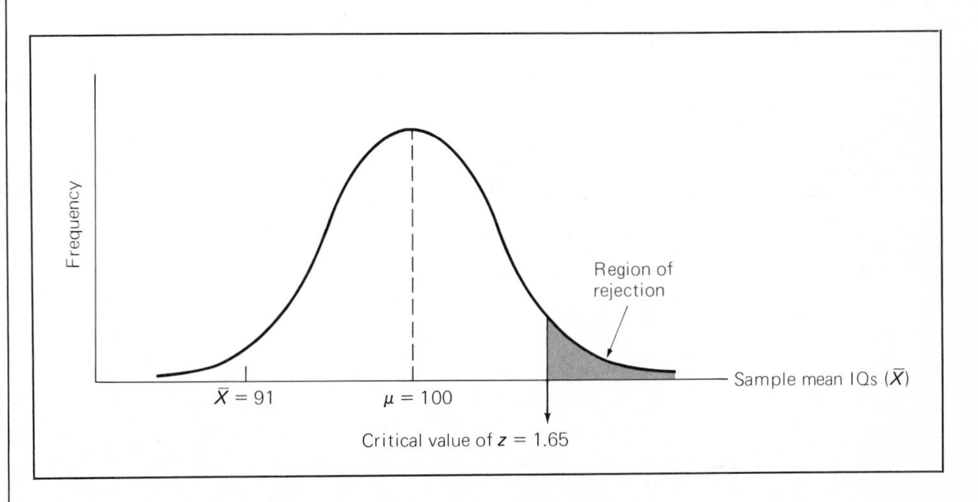

FIGURE 7-8
To reject H_0 in a one-tailed test, the sample mean must be in the direction specified in H_A. If H_A is $\mu > 100$, then your sample mean of 91 is in the opposite direction and you may retain H_0 without calculating the z score for 91.

Step 2 of the hypothesis-testing process. Then draw in the sample mean in its approximate location along the abscissa and ask, "Is the sample mean in the direction specified under H_A?" If the sample mean is *not* in the specified direction, you can stop the hypothesis test and retain H_0 before you even calculate z. Obviously, the sample mean cannot fall in the region of rejection if it is in the wrong direction. For example, you cannot conclude that the drug increases IQ if your sample mean is 91. If the sample mean *is* in the specified direction, calculate z; and if $|z| > z^*$, reject H_0. As a rule, for a one-tailed test the criterion for rejecting H_0 is as follows. If the sample mean is not in the direction specified under H_A, retain H_0. If the sample mean is in the direction specified under H_A *and* if $|z| > z^*$, reject H_0; but if $|z| \leq z^*$, retain H_0.

In hypothesis testing, the key to whether you should use a one-tailed test or a two-tailed test is to consider what you want to show. If you want to show that a treatment has *some* effect, then use a two-tailed test. Synonymous statements for "has some effect" are: "makes a difference," "influences," "changes," "alters," and so forth. Vagueness of direction is common to all these statements. Although the treatment is expected to have some effect, the direction of the effect is not specified. You use a two-tailed test when you are doing exploratory research and have no idea what effect to expect. On the other hand, if you are looking for a specific effect in one direction, then use a one-tailed test. Statements expressing a specific effect will always have a word indicating the specific direction, such as "increases," "decreases," "raises," "lowers," "more," "greater," "less," "bigger," "smaller," and so on. You use a one-tailed test if previous research, theory, or your own experience gives you a good idea in what direction the effect will be before you actually conduct the study.

Since the critical values for the z test are used so often, they are given below and in Table z^* of the Appendix for one- and two-tailed tests at the .05 and .01 levels of alpha. The z^* values are:

	One-tailed test	*Two-tailed test*
for $\alpha = .05$	1.65	1.96
for $\alpha = .01$	2.33	2.58

The critical values are smaller for one-tailed tests than for two-tailed tests at each level of alpha. To reject H_0, your calculated z does not have to be as large with a one-tailed test. Thus when H_0 is false and your $\bar{X}$ is in the right direction, it is easier to reject H_0 with a one-tailed test than with a two-tailed test. Because it is easier to reject H_0 with a one-tailed test, you are less likely to make a beta error (incorrectly retaining a false H_0), and, consequently, power is higher. For a given level of alpha, a one-tailed test has more power than a two-tailed test, so you are more likely to detect a real effect with a one-tailed test. Therefore it is to your advantage to use a one-tailed test whenever possible. We caution you, however, to use a one-tailed test only when you have a very good idea before conducting the study of the direction of the treatment effect. This note of caution stems from the fact that once you have decided on a one-tailed test, you cannot reject H_0 if you get a sample mean that is extreme but in the wrong direction. Once the results are obtained, you cannot change your mind and with the same data decide to do a two-tailed test, or a one-tailed test in the other direction. You would have to conduct the whole experiment over and change H_0 and H_A in light of these results!

The reason you cannot change your mind and alter your hypotheses or alpha level to suit the results is because the hypothesis-testing procedure is a predetermined set of rules to help you make a decision. You decide on the rules and then abide by them. If you abide by them, you know the chance of making an alpha error is the level of alpha you set. If you change the rules in the middle of the decision process you will change the probability of making an erroneous decision. *Before you gather any data* in a study, you should specify the following decision criteria and maintain them:

1. Whether you are conducting a one-tailed test or a two-tailed test.
2. The level of alpha.

Should you change either of these rules while gathering the data or after analyzing the data, you are allowing the data to influence your decision. To properly make a decision on the data itself, you must lay down the rules beforehand and stick to them. This is the purpose of having decision rules in the first place.

If you change the rules in the middle of the hypothesis-testing procedure, you cheat yourself and anyone who might read the report of the research. For example, if you start with a one-tailed test and $\alpha = .05$, and then switch to a two-tailed test because your sample mean turned out extreme but in the direction opposite from what you expected, then in reality you have added to the region of rejection, as shown in Figure 7-9. The

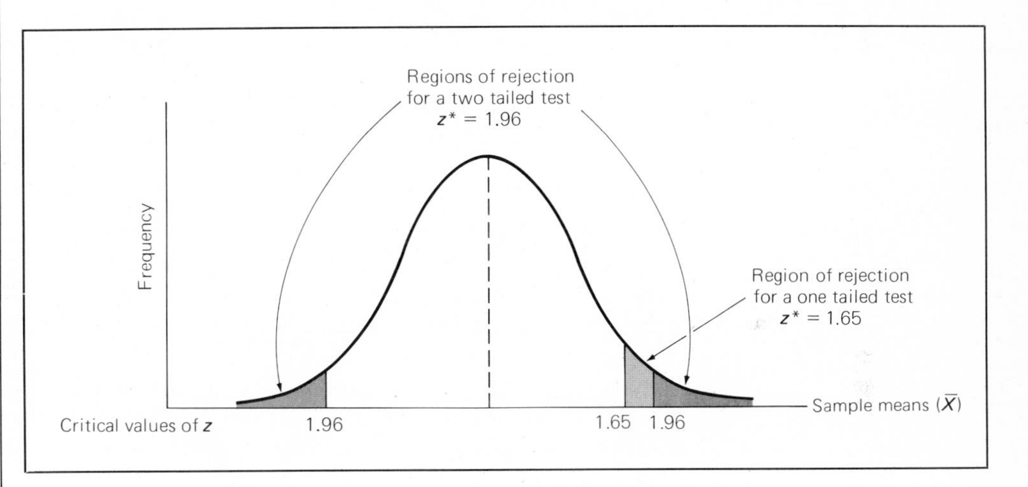

FIGURE 7-9

If a two-tailed test is performed on the same data after a one-tailed test, or vice versa, then the region of rejection is expanded to include the entire shaded area under the curve.

consequence of expanding the region of rejection is that alpha is now greater than .05, and so the probability of making an alpha error is increased. If you eventually report that you reject H_0 with a two-tailed test and $\alpha = .05$, you are misrepresenting the facts because your alpha was ultimately greater than .05.

Consider another example of misrepresentation. Suppose you are conducting a two-tailed test with $\alpha = .05$, and you observe a sample mean with a calculated z score of 1.8. Since $|1.8|$ is less than the critical value of 1.96, you must retain H_0. You might be tempted to switch and do a one-tailed test, reasoning that if you did, you could reject H_0 because the sample mean is in the correct direction and $|1.8| > 1.65$. However, if you switch, you again are altering the region of rejection. The true region of rejection is the entire shaded area of Figure 7-9, which indicates that alpha is greater than .05.

It is very frustrating to conduct a study only to find that your results fall just short of significance. Since you cannot change from a two-tailed test to a one-tailed test or change the level of alpha, what can you do? If H_0 is actually false, the failure to find a significant result is a beta error. The test you used might have had insufficient power to allow you to reject H_0. Your only recourse is to select a new sample and conduct the entire research study over again using a more powerful hypothesis test. There are only three practical ways for you to increase the power of a given hypothesis test. One way is to use a lower alpha level. This may not be feasible if alpha is already set at .05 or if the cost of an alpha error is high. Another way is to use a one-tailed test instead of a two-tailed test. While it is legitimate to do a one-tailed test on a new set of data, many professionals would question the appropriateness of using a previous non-significant research result as the sole basis for predicting the direction of an effect so that a one-tailed test can be used. Before using a one-tailed test, you should have a very good idea which direction the effect will be. The third way of increasing power, and probably the most practical way, is to re-perform your research study with a new, larger sample of subjects. In Formula 7.1 for the z test, you can increase the calculated z by decreasing $\sigma_{\bar{x}}$. Since $\sigma_{\bar{x}}$ is equal to $\sigma/\sqrt{N}$, as N increases, $\sigma_{\bar{x}}$ decreases. Therefore, for a given difference between $\bar{X}$ and μ, z will be larger for larger values of N. For example, consider the situations in the drug study when you have a two-tailed test and $\alpha = .01$. You selected 16 subjects and found their mean IQ to be 107.75 after taking the drug. Since the calculated value of $z = 2.07$ is less than the critical value of 2.58, you cannot reject H_0. Suppose you randomly select twenty-five new subjects, administer the drug, and measure the IQs. Suppose the mean IQ of this sample also happens to be $\bar{X} = 107.75$. When you calculate z for this new sample of $N = 25$, z is 2.583:

$$z = \frac{\bar{X} - \mu}{\dfrac{\sigma}{\sqrt{N}}} = \frac{107.75 - 100}{\dfrac{15}{\sqrt{25}}} = \frac{7.75}{3} = 2.583$$

Since 2.583 is greater than the critical value of 2.58, you can now reject H_0, even though you got the same sample mean of 107.75 in each case. Notice that by increasing the sample size from 16 to 25, you have decreased $\sigma_{\bar{x}}$ from 3.75 to 3.0. This is one aspect inferred by the central limit theorem. When N increases, the sample means will be less variable and more tightly clustered about the mean of the sampling distribution. The larger the sample, the more stable the results. You can decrease $\sigma_{\bar{x}}$ to extremely small values by simply increasing the number of subjects. This consequently increases the power of your test. Figure 7-10 shows that when there is less variability in the sample means, the mean of your sample need not be as extreme to fall into the region of rejection.

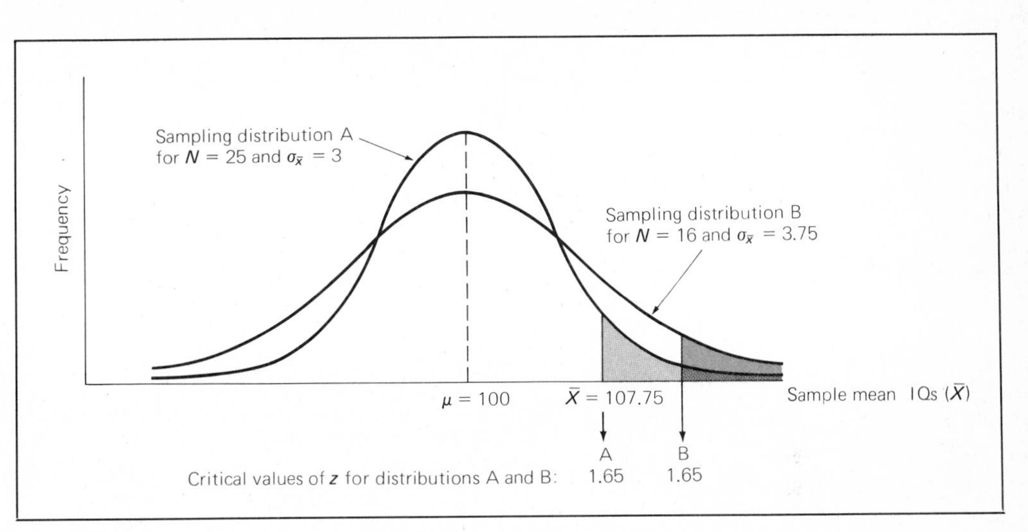

FIGURE 7-10

The effect of increasing the sample size is to decrease the variability of the sample means and thereby decrease $\sigma_{\bar{X}}$. With a larger N and smaller $\sigma_{\bar{X}}$, the sample means are more tightly clustered around μ. If your sample mean falls just short of significance, you may be able to reach significance by re-performing your research study with a new, larger sample.

One final comment is in order here. Increasing the power of a test is only helpful when H_0 is actually false. When H_0 is true, re-performing your research study with a larger sample will be to no avail. Thus it is to your advantage to use as large a sample size as feasible the first time you conduct your research study.

When σ is not known—using s to estimate σ, and the one-sample t test

The z test can be used when you have one sample, when the central limit theorem applies (or the sampling distribution is normal with a known mean), and when the population standard deviation, σ, is known. Since the only requirement for applying the central limit theorem is that the sample size is sufficiently large, there is not much difficulty in using the z test. However, σ is not known in most research situations. Since Formula 7.1 for the z test requires that you know σ, how do you calculate the probability of getting your results by chance when you do not know σ?

As an example of this type of problem, suppose Dr. Slimwell comes out with a new low-carbohydrate, fast weight-loss diet. He claims that if you follow his diet plan and eliminate virtually all sources of carbohydrate from your menu, you will lose at least one pound a day in the first week of the diet. Feeling skeptical, you want to test his diet program. You begin by finding a sample of ten volunteers who agree to adhere to Dr. Slimwell's diet, and then you weigh them. After one week on the diet, you weigh them again and record the following weight losses:

4.3, 4.4, 3.5, 6.7, 6.9
5.2, 9.3, 5.8, 4.1, 1.5

The mean weight loss is 5.17 pounds. Since this is less than Dr. Slimwell's claimed 7 pounds per week, should you decide that Dr. Slimwell is a charlatan, or should you conclude that the smaller mean weight loss is due to sampling error? Since Dr. Slimwell did not publish the standard deviation of the weight loss for his claimed population, you cannot use a z test. However, all is not lost. You can *estimate* the value of σ from your sample.

Since you do not know σ, the standard deviation of the population, you might think you can plug in the standard deviation of your sample as an estimate for σ in the z test formula. This will cause a problem because samples generally have smaller standard deviations than the populations from which they come. Since the sample standard deviation generally underestimates the population standard deviation, the sample standard deviation is referred to as a biased estimate of σ. The smaller the sample, the more biased the estimate of σ. Fortunately, it is possible to determine mathematically how much the sample standard deviation underestimates σ, and to correct for the bias. It can be shown that, in general:

$$\text{estimate of } \sigma = \left(\begin{array}{c} \text{sample} \\ \text{standard deviation} \end{array} \right) \sqrt{\frac{N}{N-1}}$$

correction factor

where N is the sample size, and $\sqrt{N/(N-1)}$ is the **correction factor.** Although we shall not prove this mathematically derived estimate of σ, we shall give its implications. Notice that with a small N the correction factor is relatively large, but with a large N the correction factor is relatively small. This simply means that larger samples are more representative of the population than smaller samples. For example, if $N = 2$, the correction factor is 1.4, whereas if $N = 100$, the correction factor is 1.005. Since the standard deviation of a small sample underestimates σ more than that of a large sample, a smaller sample's bigger correction term will raise it more to the approximate value of σ. Therefore, when you multiply the sample standard deviation by the correction term, you get a better estimate of σ. This estimate of σ is abbreviated with a small s in most texts.[4]

There is a more convenient formula for the estimate of σ that blends the correction term into the formula for the sample standard deviation. Formula 3.7 for the standard deviation was:

[4]Although the corrected version of the sample variance is an unbiased estimate of σ^2, the corrected version of the sample standard deviation, s, is not necessarily an unbiased estimate of σ. However, this does not matter for the purpose of the t test because this bias is compensated for.

$$\sigma = \sqrt{\frac{\Sigma(X - \mu)^2}{N}}$$

Since you are applying this formula to a sample, you cannot use the symbol σ anymore since σ refers only to the population standard deviation. Furthermore, in applying Formula 3.7 to a sample, you must replace μ with the sample mean, $\bar{X}$. Thus the formula becomes

$$\text{sample standard deviation} = \sqrt{\frac{\Sigma(X - \bar{X})^2}{N}}$$

You can take this formula and multiply it by the correction factor $\sqrt{N/(N - 1)}$ and simplify as follows:

$$\sqrt{\frac{\Sigma(X - \bar{X})^2}{N}}\sqrt{\frac{N}{N-1}} = \sqrt{\frac{N\Sigma(X - \bar{X})^2}{(N-1)N}} = \sqrt{\frac{\Sigma(X - \bar{X})^2}{N-1}}$$

Thus, the **definitional formula for *s***, the estimate of σ, is:

FORMULA 7.2

definitional formula for *s*

$$s = \sqrt{\frac{\Sigma(X - \bar{X})^2}{N - 1}}$$

The major difference between this formula and Formula 3.7 for σ is that the formula for s has $N - 1$ in the denominator instead of simply N. Remember that s is to be used only as an estimate, when it is necessary to infer σ based on a sample. If you are interested in simply describing the standard deviation of a sample and have no interest in making inferences about the characteristics of the population, then you should use Formula 3.7 properly modified, using $\bar{X}$ instead of μ, or Formula 3.9. For example, if you are converting some raw scores to z scores, then use either Formula 3.7 or 3.9 to calculate the standard deviation.

Now that you know how to estimate σ with s, you might reason that you can complete the hypothesis-testing process by substituting s for σ in the z test as follows:

$$z = \frac{\bar{X} - \mu}{\frac{s}{\sqrt{N}}}$$

This is an excellent idea. However, some modifications are necessary because of an additional problem. Since s is calculated from a sample and is subject to random sampling error, s varies from sample to sample. You might select several different samples and get several different values for s. Since different samples give you different values of s, you will get different values of z depending on the sample you take. Thus when you substitute s for σ in the z test, you introduce a new source of variability into the hypothesis-testing process.

Fortunately, it is possible to mathematically determine exactly how much variability you can expect to introduce when you estimate σ with s. This extra variability can be compensated for. For example, if you know the value of σ and you calculate z for your sample mean, you are calculating the exact number of standard deviation units your sample mean lies away from the mean of the sampling distribution. On the other hand, if you do not know σ and you have to estimate it with s, and you go ahead and calculate z by substituting s for σ in the z test formula, then you are calculating the *estimated* number of standard deviation units that your sample mean lies away from the mean of the sampling distribution. Because estimating σ by s introduces a new source of variability, the probability that a sample mean will lie a certain number of estimated standard deviations away from the mean of the sampling distribution differs from the probability that it will lie a certain number of actual standard deviations away. Thus estimating σ by s alters the appropriate critical values that should be used in the z test. Since you can mathematically determine the variability introduced by estimating σ, it is possible to set new critical values that adjust for using s instead of σ in the z test formula. To keep things straight, when z is calculated in this fashion by using s, it is called a **one-sample *t* test**, and the calculated value is called a ***t score.*** The **definitional formula for the one-sample *t* test** is:

one-sample *t* test
***t* score**

FORMULA 7.3

definitional formula for the one-sample *t* test

$$t = \frac{\bar{X} - \mu}{\frac{s}{\sqrt{N}}}$$

The only difference between the one-sample t test and the z test is that for the t test σ must be estimated with s, and the critical values, symbolized t^*, are different.

*t**

The critical values for t are bigger than the critical values for z to compensate for the variability introduced by estimating σ with s. This means that for a sample mean to be significant (to have a probability of occurrence that is less than alpha) it must be more (estimated) standard deviation units away from the mean of the H_0 sampling distribution than when the population standard deviation is known. Despite the correction factor incorporated in Formula 7.2, s is a poorer estimate of σ with small samples than with larger samples. Smaller samples are less representative of their parent populations than are larger samples and hence are more prone to variability and sampling error. Consequently, for small samples the t^* values are relatively large compared to the corresponding z^* values. With larger samples s becomes a better estimate of σ and the t^* values come closer to z^* values.

Since t^* values vary depending on the sample size, alpha, and whether you are conducting a one- or a two-tailed test, the critical values of t are listed for each combination of these three factors. Table t^* in the Appendix lists t^* values according to alpha, whether you are doing a one-tailed test or a two-tailed test, and as a function of the sample size referred to as **degrees of freedom** and abbreviated ***df***. The term *df* is used instead of N because two other types of t tests to be discussed in the next chapter also use Table t^* to find critical values. Since these other two t tests involve two samples instead of one, the proper t value to select depends a little differently on the sample size than the one-sample t test. So that the same table of critical values can be used for all three tests, it is more convenient to list the t^* values in terms of *df* for the particular test used. For the one-sample t test, *df* is equal to $N - 1$, where N is the sample size.

degrees of freedom
df

If your exact value of *df* cannot be found in Table t^*, a conservative procedure is to use the closest *smaller* value of *df* that is listed. This will give you a larger value of t^*, thus making it harder to reject H_0 and reducing the chances of an alpha error. For example, if your actual $df = 50$, choose the t^* for the closest smaller listed *df* of 40.

The hypothesis-testing procedure using the one-sample t test is:

1. Determine the appropriate H_0 and H_A. Assume H_0 is true.
2. Set alpha and calculate $df = N - 1$. Determine the critical value of t from Table t^* according to your alpha, *df*, and whether you are conducting a one- or two-tailed test.

   If you are conducting a two-tailed test, proceed with Step 3.

   If you are conducting a one-tailed test, draw a diagram of the sampling distribution, and mark the region of rejection to see if the sample mean is in the direction expected with H_A.

   If it is in the expected direction, proceed with Step 3.

   If it is in the opposite direction, retain H_0 and conclude that there is not sufficient evidence to say that the population from which your sample was drawn is different from the population specified in H_0.
3. Calculate a t score for your sample mean by:

$$t = \frac{\bar{X} - \mu}{\dfrac{s}{\sqrt{N}}}$$

where

$$s = \sqrt{\frac{\Sigma(X - \bar{X})^2}{N - 1}}$$

Refer to the value of t obtained from this formula as the calculated value of t, and symbolize its absolute value as $|t|$.

4. If $|t| > t^*$, reject H_0 and conclude that the mean of the population from which your sample was drawn is not the value specified in H_0.

   If $|t| \leq t^*$, retain H_0 and conclude that there is not sufficient evidence to say that the population from which your sample was drawn is different from the population specified in H_0.

Apply this process to the problem of deciding about Dr. Slimwell's diet plan. You want to see whether the diet is as effective as claimed. This is a difficult example and you must be careful when you specify H_0 and H_A. If you obtain a mean weight loss that is very much *less* than 7 pounds, you can conclude that the diet is not as effective as claimed. Therefore, this is a one-tailed test and H_A is that the diet is not as effective as claimed. H_0 is then that the diet is at least as effective as Dr. Slimwell claims. That is, if Dr. Slimwell is being honest, the mean weight loss per week will be 7 pounds or more. H_0 implies that the low sample mean of $\bar{X} = 5.17$ is due to sampling error. Notice you cannot "prove" the diet to be at least as effective as claimed because that is an attempt to "prove" the null hypothesis. Remember, H_0 is the hypothesis you can use to calculate the probability of getting your results. The hypothesis that specifies some exact value for the mean of a population must therefore be H_0. The steps you take to decide between these two competing hypotheses are:

1. H_0: $\mu \geq 7$ (The diet is effective)
   H_A: $\mu < 7$ (The diet is not effective)
2. Let $\alpha = .01$. The .01 level is chosen because you do not want to indict Dr. Slimwell without good cause, and this more stringent level will reduce the probability of an alpha error.

   Since you have ten subjects, $df = N - 1 = 10 - 1 = 9$. Use Table t^* to find that for a one-tailed test at $\alpha = .01$ and $df = 9$, $t^* = 2.821$.

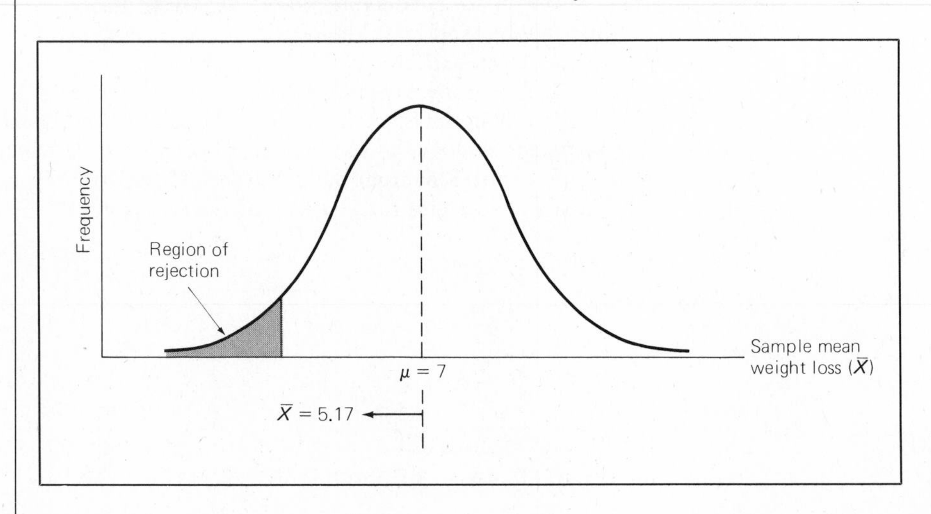

Since this is a one-tailed test and you are looking for an extremely low sample mean to reject H_0, you draw a diagram of the sampling distribution (as shown) and see whether your sample mean s is in the direction specified by H_A. The region of rejection is in the left tail.

3. Since the sample mean of 5.17 lies in the direction specified by H_A, you calculate t:

$$t = \frac{\bar{X} - \mu}{\frac{s}{\sqrt{N}}} = \frac{5.17 - 7.0}{\frac{2.148}{\sqrt{10}}} = \frac{-1.83}{\frac{2.148}{3.16}}$$

$$= \frac{-1.83}{.679} = -2.695$$

where

$$s = \sqrt{\frac{\Sigma(X - \bar{X})^2}{N - 1}} = \sqrt{\frac{41.532}{9}}$$

$$= \sqrt{4.614} = 2.148$$

4. Since $|-2.695| < 2.821$, retain H_0 and conclude that there is not sufficient evidence to say Dr. Slimwell's diet is not as effective as claimed.

Comparison of the z test and the one-sample t test

Essentially the same hypothesis-testing process is followed for both the z test and the one-sample t test. The main difference is that with the one-sample t test you do not know σ and must estimate it from your sample. Since estimating introduces additional variability, you compensate by using larger critical values for the one-sample t test than for the z test. That is, you adopt a more stringent criterion for rejecting H_0 by requiring your sample mean to be more estimated standard deviations away from the mean of the sampling distribution before you judge it "too unlikely to occur by chance." Comparing the definitional formulas for the two tests shows their similarity; that is,

$$z = \frac{\bar{X} - \mu}{\sigma_{\bar{x}}} = \frac{\bar{X} - \mu}{\frac{\sigma}{\sqrt{N}}}$$

$$t = \frac{\bar{X} - \mu}{\text{estimate of } \sigma_{\bar{x}}} = \frac{\bar{X} - \mu}{\frac{s}{\sqrt{N}}}$$

In the z test you calculate the number of standard deviation units that your sample mean lies away from the mean of the sampling distribution. In the one-sample t test you have to use s to estimate σ, and hence the denominator is an estimate of $\sigma_{\bar{x}}$ instead of the actual value of $\sigma_{\bar{x}}$ that appears in the z test formula. Thus the one-sample t formula calculates the *estimated* number of standard deviation units that your sample lies away from the mean of the sampling distribution. As a rule, apply the z test when you know σ, and apply the one-sample t test when you do not know σ and must estimate it using s from your sample.[5]

Computational formulas for s and the one-sample t test

The computational formula for s is very similar to the computational formula for σ (Formula 3.9) except that the denominator in the formula for s is $N - 1$ instead of N. Here is the **computational formula for s**:

FORMULA 7.4

computational formula for s

$$s = \sqrt{\frac{\Sigma X^2 - \frac{(\Sigma X)^2}{N}}{N - 1}}$$

This computational formula is generally easier to use than the definitional formula and yields the same value. For example, using your sample of ten volunteers in Dr. Slimwell's sample, Formula 7.4 also gives $s = 2.15$.

	X	X^2
	4.3	18.49
	4.4	19.36
	3.5	12.25
	6.7	44.89
$N = 10$	6.9	47.61
	5.2	27.04
	9.3	86.49
	5.8	33.64
	4.1	16.81
	1.5	2.25

$$\Sigma X = 51.7 \qquad \Sigma X^2 = 308.83$$

$$(\Sigma X)^2 = (51.7)^2 = 2672.89$$

$$s = \sqrt{\frac{\Sigma X^2 - \frac{(\Sigma X)^2}{N}}{N - 1}} = \sqrt{\frac{308.83 - \frac{2672.89}{10}}{10 - 1}}$$

[5]In older books and professional journals you might find a z or CR (Critical Ratio) used instead of t in the formulas for the t test when samples are large. This reflects the fact that the critical values for z and t are very similar with large samples ($N \geq 30$).

$$= \sqrt{\frac{308.83 - 267.289}{9}}$$

$$= \sqrt{\frac{41.541}{9}} = \sqrt{4.616} = 2.148$$

After solving for s you can plug the value into Formula 7.3 for the one-sample t test and derive the t score for your sample mean. In fact, with a little mathematical finesse you can combine the computational formula for s with the definitional formula for t to derive an easier way to calculate the t score:

$$t = \frac{\bar{X} - \mu}{\frac{s}{\sqrt{N}}} = \frac{\bar{X} - \mu}{\sqrt{\frac{\frac{\Sigma X^2 - \frac{(\Sigma X)^2}{N}}{N - 1}}{\sqrt{N}}}}$$

Further simplification gives you the **computational formula for the one-sample t test:**

FORMULA 7.5

computational formula for the one-sample t test

$$t = \frac{\bar{X} - \mu}{\sqrt{\frac{\Sigma X^2 - \frac{(\Sigma X)^2}{N}}{N(N - 1)}}}$$

With this formula you can conduct a one-sample t test without separately calculating s beforehand. The computational Formula 7.5 is the easiest to use because it is a one-step process, whereas the definitional Formula 7.3 involves the two steps of first estimating σ and then calculating t. If you have already calculated s, or are given s in a problem, then Formula 7.3 is the most appropriate. To test the effectiveness of Dr. Slimwell's diet, the computational Formula 7.5 can be applied in Step 3 of the hypothesis-testing process as follows:

X	X^2
4.3	18.49
4.4	19.36
3.5	12.25

X	X^2
6.7	44.89
6.9	47.61
5.2	27.04
9.3	86.49
5.8	33.64
4.1	16.81
1.5	2.25

$$\Sigma X = 51.7 \qquad \Sigma X^2 = 308.83$$

$$(\Sigma X)^2 = (51.7)^2 = 2672.89$$

$$t = \frac{\bar{X} - \mu}{\sqrt{\dfrac{\Sigma X^2 - \dfrac{(\Sigma X)^2}{N}}{N(N-1)}}} = \frac{5.17 - 7.0}{\sqrt{\dfrac{308.83 - \dfrac{(2672.89)^2}{10}}{10(10-1)}}}$$

$$= \frac{-1.83}{\sqrt{\dfrac{308.83 - 267.289}{10(9)}}}$$

$$= \frac{-1.83}{\sqrt{\dfrac{41.541}{90}}}$$

$$= \frac{-1.83}{\sqrt{.4616}} = \frac{-1.83}{.679} = -2.695$$

Regardless of whether you use Formula 7.3 or Formula 7.5 to calculate t, you can check to make sure you are on the right track by verifying that the ultimate value in the denominator is a positive number. The denominator should always be positive. Your calculated t score may be negative, but this would be due to a negative value in the numerator, which will occur when μ is greater than $\bar{X}$.

Assumptions for the z and one-sample t test

When you use either the z or one-sample t test you must assume that your sample is a random sample from the population. This is not just an assumption for these particular tests, but it is an assumption for inferential statistics in general. You can make inferences from samples to populations when you have random samples.

In addition, for the one-sample t test you must assume the population distribution is normal. This is necessary so that the t test can be used with small samples, where the normality aspect of the central limit theorem does not apply, and also so that the t^* tables can compensate for the variability introduced by estimating σ from s. With the z test you do not have to assume normality of the population unless the sample size is small ($N < 25$). If $N \geq 25$, the sampling distribution will be normal even if the population distribution is not, so you can employ the z test without assuming population normality.

robustness

An important aspect of the t test is **robustness.** Robustness means that the t test will give accurate results even when the assumption of population normality is somewhat violated. Thus you do not have to be overly meticulous in assuring that the population is normally distributed. Anything reasonably close to a normal distribution will do.

Journal form

journal form

One of the practical benefits derived from a knowledge of statistics is that it enables you to understand the statistical terminology and critically assess the statistical analysis presented in many books and professional journal articles. The results of hypothesis-testing procedures are generally put in an abbreviated form for journals. We will refer to this form as **journal form.** For example, you may read, "The effect of the 'enrichment' treatment was significant, $z = 2.98$, $p < .05$," or "Administration of the drug had no effect on performance, $t\ (19) = 1.35$, $p > .05$." Knowing what the journal forms "$z = 2.98$, $p < .05$" and "$t\ (19) = 1.35$, $p > .05$" mean will facilitate your understanding of research results. The general journal form proceeds in the following order:

First: Give the abbreviated name of the hypothesis test statistic you calculated, such as the letters z or t.

Second: Give the degrees of freedom, if any, in parentheses, followed by an equal sign.

Third: Give the calculated value of your test statistic.

N.S.

Fourth: Give the probability of your results occurring by chance. This is done by writing a p, then $<$ or $>$, and the level of alpha used in the table of critical values. The p means "probability" and the symbol $<$ reads "less than" and $>$ reads "greater than." If the probability of a chance occurrence is less than alpha, use $<$; if the probability is greater than alpha, use $>$. For example, $p < .05$ means that the probability of your results occurring by chance is less than $\alpha = .05$, whereas $p > .05$ means that the results are not significant at $\alpha = .05$. You can also write **N.S.** for "not significant" when $p > \alpha$.

Consider the following results of hypothesis tests written in journal form:

Journal form	*Meaning*
$z = 4.01$, $p < .05$	A z test was used and the calculated value of $z = 4.01$. The results have a probability less than .05 of occurring by chance, and hence are significant at the .05 level.

$z = 4.01, p < .01$ — A z test was used and the calculated value of $z = 4.01$. The results have a probability less than .01 of occurring by chance, and are hence significant at the .01 level. Notice that in the example above $\alpha = .05$, whereas $\alpha = .01$ here.

$z = .96, p > .05$ — A z test was used and the calculated value of $z = .96$, which is not significant at the .05 level.

$t(15) = 1.33, p > .05$ or $t(15) = 1.33$, N.S. — A t test was used and the calculated value of $t = 1.33$. With $df = 15$, the results are not significant at the .05 level.

$t(41) = 5.21, p < .05$ — A t test was used and the calculated value of $t = 5.21$. With $df = 41$, the results are significant at the .05 level. (These results would also have been significant at the .01 level.)

Journal form does not indicate whether a one- or two-tailed test is used. This information is often stated verbally.

Remember that a result that is significant at $\alpha = .05$ indicates that the probability of the result occurring by chance is less than .05, and significance at $\alpha = .01$ means that the probability of the result occurring by chance is less than .01. When the probability of a result occurring by chance is less than .05, researchers often say it is significant "beyond the .05 level." Likewise, when the probability is less than .01, the result is significant "beyond the .01 level." Since a sample mean that is significant at the .01 level will also be significant at the .05 level, $\alpha = .01$ is a *higher level of significance* than $\alpha = .05$. You can also say that a result that is significant at the .05 level of significance is "more significant" than a result that is significant at the .05 level, where "more significant" means only that the probability of an alpha error is lower.

KEY CONCEPTS

1. The sampling distribution of the mean is a frequency distribution of the means of samples of a given size drawn randomly from a population.

2. The central limit theorem assures that if the sample size is sufficiently large ($N \geq 25$), the sampling distribution will be approximately normal, with a mean equal to the population mean, μ, and with a standard deviation equal to σ / N, where N is the sample size. The standard deviation of the sampling distribution is called the standard error of the mean, and is symbolized $\sigma_{\bar{x}}$. Thus $\sigma_{\bar{x}} = \sigma / \sqrt{N}$.

3. The z test is used in hypothesis testing to determine whether a sample mean resulted by chance from a population with a given mean. You can specify a critical value of z, symbolized z^*, such that in the sampling distribution specified in H_0, any sample mean with a calculated z score

greater in absolute value than z^* is less likely than alpha. The critical values of z define the regions of rejection at one or both tails of the sampling distribution. If a sample mean lies in the region of rejection, it is said to be significant because you can reject H_0.

4 Using the z test, the hypothesis-testing procedure is as follows:

a. State H_0 and H_A. Assume H_0 is true.

b. Set alpha. Determine z^* from Table z^* for your level of alpha and whether you are conducting a one- or two-tailed test. If it is a two-tailed test, proceed with Step 3. If it is a one-tailed test, draw a diagram of the sampling distribution and mark the region of rejection to see whether $\bar{X}$ is in the direction expected with H_A. If $\bar{X}$ is in the expected direction, proceed with Step 3. If $\bar{X}$ is in the opposite direction, retain H_0.

c. Calculate z for $\bar{X}$ with the formula:

$$z = \frac{\bar{X} - \mu}{\dfrac{\sigma}{\sqrt{N}}}$$

d. If $|z| > z^*$, reject H_0. If $|z| \leq z^*$, retain H_0.

5 Use a two-tailed test when you want to see whether a treatment has an effect in either direction. Use a one-tailed test when you want to show that a treatment has an effect in one specified direction.

6 Often you will not know σ, so you cannot use the z test. You can estimate σ from the sample, with either the definitional formula on the left or the computational formula on the right.

$$s = \sqrt{\frac{\Sigma(X - \bar{X})^2}{N - 1}} \qquad s = \sqrt{\frac{\Sigma X^2 - \dfrac{(\Sigma X)^2}{N}}{N - 1}}$$

7 You can use s in the expression $s/\sqrt{N}$ to estimate $\sigma_{\bar{x}}$.

8 The one-sample t test is used when you have to estimate σ with s. The critical values of t, symbolized t^*, are larger than the critical values of z because of the variability introduced by estimating σ.

9 Using the one-sample t test, the hypothesis-testing process is as follows:

a. State H_0 and H_A. Assume H_0 is true.

b. Set alpha and calculate $df = N - 1$. Determine t^* from Table t^* for your level of alpha, df, and whether you are conducting a one- or two-tailed test. If it is a two-tailed test, proceed to the next step. If it is a one-tailed test, draw a diagram of the sampling distribution and mark the region of rejection to see whether $\bar{X}$ is in the direction expected with H_A. If $\bar{X}$ is in the expected direction, proceed with the next step. If it is in the opposite direction, retain H_0.

c. Calculate t for $\bar{X}$ with either the definitional formula shown below on the left or the computational formula on the right:

$$t = \frac{\bar{X} - \mu}{\frac{s}{\sqrt{N}}} \qquad t = \frac{\bar{X} - \mu}{\sqrt{\frac{\Sigma X^2 - \frac{(\Sigma X)^2}{N}}{N(N-1)}}}$$

d. If $|t| > t^*$, reject H_0. If $|t| \leq t^*$, retain H_0.

10 If your results are not significant, you must not switch from a one-tailed test to a two-tailed test or vice versa, nor can you alter alpha in an attempt to reach significance. You may repeat the research study with a larger sample size if your results are close to being significant, in the hope that the increased power resulting from a decrease in $\sigma_{\bar{x}}$ will enable you to reject H_0.

11 The assumptions for the one-sample t test are that the population distribution is normal, and the sample is random. The t test is robust with respect to violation of the normality assumption and will give accurate results even when the population distribution departs from normality.

12 Journal form is an abbreviated way to list the results of hypothesis tests. The abbreviated sequence is: give the hypothesis test (usually by letter); give the degrees of freedom in parentheses (if relevant); give the calculated value of your test; and give the probability that your observed results may have occurred by chance.

EXERCISES

1 Given the following finite population of 50 scores:

5	3	11	7	5	9	5	9	13	7
3	3	7	5	13	7	7	11	7	13
7	1	9	5	7	11	9	7	11	3
1	7	3	11	1	5	13	9	9	5
3	9	7	9	5	7	1	5	9	11

a. Calculate the population mean, μ.

b. Construct a sampling distribution as follows: select 12 random samples of two scores each from the population (replacing each sample's scores before drawing the next sample), and calculate the mean of each sample. Construct a frequency distribution showing how often each sample mean occurs.

c. Construct another sampling distribution as follows: select 12 random samples of ten scores each from the population (replacing each sample's scores before drawing the next sample), and calculate the mean of each sample. Construct a frequency distribution showing how often each sample mean occurs.

d. Describe the effect that increasing sample size has on the variability among the sample means in the sampling distributions in (b) and (c).

2 Identify the appropriate H_0 and H_A in the following research situations, and identify whether it is a one- or a two-tailed test.

a. Alpha Epsilon sorority wanted to know whether giving Valentine's Day cards to their male statistics instructor would influence the sorority members' final scores. The previous mean for all Alpha Epsilons taking statistics from this instructor was 79. The seven members currently enrolled in the class mailed valentines. The final scores were later obtained.

b. Alpha Epsilon sorority wanted to know whether giving Valentine's Day cards to their male statistics instructor would increase the sorority members' final scores. The previous mean of all Alpha Epsilons taking statistics from this instructor was 79. The seven members currently enrolled in the class mailed valentines. The final scores were later obtained.

3 You are interested in the time it takes 6-year-old children to put together a design made of colored blocks so as to match the design shown in a picture. Such a matching task is part of a general test of intelligence. You have norms showing that it takes children this age an average of 45 seconds to complete the matching task, with a standard deviation of 16 seconds. This population of times is known to be approximately normally distributed. You believe that practice in putting jigsaw puzzles together will have an effect on the time it takes the child to complete the matching task with the blocks. You select a sample of 20 children 6 years of age with average intelligence (as measured by a different intelligence test) and give them the five jigsaw puzzles to put together at home. After the jigsaw puzzles were put together, you give the child the matching task with the blocks. You obtain the following scores (in seconds):

28	78	30	30	32
44	33	52	41	33
29	17	35	32	38
32	57	39	36	34

a. Setting $\alpha = .05$, use the z test and complete the four steps of the hypothesis-testing procedure. What can you conclude about the effect that practice in putting jigsaw puzzles together has on performance in the matching task with blocks?

b. Put your results in journal form.

4 Each of the data sets below are random samples of scores drawn from the population given in Exercise 1. Assume that you do not know the population standard deviation, nor do you have any way of calculating it exactly. For each of the samples in (a) to (d), use both the definitional and computational formulas and solve for s (Formulas 7.2 and 7.4) to estimate σ. Does one method of calculating s appear easier than the other?

a. 1, 1, 9

b. 11, 9, 1, 3, 11, 7

c. 5, 9, 1, 13, 9, 7, 7, 3, 7, 3

d. 7, 13, 11, 9, 5, 7, 3, 13, 5, 9, 7, 9, 7, 3, 1

e. The actual standard deviation, σ, of the population of 50 scores given in

Exercise 1 is 3.323. Compare your estimates in (a) through (c) with the actual value of σ, and describe the general effect that increasing sample size has on how closely s, your estimate of σ, comes to the actual value of σ.

5 As part of a committee reviewing applicants for entrance in a master's degree program in early childhood education, you find that the 45 applicants obtained a mean of 1073 on the combined GRE aptitude tests of quantitative ability and verbal ability. You know from years of experience that the distribution of previous applicants is normal with a mean of 1100 with a standard deviation of 80. You assert, "This is an inferior group of applicants."

a. Which is the appropriate hypothesis test to employ here, the z test (Formula 7.1) or the one-sample t test (Formula 7.3)?

b. Setting $\alpha = .01$, complete the four steps of the hypothesis-testing procedure. What can you conclude about the inferiority of these applicants relative to previous applicants?

c. Put your results in journal form.

6 Absenteeism due to common cold averages a total of 8 days per year in elementary school children in a large New England school district. You wish to determine the effect, if any, daily doses of vitamin C have on the total number of days a child in the district is absent due to colds. A random sample of 70 elementary school children was obtained from throughout the district and you arranged that they daily received 250-milligram tablets of vitamin C throughout the year. The results at the end of the year revealed a sample mean of 6.55 total days absent ($\bar{X} = 6.55$), with a standard deviation of 3.21 days ($s = 3.21$).

a. Which is the appropriate hypothesis test to employ here, the z test (Formula 7.1) or the one-sample t test (Formula 7.3)?

b. Setting $\alpha = .01$, use the one-sample t test (Formula 7.3) and complete the four steps of the hypothesis-testing procedure. What can you conclude about the effect of vitamin C on absenteeism?

c. Put your results in journal form.

7 In a large introductory psychology course 800 students took a final exam, but one section of 15 students was accidently given 20 extra minutes to complete the final. Consider the class of 800 students as the population, and the section of 15 students as a sample drawn from this population. The instructor reasoned that the extra time may have raised these 15 students' scores, and wants them to take the final (another version) over again. You know that the population is normally distributed with $\mu = 70$ and $\sigma = 12.5$. The students given the extra time made the following scores on the original exam:

78	93	69
59	75	71
77	51	76
64	81	60
62	77	55

a. Which is the most appropriate hypothesis test to use in this study, the z test or the one-sample t test?

b. Set alpha at .05 and complete the hypothesis-testing procedure. What can you conclude about any facilitative effect extra time had on the final exam scores?

8 You conduct a study setting alpha at .01 and complete a hypothesis-testing procedure using a two-tailed test. Your results are *almost* significant: $t(19) = 2.693$, $p > .01$. Which one of the following could you do?

a. Say the results are significant.
b. Change alpha to .05.
c. Change to a one-tailed test.
d. Repeat the study with a larger sample.

9 **a.** Is the region of rejection larger for a one- or a two-tailed test?
b. Is the region of rejection larger for $\alpha = .01$ or for $\alpha = .05$?

10 Assume that, on the average, normal rhesus monkeys spend approximately 60 minutes dreaming while sleeping at night. You are interested in the effects a new appetite-suppressing drug has on dreaming since a reduction in the time spent dreaming is known to cause irritability in humans. Thus, you do not want to pursue marketing the drug for human consumption if there is any evidence that it significantly decreases nightly dreaming time in nonhuman primates. Ten rhesus monkeys are administered appetite-suppressing doses of the drug in the morning and their dream time (in minutes) is monitored that night. You obtain the following times:

42, 19, 64, 60, 57, 35, 59, 15, 23, 58

a. Which level of alpha would you select to have the most power?
b. Use the definitional Formula 7.3 for the one-sample t test and complete the hypothesis-testing procedure. What can you conclude about the desirability of marketing the drug for human consumption?
c. Put your results in journal form.
d. Use the computational Formula 7.5 for the one-sample t test. Does one method for calculating t appear easier than the other?

11 As a diagnostician in a district of eight high schools, adolescents are often referred to you because of disruptive classroom behavior. On the basis of your experience in obtaining numerous personality profiles on such adolescents, you believe that they actually have higher than average "ergic" tension (that is, they are tense, frustrated, and driven individuals), and that their disruptive behavior is merely a symptom of high ergic tension. You know that average adolescents obtain a mean of 36 on a personality test designed to measure ergic tension. You assume that the population distribution of ergic tension is normally distributed. To test your speculation, you randomly select 20 adolescents who had been previously diagnosed as conduct disorders and administer the personality test. The scores were as follows (high scores reflect high ergic tension):

28	32	53	41
59	25	37	41
41	47	43	37
38	40	35	39
40	42	34	48

a. What is the appropriate hypothesis test to employ in this study, the z test or the one-sample t test?
b. Setting alpha at .05, complete the hypothesis-testing procedure. What can you conclude about heightened ergic tension in disruptive adolescents?
c. Put your results in journal form.

8

TWO-SAMPLE HYPOTHESIS TESTING

CHAPTER GOALS

After studying this chapter, you should be able to:

1. Define the sampling distribution of the difference for the two-sample independent t test, and state the mean and estimated standard deviation of this sampling distribution.
2. Perform the two-sample independent and dependent t tests using both the definitional and computational formulas.
3. State the assumptions for the two-sample independent and dependent t tests.
4. Identify the observed result, the mean of the sampling distribution, and the (estimated) standard deviation of the sampling distribution in the definitional formulas for each of the following hypothesis tests: z test; one-sample t; two-sample independent t; and two-sample dependent t.
5. Specify when to use each of the four hypothesis tests mentioned in Goal 4 above.

CHAPTER IN A NUTSHELL

A very common research problem is whether an observed difference between the means of two samples is the result of an actual difference in population means or is the chance result of sampling error. The general hypothesis-testing procedure can be employed to help you decide. If the subjects comprising each sample are independent, you subtract the mean of one sample from that of the other and compare this difference with a distribution of differences you expect to get by chance, assuming that the means of the populations from which the samples were drawn are equal. This distribution of differences between the means is called the sampling distribution of the difference. From your samples you can estimate the standard deviation of the sampling distribution of the difference. You then use the two-sample independent *t* test to determine the probability that the difference between your group means arose by chance. If the samples are matched (such that a subject tested under one condition has a paired subject tested under the other condition), or if the same subjects are tested and retested under different conditions, such as "before" and "after" a given treatment, then for each pair of subjects' scores you subtract one score from the other to get a difference score. You compute the mean of these difference scores for all subject pairs and then perform a one-sample *t* test to decide if a sample with your given mean difference score could likely have occurred by chance from a population with a mean difference score of zero. If not likely, then this implies that your treatment has an effect since the mean difference between "before" and "after" scores in the population is not zero.

Introduction to two-sample hypothesis tests

In research situations in which the population mean, μ, is known or assumed to be a given value, researchers often use this value as a norm with which to compare some other specific group. One sample may be selected from that specific group, and the one-sample hypothesis tests introduced in Chapter 7 can help you to decide whether your sample mean of $\bar{X}$ came from the population with the given mean of μ. However, in numerous situations there is no previously established norm or standard available for comparison. In this case you may select two samples that differ in some naturally occurring characteristic in a naturalistic observation (such as "high socioeconomic status" and "low socioeconomic status") or level of the independent variable in an experiment (such as "zero" delay and "one-second" delay of reinforcement). The dependent variable is then measured and the sample means can be compared to each other. The *two-sample* hypothesis tests described in this chapter help you decide whether the two samples could have been drawn from populations with the same mean on the dependent variable.

Samples representing two different naturally occurring conditions or levels of the independent variable can have different means because they are drawn from populations with different means. Alternatively, two samples can have different means because of sampling error. The populations may actually be the same on the dependent variable, meaning that the samples are drawn from populations with equal means. However, due to chance factors, the sample means differ. These two possible explanations for why two sample means may differ are outlined in Figure 8-1. The probability that the sample means differ simply by chance can be calculated using two-sample hypothesis tests such as the independent t test or the dependent t test.

The two-sample independent t test

Suppose you are interested in determining which of two study techniques is more effective, cramming right before a test, or spacing out the study time over a longer period. You randomly assign each of 24 subjects to one of two different treatment conditions: massed study, in which the subjects study for five hours and then immediately take a test on the material; and spaced study, in which the subjects study for one hour a day for five consecutive days and then take a test immediately after the fifth study session. Although data were collected for 12 subjects in each condition, you accidentally dropped the data sheets and the data from two random subjects in the spaced condition blew away and were lost, leaving only 10 subjects in that condition.[1] The scores (number correct) for your remaining 22 subjects are as follows:

[1] Since the data loss was random, you still have a random sample from the population. The two-sample independent t test can still be performed even though the sample sizes are not equal.

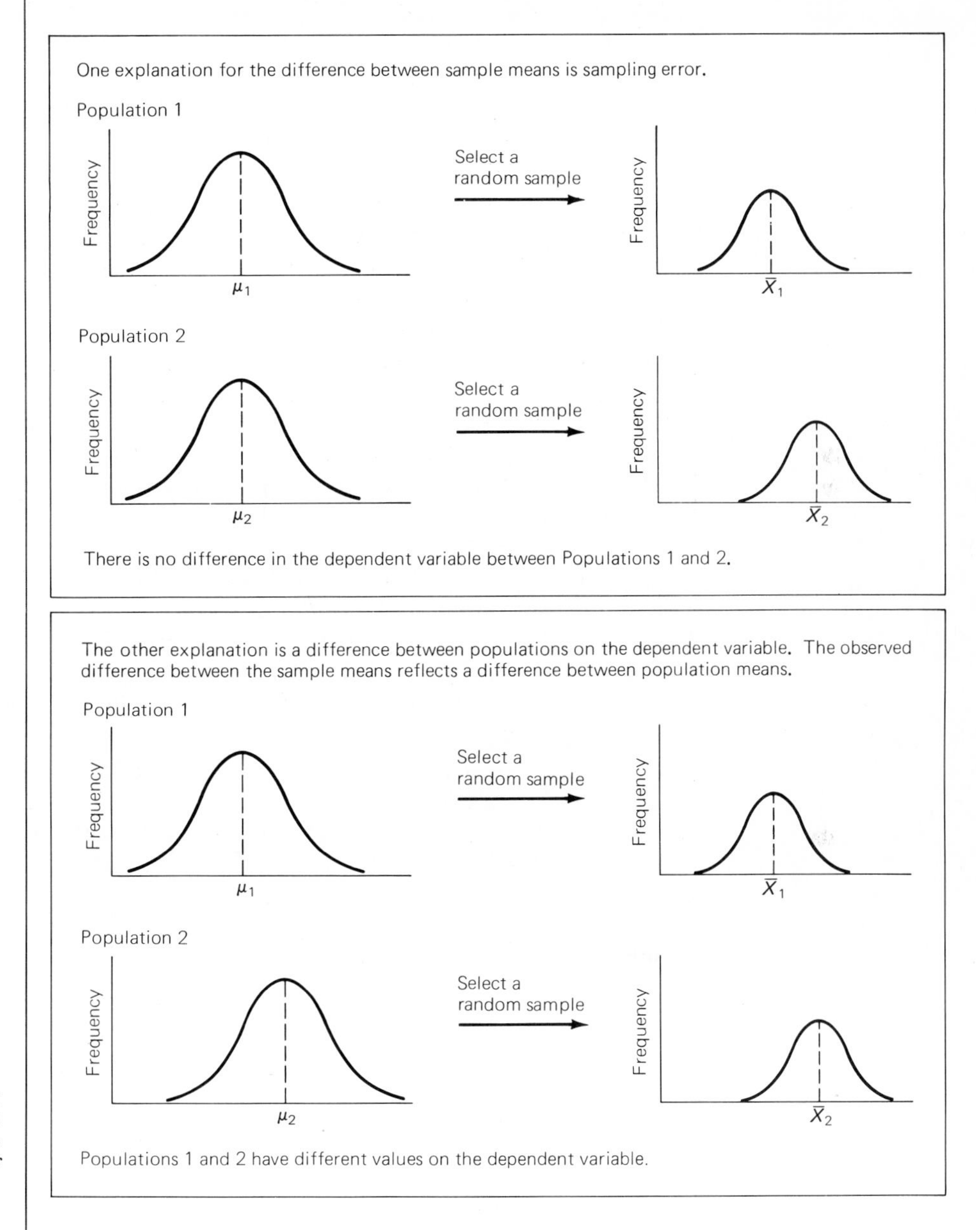

FIGURE 8-1

The two competing explanations for an observed difference between the means of two samples.

Condition 1: Spaced study subject	X_1	*Condition 2: Massed study subject*	X_2
S_1	26	S_{11}	20
S_2	62	S_{12}	57
S_3	68	S_{13}	64
S_4	31	S_{14}	42
S_5	10	S_{15}	39
S_6	44	S_{16}	26
S_7	84	S_{17}	38
S_8	61	S_{18}	58
S_9	62	S_{19}	16
S_{10}	51	S_{20}	50
	$\bar{X}_1 = 49.9$	S_{21}	49
		S_{22}	33
			$\bar{X}_2 = 41$

A comparison of the sample means reveals that the subjects in the spaced-study condition did better on the test. However, there are two possible explanations for this difference. One is that spaced study produces superior performance, and the second is that the difference between the sample means is due to some other random factor. For example, the subjects randomly assigned to the spaced condition might be smarter, more motivated, or more interested in the study material to begin with.

You have observed a difference of 8.9 points between the means of the test scores and you need to perform some sort of hypothesis test to decide whether to attribute the difference to the study conditions or to sampling error. How do you decide? The z test and the one-sample t test are not appropriate here for several reasons. Among these reasons are the simple mechanical ones that you do not have the appropriate data to "plug into" the formulas. For example, you are not given any normative value for μ to plug into either the z test formula or the one-sample t test formula, nor are you given the value of σ to plug into the z test formula. Furthermore, you have two samples, and hence two *sample* means, $\bar{X}_1$ and $\bar{X}_2$. Where can you plug these values into the z test or the one-sample t test formulas? The context is simply not appropriate to use these formulas.

To help you make a decision it is possible to apply the same hypothesis-testing logic to which you were introduced in Chapters 6 and 7. You begin by assuming that the difference between the sample means is due to a chance sampling error. If chance alone is operating, H_0 is that the means of the populations from which each sample is drawn are equal—or, symbolically, $\mu_1 = \mu_2$. This says there is no difference in the effects of the two study conditions and the difference in the sample means is due to sampling error. Then, assuming H_0 is true, you need to calculate the probability of getting a difference between sample means as extreme as 8.9. If this difference of 8.9 is very unlikely, you can reject H_0 and conclude that the means of the populations are different. In other words, the two study conditions produce different effects. How do you calculate the probability of getting a chance difference between sample means?

One way to calculate the probability of getting a chance difference between two sample means is to proceed with the following four steps: (1) assume you have two populations with the same mean; (2) draw a random sample from one population and then draw a random sample from the other, find the means of both samples and then subtract one mean from the other; (3) repeat this procedure of drawing pairs of samples many many times; and (4) construct an empirical frequency distribution of the differences between the pairs of sample means and depict this with a frequency polygon. This procedure is illustrated in Figure 8-2. The resulting distribution of the differences between the pairs of sample means is another type of a sampling distribution. To distinguish it from the sampling distribution of the mean introduced in Chapter 7, it is called the **sampling distribution of the difference between means,** or **sampling distribution of the difference** for short.

sampling distribution of the difference

If you actually go out and randomly draw all those pairs of samples, calculate the difference between means, and so forth, the resulting sampling distribution is called an empirical sampling distribution of the difference. However, as you discovered in Chapter 7, constructing empirical sampling distributions is an exceedingly impractical way to go about calculating the probability of observed results. Fortunately, as in the case with the sampling distribution of the mean, it is possible to mathematically infer the characteristics of a theoretical sampling distribution of the difference. This theoretical distribution is a hypothetical entity and never occurs exactly in nature. However, it is an excellent approximation to the average empirical sampling distribution of the difference. In fact, under ideal conditions you expect a well-constructed empirical sampling distribution of the difference to look just like the theoretical distribution, provided that enough paired samples are actually drawn. If the populations from which the pairs of samples are hypothetically drawn are normal and have equal means, the theoretical sampling distribution of the difference has the following inferred characteristics.

---

characteristics of the sampling distribution of the difference for independent samples

The sampling distribution of the difference for independent samples is an approximately normal distribution with a mean of 0 and a standard deviation (symbolized $\sigma_{\bar{x}_1-\bar{x}_2}$) equal to

$$\sqrt{\sigma^2\left(\frac{1}{N_1}+\frac{1}{N_2}\right)}$$

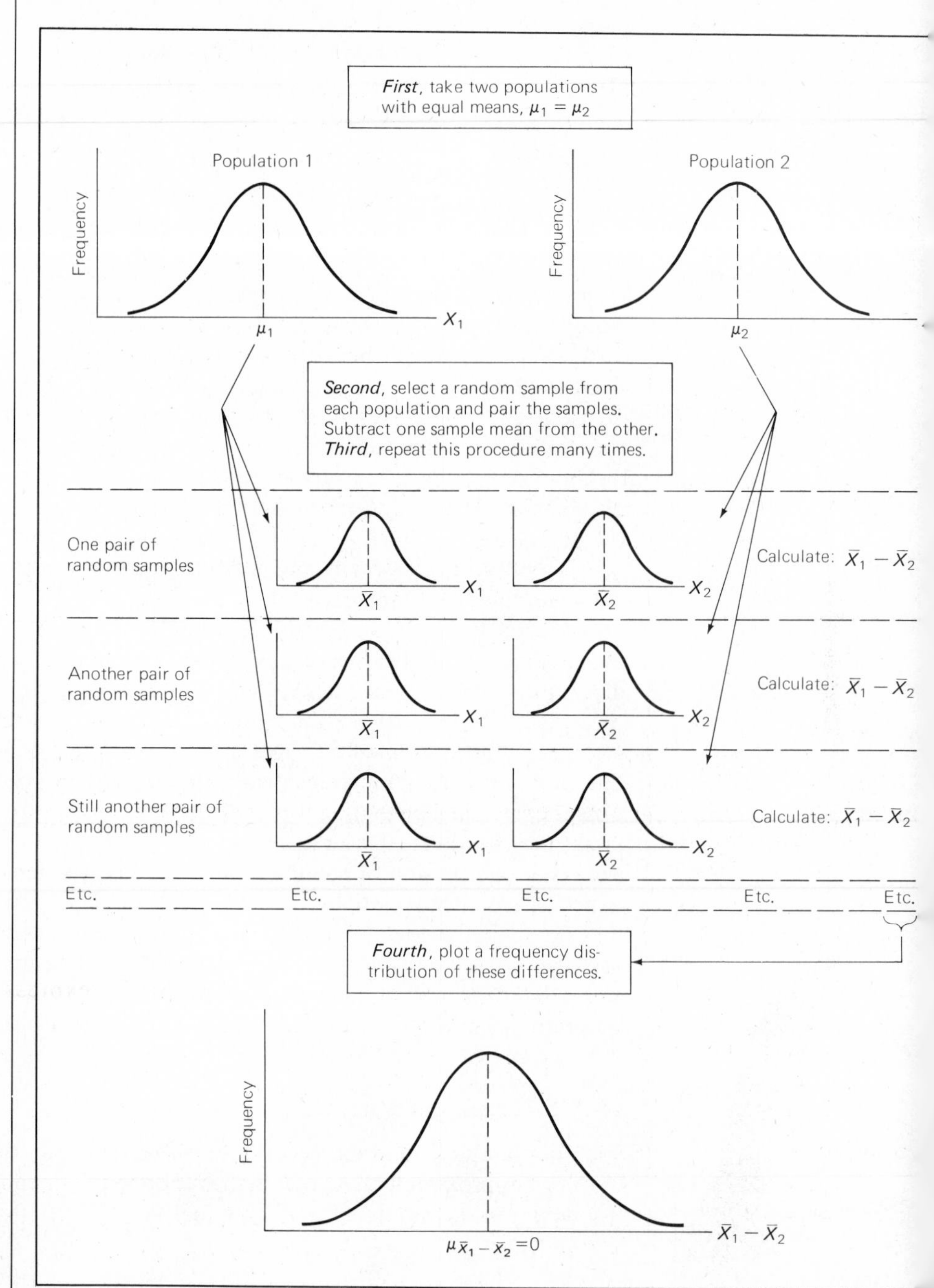

FIGURE 8-2
Constructing an empirical sampling distribution of the difference (between means) for independent samples.

In this statement, σ^2 is the variance of each of the two populations from which the samples are hypothetically drawn (assuming that the two populations have equal variances and equal means), and N_1 and N_2 are the respective sizes of each sample in a pair of samples. The standard deviation of the sampling distribution, $\sigma_{\bar{x}_1-\bar{x}_2}$, is also called the **standard error of the difference.** Figure 8-3 depicts the characteristics of the sampling distribution of the difference.

standard error of the difference $\sigma_{\bar{x}_1-\bar{x}_2}$

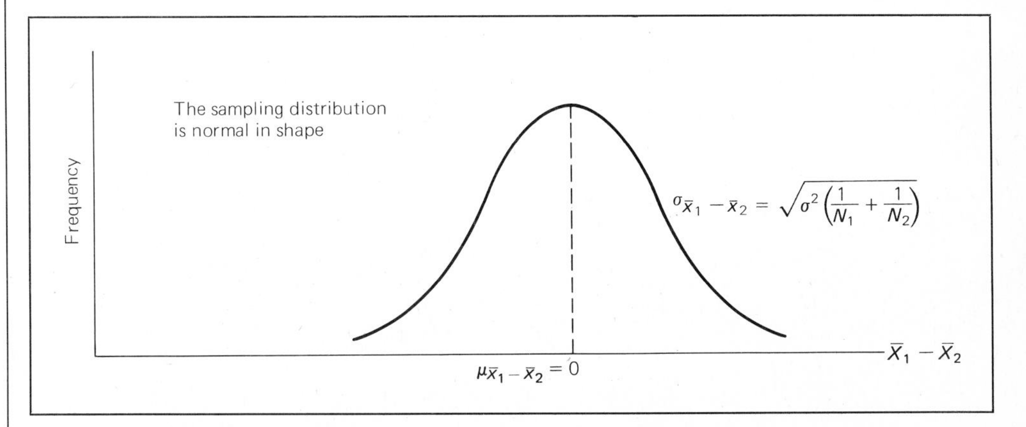

FIGURE 8-3
Characteristics of the sampling distribution of the difference for independent samples.

$\mu_{\bar{x}_1-\bar{x}_2}$

The mean of the sampling distribution of the difference, $\mu_{\bar{x}_1-\bar{x}_2}$, is zero because H_0 specifies that the populations from which the samples are drawn have the same mean. When you randomly select pairs of samples, you expect on the average to get a zero difference between the pairs of means. Although the formula for the standard deviation of the sampling distribution of the difference will not be mathematically proved here, you can get an intuitive feeling for why

$$\sigma_{\bar{x}_1-\bar{x}_2} = \sqrt{\sigma^2\left(\frac{1}{N_1}+\frac{1}{N_2}\right)}$$

Take this formula and distribute σ^2 into the expression in parentheses, giving you the right-hand formula:

$$\text{standard error of the difference} = \sigma_{\bar{x}_1-\bar{x}_2} = \sqrt{\sigma^2\left(\frac{1}{N_1}+\frac{1}{N_2}\right)} = \sqrt{\frac{\sigma^2}{N_1}+\frac{\sigma^2}{N_2}}$$

Notice the similarity between the individual terms under the radical in the right-hand formula above for $\sigma_{\bar{x}_1-\bar{x}_2}$, and the expression below on the right derived from the formula for the standard error of the mean, $\sigma_{\bar{x}}$, introduced in Chapter 7:

$$\text{standard error of the mean} = \sigma_{\bar{x}} = \frac{\sigma}{\sqrt{N}} = \sqrt{\frac{\sigma^2}{N}}$$

Each term under the radical for the standard error of the difference is analogous to the single term under the radical for the standard error of the mean. The formula for the standard error of the difference has two terms under the radical because the standard error of the difference reflects two sources of variability, one source of variability for each of the two samples drawn. Thus the sampling distribution of the difference has greater variability than either of the sampling distributions of the mean for each population alone. This makes sense when you consider the many ways in which large differences between pairs of sample means can arise. It is possible to pick a sample with a high mean from the first population and pick a sample with a low mean from the second population. When you subtract the second mean from the first, you will get a large positive difference. It is also possible to pick a sample with a low mean from the first population and pick a sample with a high mean from the second population. When you subtract the second mean from the first, you will get a large negative difference.

Now that you know the sampling distribution of the difference is normal, has a mean of 0, and a standard deviation of

$$\sqrt{\sigma^2\left(\frac{1}{N_1}+\frac{1}{N_2}\right)}$$

you might consider setting alpha and establishing regions of rejection by marking the critical values of z on the sampling distribution of the difference. Then perhaps you can transform your observed difference between sample means into a z score in the sampling distribution of the difference by using the z test formula:

$$z = \frac{(\bar{X}_1 - \bar{X}_2) - \mu_{\bar{x}_1-\bar{x}_2}}{\sigma_{\bar{x}_1-\bar{x}_2}} = \frac{(\bar{X}_1 - \bar{X}_2) - 0}{\sqrt{\sigma^2\left(\frac{1}{N_1}+\frac{1}{N_2}\right)}}$$

It is appropriate to use the z test formula here if the sampling distribution is normal, which it is, and if you know the value of the standard deviation, $\sigma_{\bar{x}_1-\bar{x}_2}$. This is the proper way to proceed in the hypothesis-testing process except for the fact that you rarely know the exact value of $\sigma_{\bar{x}_1-\bar{x}_2}$. This is because the variance of the two populations, σ^2, is usually unknown.

The fact that σ^2 is usually unknown can be overcome by using the same procedure presented in Chapter 7 for the one-sample t test when σ was unknown. That is, σ^2 can be estimated from the samples. Specifically, each sample can be used to get an independent estimate of σ^2 by using Formula 7.2. Formula 7.2 for estimating σ is:

$$s = \sqrt{\frac{\Sigma(X - \bar{X})^2}{N - 1}}$$

Squaring this gives the definitional formula for s^2, which is an estimate of σ^2:

FORMULA 8.1
definitional formula for s^2

$$s^2 = \frac{\Sigma(X - \bar{X})^2}{N - 1}$$

Also, recall that the computational counterpart to Formula 7.2 was given by Formula 7.4. Squaring Formula 7.4 gives:

FORMULA 8.2
computational formula for s^2

$$s^2 = \frac{\Sigma X^2 - \dfrac{(\Sigma X)^2}{N}}{N - 1}$$

Formula 8.2 is thus the computational formula for s^2.

Since you have two independent samples, you have two independent estimates of σ^2. You can calculate s_1^2 from the subjects in your first treatment condition, and s_2^2 from the subjects in your second treatment condition. Now s_1^2 and s_2^2 should have values that are close to each other, but due to sampling error they will probably not be equal. Which one should you choose? If the sample sizes are different, you might suggest using the variance estimate from the larger sample because it will generally give you a better estimate. But which estimate will you choose when the sample sizes are equal? Actually, the best estimate of σ^2 is obtained by combining the variance estimates from both samples so as to use as many subjects as possible on which to base your estimate. An adjustment can be made so that more weight is given to the larger sample whenever the sample sizes are different. This is done by forming a weighted average of the variance estimates from each sample. The weighting can be seen in the numerator of Formula 8.3, where each sample estimate is multiplied (weighted) by its associated degrees of freedom. If the sample sizes are equal, the associated degrees of freedom $(N - 1)$ are equal, and consequently both variance estimates are multiplied by the same value and are equally weighted. However, if the sample sizes are different, the variance estimate from the larger of the two is given more weight since its degrees of freedom are greater. The denominator contains the total degrees of freedom of both samples, $(N_1 - 1) + (N_2 - 1)$, which can be simplified to $N_1 + N_2 - 2$. Formula 8.3 for estimating σ^2 is called the **pooled variance estimate,** or s^2_{pooled}:

pooled variance estimate

FORMULA 8.3

s^2_{pooled}

$$s^2_{pooled} = \frac{(N_1 - 1)s_1^2 + (N_2 - 1)s_2^2}{N_1 + N_2 - 2}$$

This s^2_{pooled} may be substituted for s^2 in the formula for the standard error of the difference to give an estimate[2] of $\sigma_{\bar{x}_1 - \bar{x}_2}$ as in Formula 8.4:

FORMULA 8.4

$$\text{estimate of } \sigma_{\bar{x}_1 - \bar{x}_2} = \sqrt{s^2_{pooled}\left(\frac{1}{N_1} + \frac{1}{N_2}\right)}$$

Since you can estimate $\sigma_{\bar{x}_1 - \bar{x}_2}$, you might reason that you can complete the hypothesis-testing process by substituting the estimate of $\sigma_{\bar{x}_1 - \bar{x}_2}$ into the z test formula that was rewritten earlier on page 154.

$$z = \frac{(\bar{X}_1 - \bar{X}_2) - \mu_{\bar{x}_1 - \bar{x}_2}}{\sigma_{\bar{x}_1 - \bar{x}_2}} = \frac{(\bar{X}_1 - \bar{X}_2) - 0}{\sqrt{s^2_{pooled}\left(\frac{1}{N_1} + \frac{1}{N_2}\right)}}$$

Knowing how to estimate $\sigma_{\bar{x}_1 - \bar{x}_2}$ thus allows you to calculate the number of *estimated* standard deviations that an observed difference between a pair of sample means, $\bar{X}_1 - \bar{X}_2$, lies away from the mean of the sampling distribution of the difference. The probability of such an event occurring by chance can then be deduced from this information. However, it is important to realize that estimating $\sigma_{\bar{x}_1 - \bar{x}_2}$ introduces a new source of variability, which must be neutralized. The added variability can be compensated for by setting new critical values as in the one-sample research situation when a t score was derived instead of a z score. The same procedure is followed here with two samples. To keep things straight, when z is calculated using the estimate of $\sigma_{\bar{x}_1 - \bar{x}_2}$, it is called a **two-sample independent *t* test** or **independent *t* test.** The definitional formula for the independent t test is:

independent *t* test

[2] s^2_{pooled} is an unbiased estimate of σ^2, but s_{pooled} is not necessarily an unbiased estimate of σ. This does not matter for the purposes of the t test, however, because this bias is compensated for.

FORMULA 8.5

definitional formula for the independent *t*

$$\text{independent } t = \frac{(\bar{X}_1 - \bar{X}_2) - 0}{\sqrt{s^2_{\text{pooled}}\left(\frac{1}{N_1} + \frac{1}{N_2}\right)}}$$

$$= \frac{(\bar{X}_1 - \bar{X}_2) - 0}{\sqrt{\frac{(N_1 - 1)s_1^2 + (N_2 - 1)s_2^2}{N_1 + N_2 - 2}\left(\frac{1}{N_1} + \frac{1}{N_2}\right)}}$$

In Formula 8.5, 0 has been substituted for $\mu_{\bar{x}_1 - \bar{x}_2}$, which here is the means of the sampling distribution of the difference. Formula 8.5 demonstrates how similar the independent t is to the one-sample t in Formula 7.3. For Formula 8.5,

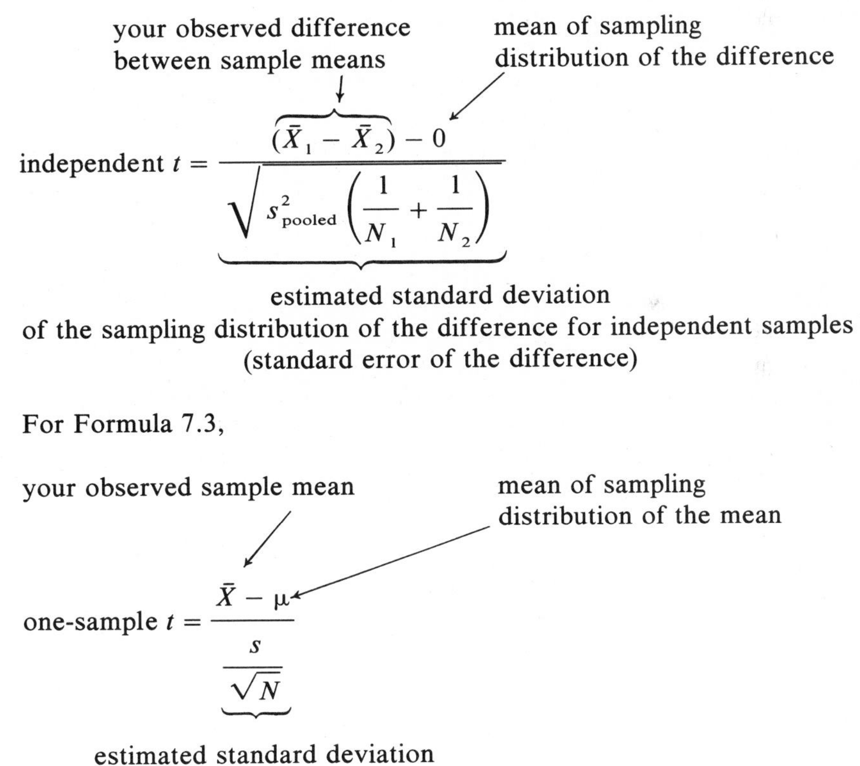

The specific symbols in the independent t test formula differ from those

in the one-sample t test formula because the tests refer to different sampling distributions.

Now you can identify the region of rejection in the distribution of the difference. Critical values of t can be specified by using Table t^*. Degrees of freedom for the independent t test are $N_1 + N_2 - 2$. For a two-tailed test, any difference between a pair of sample means that yields a calculated t score greater in absolute magnitude than t^* will be significant.

Figure 8-4 illustrates the region of rejection for one- and two-tailed tests. For a *two-tailed test* you are trying to show that the treatment conditions produce different effects. It does not matter which sample mean turns out to be greater, only that they turn out to be different. That is, when you subtract one sample mean from the other, the direction of the difference (that is, whether it is positive or negative) is immaterial. H_A is that the samples occurred from populations with different means, or $\mu_1 \neq \mu_2$. In an experiment this says that your treatment conditions have different effects. H_0 is the logical complement of H_A and states that the samples occurred by chance from populations with equal means, or $\mu_1 = \mu_2$. The two-tailed case is shown in the top sampling distribution in Figure 8-4. For a *one-tailed test,* the tail in which the region of rejection is located depends on the specific conclusion you want to reach. If you want to conclude that Treatment 1 increases scores over that of treatment 2, H_A is that the mean of the population from which sample 1 was selected is greater than that of the population from which sample 2 was selected: $\mu_1 > \mu_2$. H_0 is $\mu_1 \leq \mu_2$. On the other hand, if you want to conclude that treatment 1 decreases scores below that of treatment 2, H_A is $\mu_1 < \mu_2$, and H_0 is $\mu_1 \geq \mu_2$. These one-tailed cases are depicted with the middle and bottom sampling distributions in Figure 8-4.

The hypothesis-testing procedure using the independent t test is:

1. Determine the appropriate H_0 and H_A. (H_0 for a two-tailed test is that your samples are drawn from populations having the same means.) Assume H_0 is true.
2. Set alpha and calculate $df = N_1 + N_2 - 2$. Determine the critical value of t from Table t^* for your alpha, *df*, and whether you are conducting a one- or two-tailed test.

   If you are conducting a two-tailed test, then proceed to step 3.

   If you are conducting a one-tailed test: if you predict $\mu_1 > \mu_2$, subtract $\bar{X}_2$ from $\bar{X}_1$ and expect a positive difference; if you predict $\mu_1 < \mu_2$, subtract $\bar{X}_2$ from $\bar{X}_1$ and expect a negative difference. Draw a diagram of the sampling distribution of the difference and mark the region of rejection to see whether the difference between your sample means is in the direction expected with H_A.

   If $\bar{X}_1 - \bar{X}_2$ is in the expected direction, proceed to Step 3.

   If $\bar{X}_1 - \bar{X}_2$ is in the opposite direction, retain H_0 and conclude that there is not sufficient evidence to say your samples were drawn from populations

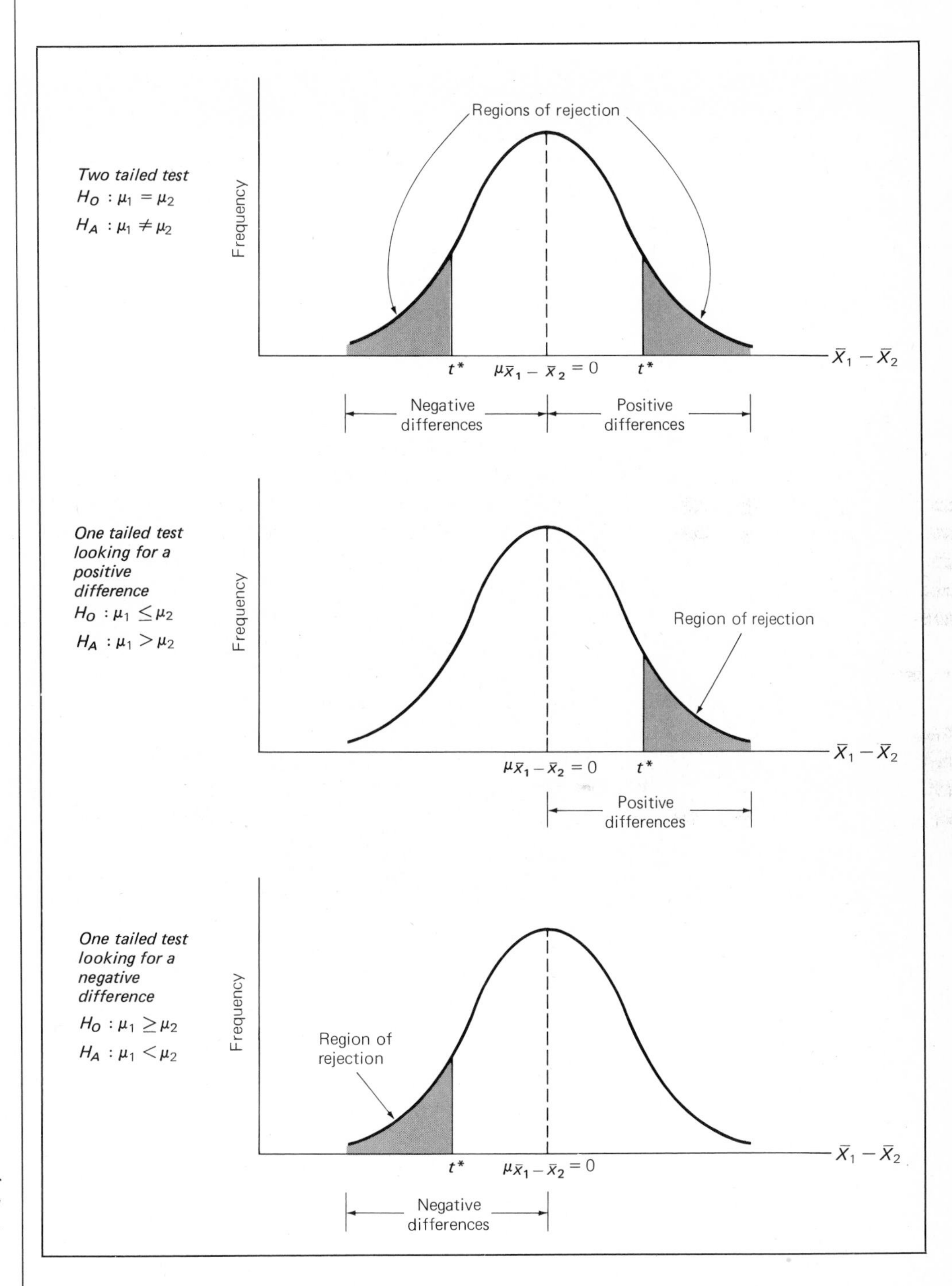

FIGURE 8-4

The above three sampling distributions of the difference (between means) illustrate the regions of rejection for one- and two-tailed tests using the independent t test.

having means that differ in the direction specified in H_A.

3. Calculate a t score for the difference between your sample means by Formula 8.5:

$$t = \frac{(\bar{X}_1 - \bar{X}_2) - 0}{\sqrt{\frac{(N_1 - 1)s_1^2 + (N_2 - 1)s_2^2}{N_1 + N_2 - 2}\left(\frac{1}{N_1} + \frac{1}{N_2}\right)}}$$

where s_1^2 for Treatment 1 and s_2^2 for Treatment 2 are calculated by either Formulas 8.1 or 8.2:

$$s^2 = \frac{\Sigma(X - \bar{X})^2}{N - 1} = \frac{\Sigma X^2 - \frac{(\Sigma X)^2}{N}}{N - 1}$$

Refer to the value of t obtained by Formula 8.5 as the *calculated value of t*, and symbolize its absolute value as $|t|$. This is the number of estimated standard deviations that the difference between your pair of sample means lies from the mean of the sampling distribution of the difference.

4. If $|t| > t^*$, reject H_0 and conclude that your samples were drawn from populations having different means.

   If $|t| \leq t^*$, retain H_0 and conclude that there is not sufficient evidence to say your samples were drawn from populations having different means, as specified in H_A.

assumptions for independent *t* test

Before you apply the two-sample independent t test, you should check to see whether three **assumptions** are met:

1. The two population distributions are normal.
2. The variances of the two populations are equal: $\sigma_1^2 = \sigma_2^2$ (homogeneity of variance).
3. The subjects comprising each sample are selected randomly, and the samples are independent.

The first assumption, that the populations are normal, is necessary to infer that the sampling distribution of the difference is normal. This assumption is not necessary with the one-sample z test with larger sample sizes ($N \geq 25$), because the central limit theorem assures you the sampling distribution of the mean is normal even when the population distribution is not. However, with the one-sample t test and the independent t test you must compensate for the added variability introduced by estimating. This can be done by using the critical values of t in Table t^*, but only when the populations are approximately normally distributed.

The second assumption, that the variances of the populations are equal, allows you to use the variances of each sample and pool them to estimate $\sigma_{\bar{X}_1 - \bar{X}_2}$. This assumption of equal population variance is often referred

homogeneity of variance

random samples

independent samples

to as the assumption of **homogeneity of variance.**

The third assumption requires that subjects are selected on a random basis and that the samples are independent. In a **random sample** there must be an equal chance for every subject in the population to be included. Your sampling method must not introduce any bias by making it more likely that certain subjects rather than others will be chosen. This is a necessary assumption for all types of research involving making inferences from samples to populations. In other words, random sampling is a necessary assumption for inferential statistics in general and not merely the t tests. For the samples to be **independent,** the subjects in each group must be different subjects, and subjects from one condition cannot be "matched" or "paired" in any fashion with subjects in the other condition. The independent t test cannot be applied in a study in which you test a group of subjects under one condition and then retest them under another condition. For example, if you use the same students and test them after studying under spaced conditions, and then retest them after studying under massed conditions, you cannot use the independent t test.

Moreover, the subjects in one group cannot be "matched" or "paired" with specific subjects in the other group. You cannot apply the independent t test when you match pairs of subjects on some characteristic in an attempt to avoid possibly confounding[3] the results. For example, if you are doing a study on reaction time (RT) in two different psychiatric groups, you might want to be sure the two groups are about equal on all factors related to RT except for the difference in treatment conditions. You do not want the subjects in one condition to be much older than those in the other because RT increases as you get older. To avoid this confounding you can make sure that for every subject of a given age in one group you have a subject of the same age in the other group. This pairwise matching procedure assures the same mean age, but then you are violating the assumption of independence and cannot apply the independent t test to the RT data. You have to apply the two-sample hypothesis test described in the next section of this chapter, the dependent t test. It is usually easy to determine if the assumptions of randomness and independence are satisfied by examining the procedure used to select the samples.

It is generally possible to determine whether the assumptions of normality and equal population variance are satisfied by examining the distributions of scores in the samples. On visual inspection, if each sample distribution appears roughly normal in shape and if you have no reason to suspect the populations are not normally distributed, then it is reasonable to consider the assumption of normality to be satisfied. If the sample variances are roughly the same, then it is reasonable to conclude that the variances of the populations from which the samples were drawn are

[3] Confound means to interfere, that is, to make a clear interpretation of the results difficult.

robustness

equal, thereby satisfying the assumption of homogeneity of variance. These seemingly casual inferences about the populations are justified by a property of the t tests called **robustness.** Robustness refers to the finding that the t test leads to accurate conclusions even when the assumptions of normality and homogeneity of variance are not strictly satisfied.[4] It is therefore not necessary to worry too much about whether the population variances are exactly equal, especially if your sample sizes are equal. The assumption of homogeneity of variance is most critical when the size of the two samples differs considerably. A good rule of thumb is that if the sample sizes differ considerably, for example if the larger is over 50% greater than the smaller, consider the assumption of homogeneity of variance not satisfied when the larger sample variance is more than 50% greater than the smaller.

Now apply the independent t test to the data on spaced versus massed studying. To decide whether to attribute the difference of 8.9 points between the means to the method of study or to sampling error, use the following hypothesis-testing procedure:

1. Many previous studies similar to the one you just conducted found performance to be best with spaced studying. You therefore expect better performance in the spaced condition and conduct a one-tailed test. H_A is that spaced studying improves performance. Technically, this says the mean performance of the population of subjects who study under spaced conditions (μ_1) is greater than the mean of the population who study under massed conditions (μ_2). H_0 is that spaced studying does not improve performance; or, technically speaking, the mean of the population of subjects who study under spaced conditions is not greater than the mean of the population studying under massed conditions.

Symbolically, these hypotheses are:

H_0: $\mu_1 \leq \mu_2$ (Spaced studying does not improve performance)

H_A: $\mu_1 > \mu_2$ (Spaced studying improves performance)

2. Let $\alpha = .05$. Although several studies have demonstrated that spaced conditions produce better performance than massed conditions, your experiment does not exactly replicate the conditions of the previous studies, so you let $\alpha = .05$. Besides, should an alpha error occur in your experiment, you do not believe it will injure your professional reputation.

Since you have 10 subjects in one condition and 12 in the other, $df = N_1 + N_2 - 2 = 10 + 12 - 2 = 20$.

Using Table t^*, you find that for a one-tailed test at $\alpha = .05$ and $df = 20$, $t^* = 1.725$.

[4] There are numerous advanced statistics texts that describe precise methods for determining whether the assumptions of normality and homogeneity of variance are met. Because of the robustness of the t tests, however, these tests are rarely used.

Since this is a one-tailed test and you predict $\mu_1 > \mu_2$, you expect a positive difference when you subtract $\bar{X}_2$ from $\bar{X}_1$. If you draw a diagram (as shown) of the sampling distribution of the difference and mark the region of rejection, you will see that the positive difference of 8.9 is in the direction specified by H_A.

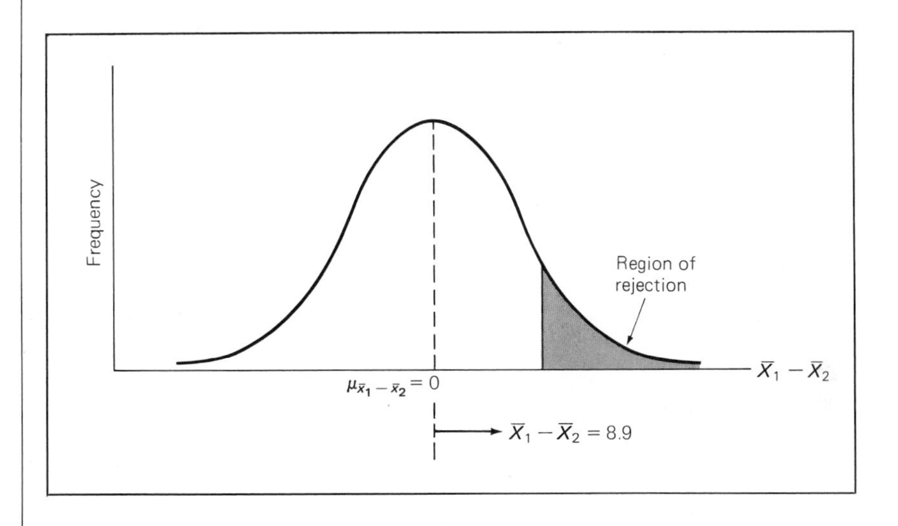

3. Since the positive difference of 8.9 is in the direction specified by H_A, you calculate a *t* score by first solving for s_1^2 and s_2^2 and then plug these two values into Formula 8.5:

	Spaced condition		*Massed condition*		
	X_1	X_1^2	X_2	X_2^2	
	26	676	20	400	
	62	3844	57	3249	
	68	4624	64	4096	
$\bar{X}_1 = 49.9$	31	961	42	1764	$\bar{X}_2 = 41$
	10	100	39	1521	
$N_1 = 10$	44	1936	26	676	$N_2 = 12$
	84	7056	38	1444	
	61	3721	58	3364	
	62	3844	16	256	
	51	2601	50	2500	
			49	2401	
			33	1089	

$\Sigma X_1 = 499$ $\quad \Sigma X_1^2 = 29{,}363$ $\quad \Sigma X_2 = 492$ $\quad \Sigma X_2^2 = 22{,}760$

$(\Sigma X_1)^2 = (499)^2 = 249{,}001$ $\quad (\Sigma X_2)^2 = (492)^2 = 242{,}064$

$$s_1^2 = \frac{\Sigma X_1^2 - \dfrac{(\Sigma X_1)^2}{N_1}}{N_1 - 1} \qquad s_2^2 = \frac{\Sigma X_2^2 - \dfrac{(\Sigma X_2)^2}{N_2}}{N_2 - 1}$$

$$= \frac{29{,}363 - \dfrac{249{,}001}{10}}{10 - 1} \qquad = \frac{22{,}760 - \dfrac{242{,}064}{12}}{12 - 1}$$

$$= 495.878 \qquad = 235.273$$

Now substitute known values into the formula for the independent t:

$$t = \frac{(\bar{X}_1 - \bar{X}_2) - 0}{\sqrt{\dfrac{(N_1 - 1)s_1^2 + (N_2 - 1)s_2^2}{N_1 + N_2 - 2}\left(\dfrac{1}{N_1} + \dfrac{1}{N_2}\right)}}$$

$$= \frac{(49.9 - 41) - 0}{\sqrt{\dfrac{(10 - 1)495.878 + (12 - 1)235.273}{10 + 12 - 2}\left(\dfrac{1}{10} + \dfrac{1}{12}\right)}}$$

$$= \frac{8.9 - 0}{\sqrt{\dfrac{9(495.878) + 11(235.273)}{20}(.1 + .083)}}$$

$$= \frac{8.9}{\sqrt{\dfrac{4462.902 + 2588.003}{20}(.183)}}$$

$$= \frac{8.9}{\sqrt{\dfrac{7050.905}{20}(.183)}}$$

$$= \frac{8.9}{\sqrt{352.545\ (.183)}}$$

$$= \frac{8.9}{\sqrt{64.515}} = \frac{8.9}{8.032} = 1.108$$

Notice that the sign (+ or −) of the calculated t is determined in the numerator and depends on which sample mean you substitute for $\bar{X}_1$ and $\bar{X}_2$. If you had arbitrarily let the mean of the massed condition (41) substitute for $\bar{X}_1$ and the mean of the spaced condition (49.9) substitute for $\bar{X}_2$, the calculated t would be -1.108. The interpretation of the t value you obtain is the same regardless of sign so long as you draw a diagram when you conduct a one-tailed test to see whether the difference between the sample means is in the direction specified by H_A.

4. The t value for the difference between your sample means is not sufficiently large to put it into the region of rejection, as you can see in the sampling distribution of the difference shown.

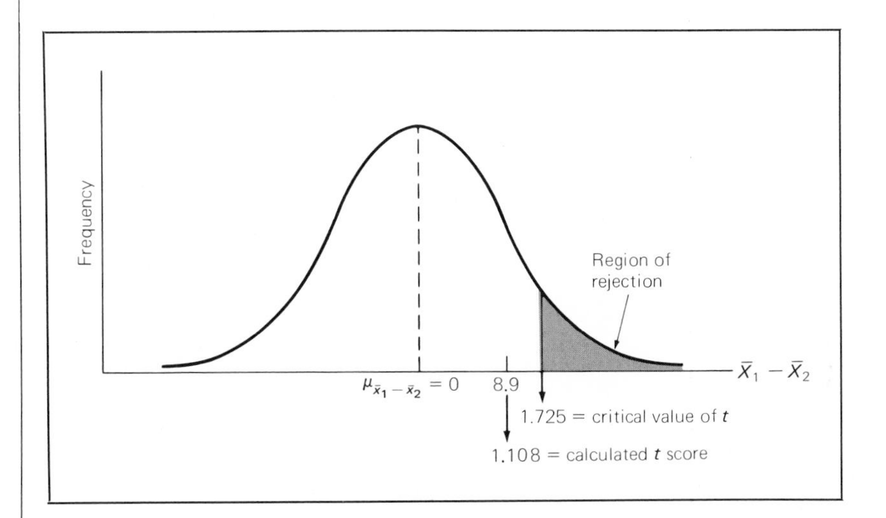

Since $|1.108| < 1.725$, retain H_0 and conclude that there is not sufficient evidence to say the spaced condition produces better performance than the massed condition.

Although the difference between sample means is in the direction you expect, the calculated t falls short of significance. You do not have sufficient evidence to corroborate previous research findings which showed spaced study conditions are superior to massed study conditions. Either the specific conditions of your study are not conducive to showing the superiority of spaced study, or your hypothesis test was not powerful enough to detect the effect if it does exist in your study. If you believe H_A is actually true but your hypothesis test was not sufficiently powerful to detect this, you might consider re-performing your experiment with a larger sample to increase power. Alternatively you might carefully re-examine the specific conditions of your experiment to see what might cause spaced study to have a weaker effect than you expected.

The two-sample dependent t test

The independent t test is appropriate whenever you have two conditions, and three assumptions are approximately met: the two populations are normal; the two population variances are equal (homogeneity of variance); and the samples are independent. However, there are many research situations in which the assumption of sample independence is not met. For example, suppose you are interested in the effect of a certain type of meditation on creativity. You select a random sample of 12 subjects, give them a test of creativity you developed, and then teach them to meditate. After two weeks of practicing meditation every day, you readminister your creativity test. The "before" and "after" scores on the creativity test are:

Condition 1: Before meditation Subject	X_1	*Condition 2: After meditation* Subject	X_2
S_1	36	S_1	41
S_2	56	S_2	61
S_3	94	S_3	83
S_4	58	S_4	33
S_5	57	S_5	43
S_6	92	S_6	96
S_7	76	S_7	69
S_8	74	S_8	69
S_9	8	S_9	26
S_{10}	53	S_{10}	33
S_{11}	49	S_{11}	18
S_{12}	83	S_{12}	79
	$\bar{X}_1 = 61.33$		$\bar{X}_2 = 54.25$

You can compare the sample means and see that subjects did better on the test of creativity before they practiced meditation. But is this difference due to some inhibiting effect of meditation or is it due to sampling error? For example, by chance, after two weeks the subjects may have been bored, less motivated, sleepy, or the like. This might be due solely to random chance fluctuations over time and may have nothing to do with the effects of meditation (or in this experiment the extraneous confounding variable of *being forced* to meditate.) You need to decide whether to attribute the 7.08 difference between the sample means to the treatment effect or to sampling error. Although you have two treatment conditions, the independent t test is not appropriate because the same subjects participated in both conditions and therefore the samples are not independent. Fortunately, it is possible to create a single new sample of independent scores on which you can apply a variation of the one-sample t test. This is done by subtracting each subject's creativity score "after" meditation from the score "before" meditation. This renders a new sample of independent scores called difference scores, or ***D* scores** for short:

D score

Subject	*Before meditation* X_1	*After meditation* X_2	*Difference (D)* $X_1 - X_2$
S_1	36	41	−5
S_2	56	61	−5
S_3	94	83	11
S_4	58	33	25
S_5	57	43	14
S_6	92	96	−4
S_7	76	69	7
S_8	74	69	5
S_9	8	26	−18
S_{10}	53	33	20
S_{11}	49	18	31
S_{12}	83	79	4
			$\Sigma D = 85$

$$\bar{D} = \frac{\Sigma D}{N} = \frac{85}{12} = 7.08$$

The D scores are independent because each individual D score comes from a different subject. The sample mean of the D scores is symbolized $\bar{D}$; in this example, $\bar{D} = 7.08$. This is the same as the difference between $\bar{X}_1$ and $\bar{X}_2$. If meditation has no effect, you expect the first creativity test score for each subject to be roughly the same as the second. Of course, there may be fluctuations in each subject's pair of scores due to chance differences in motivation or whatever. On the average, however, for each subject who does slightly poorer after meditation, there is probably one who does slightly better. Thus if meditation has no effect, you expect the individual fluctuations to balance each other out, so $\bar{D}$ will be close to zero. Conversely, if meditation has an effect, you expect $\bar{D}$ to be different from zero.

You can translate this into a consideration of H_0 and H_A. H_0 is that meditation has no effect. If meditation has no effect and you measure the creativity of everyone in the population "before meditation" and "after meditation" and then generate D scores, you should get a mean D score of zero. Since this is a population mean we symbolize it by μ_D, where the subscript D signifies that μ is the mean of a population of *difference* scores. Thus H_0 is that $\mu_D = 0$. H_A is that meditation has some effect. If meditation has some effect and you get D scores for the entire population "before" and "after" meditation, you should get a mean D score that is somewhat different from zero. Thus H_A is that $\mu_D \neq 0$. The question of whether your results are due to sampling error or to a real effect of meditation then becomes, "Is it relatively unlikely to observe a sample $\bar{D} = 7.08$ from the H_0 population where $\mu_D = 0$?" If so, you can reject H_0 and conclude that meditation has some effect on creativity. The two hypotheses you are testing are illustrated in Figure 8-5.

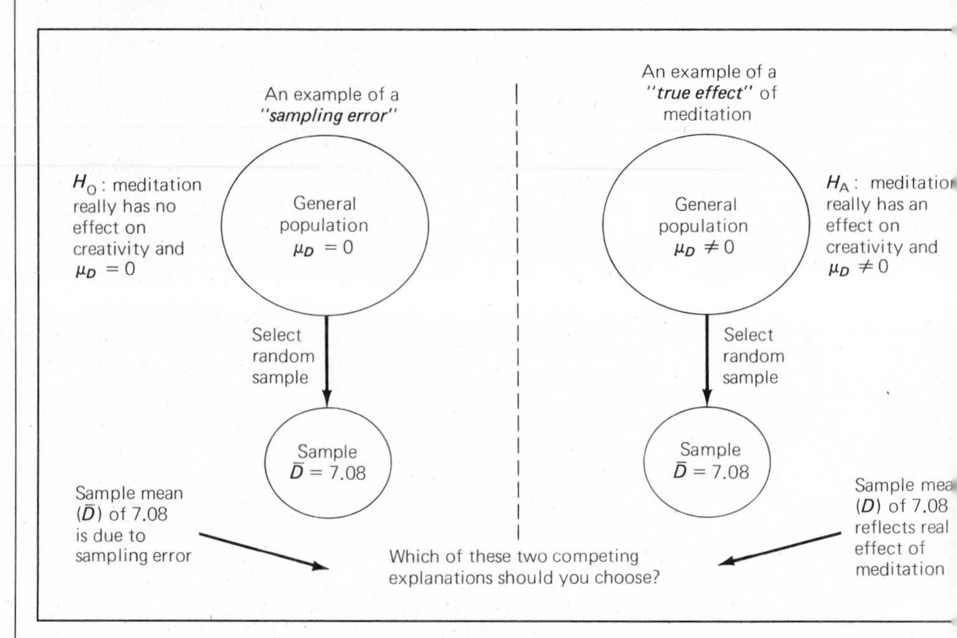

FIGURE 8-5

The two competing explanations for the results of the meditation experiment.

This is essentially the same problem you had with the one-sample hypothesis tests. Could a given sample mean have arisen by chance from a population with a mean of 0? Here you are dealing with a population and a sample of difference scores, but this does not alter the nature of the appropriate hypothesis test. Instead of X's in the formulas you will have D's, just to remind you that you are dealing with difference scores.

two-sample dependent t test

The hypothesis test you are conducting here is called the **two-sample dependent t test.** The similarity between the two-sample dependent t test and the one-sample t test is further illustrated in Figure 8-6. The population distribution for the one-sample t test is a distribution of raw scores (X's) with a mean of μ and a standard deviation of σ. For the two-sample dependent t test the population distribution is a distribution of difference scores (D's) with a mean of $\mu_D = 0$ and a standard deviation of σ_D. The subscript D reminds you that you are dealing with difference scores. To determine how unlikely a given sample mean ($\bar{X}$) is with the one-sample t test, you see where it lies in the sampling distribution of the mean, which is a distribution of sample means ($\bar{X}$'s) which would occur by chance given that H_0 is true. The mean of the sampling distribution is μ and the standard deviation (standard error) is $\sigma/\sqrt{N}$. For the two-sample dependent t test, to determine how unlikely a given sample mean difference score ($\bar{D}$) is, you need to see where it lies in a sampling distribution of $\bar{D}$ scores. The sampling distribution here is a distribution of sample $\bar{D}$'s which would occur by chance given that H_0 is true. The mean of the sampling distribution is $\mu_{\bar{D}} = 0$ and the standard deviation (standard error) is $\sigma_{\bar{D}} = \sigma_D/\sqrt{N}$.

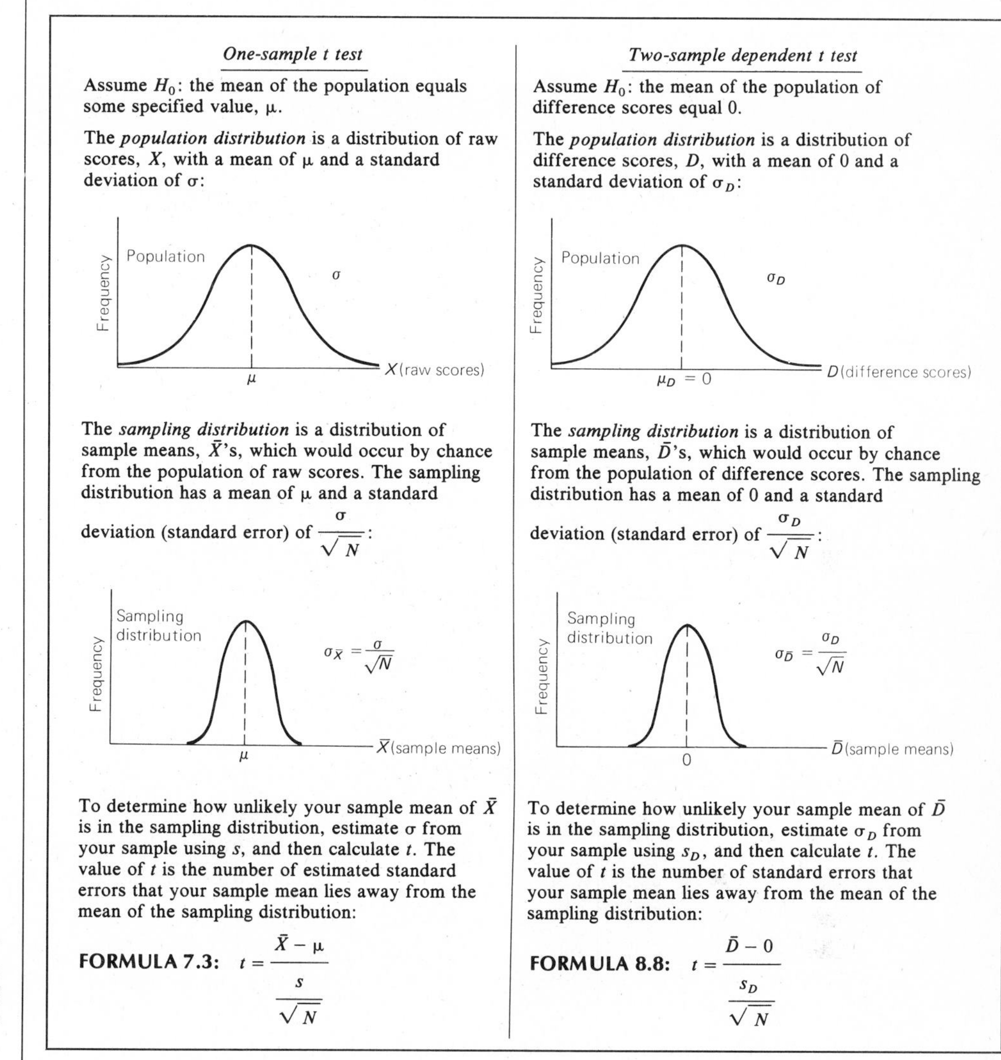

FIGURE 8-6

Comparison of the one-sample t test and the two-sample dependent t test.

If you knew what the population standard deviation (σ_D) is you could do a z test. Since you have not gone out and calculated D scores for the entire population, you obviously do not know the value of σ_D. You can, however, estimate σ_D from your sample by calculating s, and then you can follow the procedure for the one-sample t test. You will denote your calculation of s here as s_D, to remind you that you are dealing with difference scores. The formula for s_D is derived by taking Formula 7.2 for s, which is

s_D

$$s = \sqrt{\frac{\Sigma(X - \bar{X})^2}{N - 1}}$$

and substituting D for X and $\bar{D}$ for $\bar{X}$. This then gives the **definitional formula for s_D**:

FORMULA 8.6
definitional formula for s_D

$$s_D = \sqrt{\frac{\Sigma(D - \bar{D})^2}{N - 1}}$$

Similarly, you can derive a **computational formula for s_D** by taking the computational formula for s and substituting D for X as follows:

FORMULA 8.7
computational formula for s_D

$$s_D = \sqrt{\frac{\Sigma D^2 - \frac{(\Sigma D)^2}{N}}{N - 1}}$$

Formula 7.3 for the one-sample t test is:

$$t = \frac{\bar{X} - 0}{\frac{s}{\sqrt{N}}}$$

dependent t test

All you have to do to adapt this formula to the present case is to substitute $\bar{D}$ for $\bar{X}$ and s_D for s. The resulting hypothesis test is called the **two-sample dependent t test,** or **dependent t test.** The definitional formula is:

FORMULA 8.8
definitional formula for dependent t test

$$\text{dependent } t = \frac{\bar{D} - 0}{\frac{s_D}{\sqrt{N}}}$$

Here, N is the number of D scores you have, which is the same as the number of *pairs* of scores.

The hypothesis-testing procedure for the dependent t test is essentially the same, after you calculate D scores, as that for the one-sample t test. The complete hypothesis-testing procedure is:

1. Determine the appropriate H_0 and H_A. (H_0 for a two-tailed test is that the mean difference score in the population, μ_D, is zero. This says that the population of "before" scores has the same mean as the population of "after" scores, or $\mu_1 = \mu_2$.) Assume H_0 is true.

2. Set alpha and calculate $df = N - 1$, where N is the number of pairs of scores. Determine the critical value of t from Table t^* for your alpha, df, and whether you are conducting a one- or a two-tailed test.

   If you are conducting a two-tailed test, proceed to Step 3.

   Calculate a set of D scores by subtracting each subject's X_2 score ("after" score) from his or her X_1 score ("before" score). Then find the mean of the D scores.

   If you are conducting a one-tailed test: if you predict the "before" scores to be higher than the "after" scores, expect a positive $\bar{D}$; if you predict the "after" scores to be higher than the "before" scores, expect a negative $\bar{D}$.

   Draw a diagram of the sampling distribution (analogous to the sampling distribution of the mean for the one-sample t test, except that here you have a distribution of $\bar{D}$'s instead of $\bar{X}$'s) and mark the region of rejection to see whether your $\bar{D}$ is in the direction expected with H_A.

   If $\bar{D}$ is in the expected direction, proceed to Step 3.

   If $\bar{D}$ is in the opposite direction, retain H_0 and conclude that there is not sufficient evidence to say your samples were drawn from populations having means that differ in the direction specified in H_A.

3. Calculate a t score for your $\bar{D}$ with Formula 8.8:

$$t = \frac{\bar{D} - 0}{\dfrac{s_D}{\sqrt{N}}}$$

   where s_D is calculated by either Formula 8.6 or 8.7:

$$s_D = \sqrt{\frac{\Sigma(D - \bar{D})^2}{N - 1}} = \sqrt{\frac{\Sigma D^2 - \dfrac{(\Sigma D)^2}{N}}{N - 1}}$$

   Refer to the value of t obtained by Formula 8.8 as the *calculated value of t*, and symbolize its absolute value as $|t|$. This is the number of estimated standard deviations that your $\bar{D}$ lies from the mean of the sampling distribution.

4. If $|t| > t^*$, reject H_0 and conclude that your treatment has an effect in the direction specified by H_A.

   If $|t| \leq t^*$, retain H_0 and conclude that there is not sufficient evidence to say that your treatment has an effect in the direction specified by H_A.

assumptions for dependent *t* test

The two **assumptions** that should be met before applying the two-sample dependent t test are:

1. The populations of X_1 scores and X_2 scores are both normal.
2. The subjects comprising the samples of X_1 and X_2 scores are selected on a random but *dependent* basis. For example either the pairs of X_1 and X_2

scores are provided by randomly selecting subjects in specific pairs such that one member of each pair contributes the X_1 score and the other member contributes the X_2 score, or by testing and retesting the same subjects under two different conditions.

The dependent t test is very robust with respect to violations of the assumption of normality. However, you should not violate the second assumption.

You match pairs of subjects on those traits or abilities that you believe might affect the dependent variable or may mask a treatment effect. In general, the dependent t test should be used whenever the same sample of subjects is tested and then retested under different conditions, or when the subjects in two samples are deliberately paired up and then tested under different conditions.

To apply the hypothesis-testing procedure to the meditation study, proceed with the dependent t test as follows:

1. Because conflicting claims have been made, you have no strong reason to expect that the type of meditation you are investigating will either enhance or reduce creativity. You therefore conduct a two-tailed test. H_0 is that meditation has no effect on creativity. This says that the mean creativity scores of the population of subjects who are tested "before" meditation (μ_1) is equal to the mean of the population who are tested "after" meditation (μ_2). H_A is that meditation has some effect on creativity. This says that the means of the "before" meditation and "after" meditation populations are not equal:

H_0: $\mu_1 = \mu_2$ (Meditation has no effect on creativity)
H_A: $\mu_1 \neq \mu_2$ (Meditation has some effect on creativity)

2. Let $\alpha = .05$ because this is a relatively new area of research and you do not want to overlook any potentially important discoveries. Since you have 12 pairs of scores, $df = N - 1 = 12 - 1 = 11$. Using Table t^*, find that for a two-tailed test at $\alpha = .05$ and $df = 11$, $t^* = 2.201$. Create a set of D scores and calculate $\bar{D}$.

3. Calculate a t score for your $\bar{D}$ of 7.08 by first solving for s_D and then plugging these values into Formula 8.8: Here are the calculations for D, $\bar{D}$, and s_D:

	Subject	*Before* X_1	*After* X_2	D $(X_1 - X_2)$	D^2
	S_1	36	41	−5	25
	S_2	56	61	−5	25
	S_3	94	83	11	121
	S_4	58	33	25	625
	S_5	57	43	14	196
$\bar{D} = 7.08$	S_6	92	96	−4	16
$N = 12$	S_7	76	69	7	49
	S_8	74	69	5	25

Subject	*Before* X_1	*After* X_2	D $(X_1 - X_2)$	D^2
S_9	8	26	−18	324
S_{10}	53	33	20	400
S_{11}	49	18	31	961
S_{12}	83	79	4	16
			$\Sigma D = 85$	$\Sigma D^2 = 2783$

$$(\Sigma D)^2 = (85)^2 = 7225$$

$$s_D = \sqrt{\frac{\Sigma D^2 - \frac{(\Sigma D)^2}{N}}{N - 1}} = \sqrt{\frac{2783 - \frac{7225}{12}}{12 - 1}}$$

$$= \sqrt{\frac{2783 - 602.083}{11}}$$

$$= \sqrt{\frac{2180.917}{11}} = \sqrt{198.265} = 14.081$$

Now substitute known values into Formula 8.8 for the dependent t:

$$t = \frac{\bar{D} - 0}{\frac{s_D}{\sqrt{N}}}$$

$$= \frac{7.08 - 0}{\frac{14.081}{\sqrt{12}}} = \frac{7.08}{4.065} = 1.742$$

4. Since $|1.742| < 2.201$, retain H_0 and conclude that there is not sufficient evidence to say meditation affects creativity.

Comparison of the one-sample and two-sample hypothesis tests

The four hypothesis tests you have seen (z test, one-sample t test, independent t test, and dependent t test) appear very different from each other on the surface. However, it is very important to realize that basically the same procedure is employed in each. The only differences between them are that they use different sampling distributions and that it is necessary to estimate a population standard deviation in the three t tests, which is unnecessary with the z test. The underlying similarity is that they all share the same method, as can be seen in the definitional formulas below:

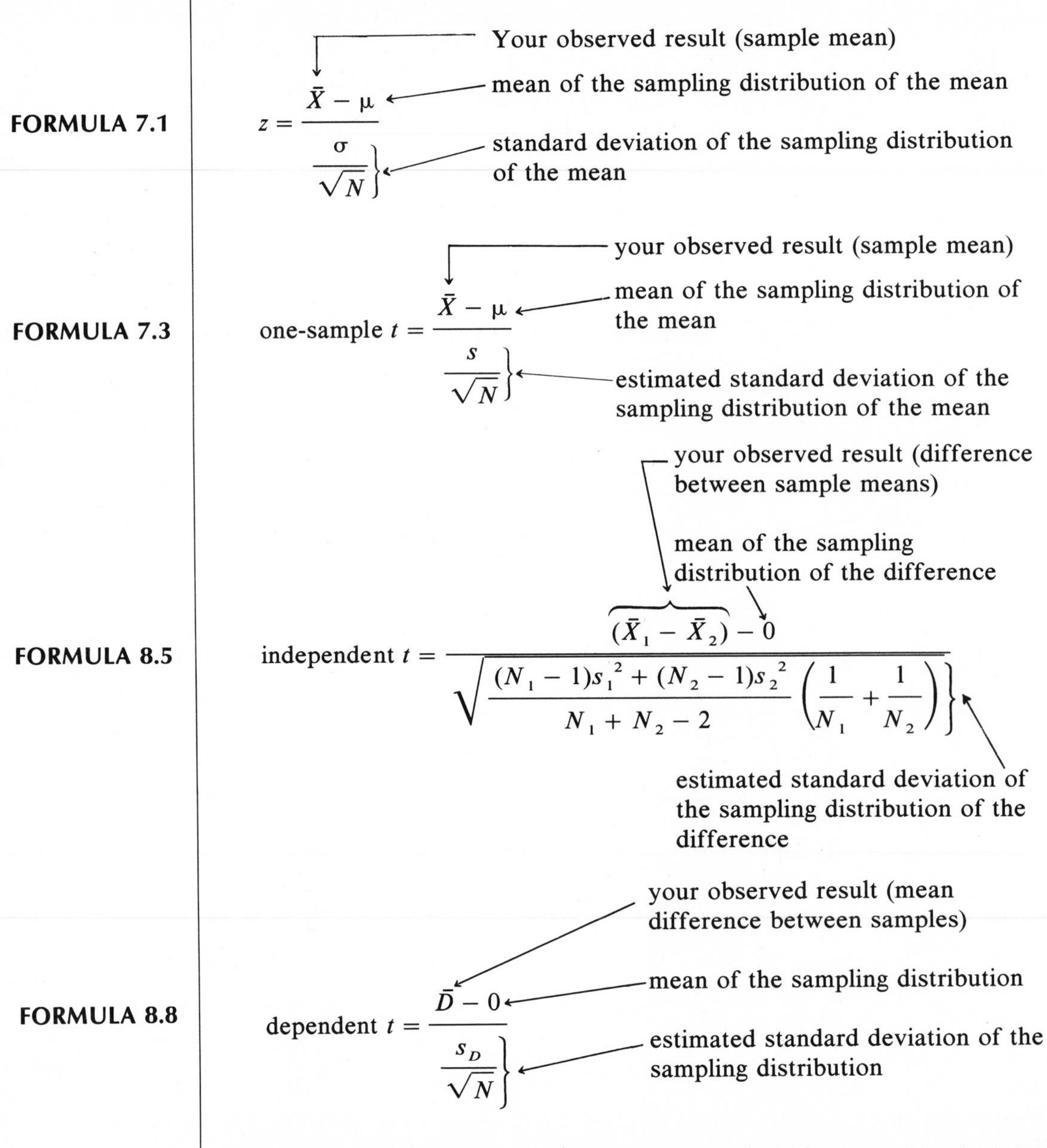

In all cases, you are finding how far away from the mean of the sampling distribution your observed result lies, with the distance measured in terms of standard deviation units (or estimated standard deviation units). The purpose is to subtract the mean of the sampling distribution from your observed result to get the distance, and then divide by the [estimated] standard deviation to find the distance in terms of standard deviation units.

When to use the one-sample and two-sample hypothesis tests

With so much material about hypothesis tests presented so rapidly, many people become confused about which hypothesis test is appropriate in a given situation. Selecting the appropriate hypothesis test is not as difficult as it first appears, however, because each of the four hypothesis tests examined so far is useful in only one specific research situation. Consequently, to decide which is the appropriate test, it is necessary to ask yourself only two questions. The first question to ask is, "Do you have *one* sample of raw scores or *two* samples of raw scores?" If the answer to the first question is "*one sample,*" then the second question to ask yourself is, "Do you know the population standard deviation (or variance)?" If the answer to this question is *yes,* use the *z* test. If the answer is *no,* you must estimate the standard deviation and use the one-sample *t* test.

If the answer to the first question is "two samples," then the second question becomes, "Are the samples composed of different subjects with no attempt to pair up the scores?" If the answer to this question is *yes,* use the independent *t*. If the answer is *no,* use the dependent *t*. The entire decision procedure is outlined below:

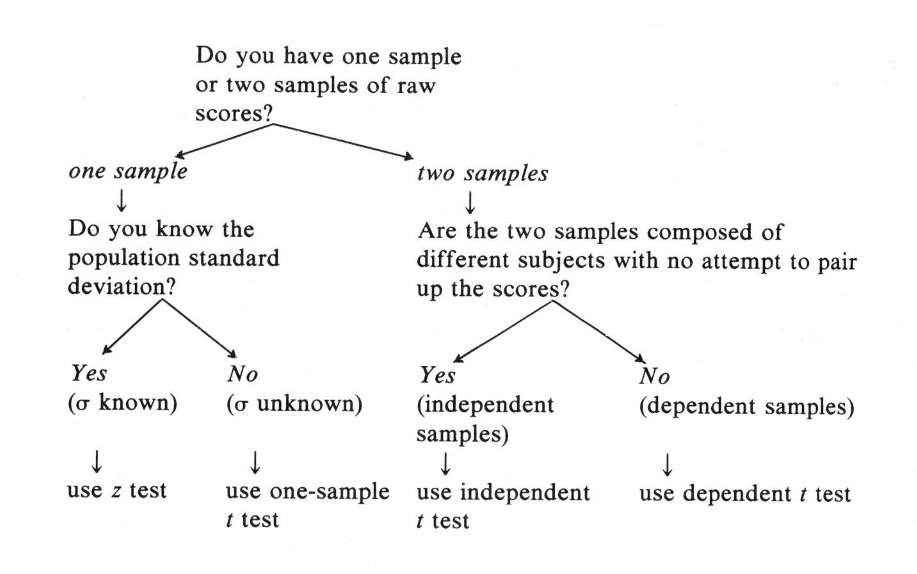

Computational formulas for the independent and dependent *t* tests

The computational formulas for the independent and dependent *t* tests are generally easier to use than the definitional formulas. For the independent *t* test, the computational formula is derived by first inserting the computational formula for s^2 (Formula 8.2) into the definitional formula for the independent *t* (Formula 8.4):

$$\text{independent } t = \frac{(\bar{X}_1 - \bar{X}_2) - 0}{\sqrt{\dfrac{(N_1 - 1)\left[\dfrac{\Sigma X_1^2 - \dfrac{(\Sigma X_1)^2}{N_1}}{N_1 - 1}\right] + (N_2 - 1)\left[\dfrac{\Sigma X_2^2 - \dfrac{(\Sigma X_2)^2}{N_2}}{N_2 - 1}\right]}{N_1 + N_2 - 2}\left(\dfrac{1}{N_1} + \dfrac{1}{N_2}\right)}}$$

To simplify things, cancel the $(N_1 - 1)$ and $(N_2 - 1)$ terms under the radical. This gives the computational formula for the independent t:

FORMULA 8.9

computational formula for the independent t

$$\text{independent } t = \frac{(\bar{X}_1 - \bar{X}_2)}{\sqrt{\dfrac{\Sigma X_1^2 - \dfrac{(\Sigma X_1)^2}{N_1} + \Sigma X_2^2 - \dfrac{(\Sigma X_2)^2}{N_2}}{N_1 + N_2 - 2}\left(\dfrac{1}{N_1} + \dfrac{1}{N_2}\right)}}$$

With Formula 8.9 you can conduct an independent t test without separately calculating s_1^2 and s_2^2 beforehand. For the spaced-study versus massed-study example, the computational Formula 8.9 also renders a calculated t of 1.108:

	X_1	X_1^2	X_2	X_2^2	
	26	676	20	400	
	62	3844	57	3249	
	68	4624	64	4096	
$\bar{X}_1 = 49.9$	31	961	42	1764	$\bar{X}_2 = 41$
	10	100	39	1521	
$N_1 = 10$	44	1936	26	676	$N_2 = 12$
	84	7056	38	1444	
	61	3721	58	3364	
	62	3844	16	256	
	51	2601	50	2500	
			49	2401	
			33	1089	

$\Sigma X_1 = 499$ $\quad \Sigma X_1^2 = 29{,}363$ $\quad \Sigma X_2 = 492$ $\quad \Sigma X_2^2 = 22{,}760$

$(\Sigma X_1)^2 = (499)^2 = 249{,}001$ $\quad (\Sigma X_2)^2 = (492)^2 = 242{,}064$

$$\text{independent } t = \frac{(\bar{X}_1 - \bar{X}_2)}{\sqrt{\dfrac{\Sigma X_1^2 - \dfrac{(\Sigma X_1)^2}{N_1} + \Sigma X_2^2 - \dfrac{(\Sigma X_2)^2}{N_2}}{N_1 + N_2 - 2}\left(\dfrac{1}{N_1} + \dfrac{1}{N_2}\right)}}$$

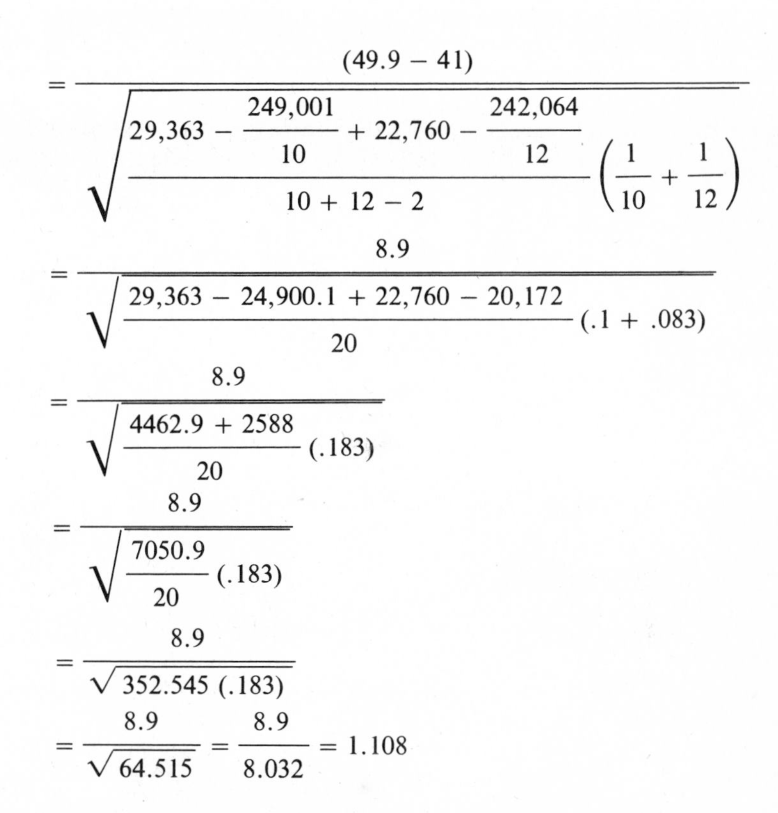

$$= \frac{(49.9 - 41)}{\sqrt{\dfrac{29{,}363 - \dfrac{249{,}001}{10} + 22{,}760 - \dfrac{242{,}064}{12}}{10 + 12 - 2}\left(\dfrac{1}{10} + \dfrac{1}{12}\right)}}$$

$$= \frac{8.9}{\sqrt{\dfrac{29{,}363 - 24{,}900.1 + 22{,}760 - 20{,}172}{20}(.1 + .083)}}$$

$$= \frac{8.9}{\sqrt{\dfrac{4462.9 + 2588}{20}(.183)}}$$

$$= \frac{8.9}{\sqrt{\dfrac{7050.9}{20}(.183)}}$$

$$= \frac{8.9}{\sqrt{352.545\ (.183)}}$$

$$= \frac{8.9}{\sqrt{64.515}} = \frac{8.9}{8.032} = 1.108$$

The computational formula for the dependent t test is derived by first substituting the computational formula for s_D (Formula 8.7) into the definitional formula for the dependent t (Formula 8.8), yielding:

$$\text{dependent } t = \frac{\bar{D} - 0}{\dfrac{\sqrt{\dfrac{\Sigma D^2 - \dfrac{(\Sigma D)^2}{N}}{N - 1}}}{\sqrt{N}}}$$

Algebraic simplification gives the computational formula for the dependent t:

FORMULA 8.10

computational formula for the dependent t

$$\text{dependent } t = \frac{\bar{D}}{\sqrt{\dfrac{\Sigma D^2 - \dfrac{(\Sigma D)^2}{N}}{N(N - 1)}}}$$

With Formula 8.10, you can conduct a dependent t test without separately calculating s_D beforehand.

To illustrate the use of Formula 8.10, consider the data from the study investigating the effect of meditation on creativity. After creating a set of D scores, you proceed as follows:

	Subject	X_1	X_2	D $(X_1 - X_2)$	D^2
	S_1	36	41	−5	25
	S_2	56	61	−5	25
	S_3	94	83	11	121
	S_4	58	33	25	625
	S_5	57	43	14	196
$\bar{D} = 7.08$	S_6	92	96	−4	16
$N = 12$	S_7	76	69	7	49
	S_8	74	69	5	25
	S_9	8	26	−18	324
	S_{10}	53	33	20	400
	S_{11}	49	18	31	961
	S_{12}	83	79	4	16
				$\Sigma D =$ 85	$\Sigma D^2 = 2783$
				$(\Sigma D)^2 = (85)^2 = 7225$	

$$\text{dependent } t = \frac{\bar{D}}{\sqrt{\dfrac{\Sigma D^2 - \dfrac{(\Sigma D)^2}{N}}{N(N-1)}}} = \frac{7.08}{\sqrt{\dfrac{2783 - \dfrac{7225}{12}}{12(12-1)}}}$$

$$= \frac{7.08}{\sqrt{\dfrac{2783 - 602.083}{12(11)}}}$$

$$= \frac{7.08}{\sqrt{\dfrac{2180.917}{132}}}$$

$$= \frac{7.08}{\sqrt{16.522}}$$

$$= \frac{7.08}{4.065} = 1.742$$

Journal form

The journal form for reporting the results of independent and dependent t tests is the same as for the one-sample t test; namely, write t, the degrees of freedom in parentheses, the calculated value of t, and then the probability that your results may have occurred by chance. For example, the journal form "$t(18) = 3.48, p < .05$" means that a t test was used and the calculated value of $t = 3.48$. With $df = 18$, the results have a probability less than .05 of occurring by chance. One might say in a journal report, "The results are significant beyond the .05 level, $t(18) = 3.48$, $p < .05$." Notice that the journal form does not reveal which of the three t tests was conducted. This information is usually gathered from reading about the design of the study.

KEY CONCEPTS

1. Many research problems use two groups (or samples) of subjects tested under different conditions. The two samples are independent when they are composed of different subjects and no deliberate attempt is made to match subjects. The two samples are dependent when the same subjects participate in both conditions, such as in a test-retest situation, or when different subjects in the two samples are deliberately matched in pairs.

2. The sampling distribution of the difference for two independent samples is a frequency distribution of the difference between pairs of sample means drawn from populations with equal means. The two-sample independent t test measures the likelihood of getting an observed difference between sample means, assuming that the samples are drawn from populations with equal means.

3. The assumptions that should be met before applying the two sample independent t test are: (a) The two population distributions are normal; (b) the two populations have equal (homogeneous) variance; and (c) the subjects in each sample are selected randomly and the samples are independent. The t tests are extremely *robust,* though, so violations of the first two assumptions usually do not matter.

4. The sampling distribution of the difference for two independent samples is approximately normal with a mean of 0 and a standard deviation of

$$\sigma_{\bar{x}_1 - \bar{x}_2} = \sqrt{\sigma^2 \left(\frac{1}{N_1} + \frac{1}{N_2} \right)}$$

where σ^2 is the population variance and N_1 and N_2 are the sample sizes.

5. You can estimate $\sigma_{\bar{x}_1 - \bar{x}_2}$ from your samples. First find s_1^2 and s_2^2 to get two estimates of the population variance, one from each sample. Then calculate the weighted average, called s^2_{pooled}, of these two estimates of the population variance:

$$s^2_{\text{pooled}} = \frac{(N_1 - 1)s_1^2 + (N_2 - 1)s_2^2}{N_1 + N_2 - 2}$$

6 The two-sample independent t test gives the number of estimated standard errors (standard deviations) your observed difference between sample means lies from the mean of the sampling distribution.

7 Using the independent t test, the hypothesis-testing procedure is:

a. State H_0 and H_A. Assume H_0 is true.

b. Set alpha and calculate $df = N_1 + N_2 - 2$. Determine t^* from Table t^* for your level of alpha, df, and whether you are conducting a one- or two-tailed test.

If it is a two-tailed test, proceed to Step 3. If it is a one-tailed test: if H_A is $\mu_1 > \mu_2$, subtract $\bar{X}_2$ from $\bar{X}_1$ and expect a positive difference; if H_A is $\mu_1 < \mu_2$, subtract $\bar{X}_2$ from $\bar{X}_1$ and expect a negative difference. Draw a diagram of the sampling distribution of the difference and mark the region of rejection to see whether $(\bar{X}_1 - \bar{X}_2)$ is in the expected direction. If $(\bar{X}_1 - \bar{X}_2)$ is in the expected direction, proceed with Step 3. If it is in the opposite direction, retain H_0.

c. Calculate an independent t for $(\bar{X}_1 - \bar{X}_2)$ with either the definitional formula shown immediately below, or the computational formula below that:

$$\text{independent } t = \frac{(\bar{X}_1 - \bar{X}_2) - 0}{\sqrt{\dfrac{(N_1 - 1)s_1^2 + (N_2 - 1)s_2^2}{N_1 + N_2 - 2}\left(\dfrac{1}{N_1} + \dfrac{1}{N_2}\right)}}$$

$$\text{independent } t = \frac{(\bar{X}_1 - \bar{X}_2)}{\sqrt{\dfrac{\Sigma X_1^2 - \dfrac{(\Sigma X_1)^2}{N_1} + \Sigma X_2^2 - \dfrac{(\Sigma X_2)^2}{N_2}}{N_1 + N_2 - 2}\left(\dfrac{1}{N_1} + \dfrac{1}{N_2}\right)}}$$

d. If $|t| > t^*$, reject H_0. If $|t| \leq t^*$, retain H_0.

8 The two-sample dependent t test is used to decide whether the two samples in a two-sample dependent research design could have come from populations having the same means. You first generate a sample of difference scores (D scores) by subtracting one score from the other for each pair of subjects. You then conduct a one-sample t test on these D scores to see if your sample could have come from a population of D scores that has $\mu_D = 0$. (If the population of "before" scores has the same mean as the population of "after" scores, then μ_D will equal zero.)

9 The assumptions that should be met before applying the two sample dependent t test are: (a) The two population distributions are normal; and

(b) the subjects in each sample are selected at random and the samples are dependent.

10 The dependent t test gives you the number of estimated standard errors (standard deviations) your observed mean difference between samples ($\bar{D}$) lies from the mean of the sampling distribution.

11 The hypothesis-testing procedure in the dependent t test is:

a. State H_0 and H_A. Assume H_0 is true.

b. Set alpha and calculate $df = N - 1$, where N is the number of pairs of scores. Determine t^* from Table t^* for your level of alpha, df, and whether you are conducting a one- or two-tailed test.

Get a set of D scores by subtracting the X_2 score from the X_1 score for each pair of raw scores. Then calculate $\bar{D}$, the mean of the D scores.

If it is a two-tailed test, proceed to Step 3. If it is a one-tailed test: if H_A is $\mu_1 > \mu_2$, expect a positive $\bar{D}$; if H_A is $\mu_1 < \mu_2$, expect a negative $\bar{D}$. Draw a diagram of the sampling distribution and mark the region of rejection to see whether $\bar{D}$ is in the expected direction. If $\bar{D}$ is in the expected direction, proceed with Step 3. If it is in the opposite direction, retain H_0.

c. Calculate a dependent t test for $\bar{D}$ with either the definitional formula on the left or the computational formula on the right:

$$\text{dependent } t = \frac{\bar{D} - 0}{\frac{s_D}{\sqrt{N}}} \qquad \text{dependent } t = \frac{\bar{D}}{\sqrt{\frac{\Sigma D^2 - \frac{(\Sigma D)^2}{N}}{N(N-1)}}}$$

d. If $|t| > t^*$, reject H_0. If $|t| \leq t^*$, retain H_0.

EXERCISES

1 You have heard it speculated that "coaching" can improve students' performance on the Law School Admissions Test (LSAT). To test this speculation, you gather 50 college seniors who had registered for the LSAT and randomly divide them into two groups of 25 subjects each. The experimental group received intensive coaching in problem-solving strategies over similar test items, whereas the control group received no coaching. Both groups subsequently took the LSAT on the same date and at the same test center. The following values are available:

Control group	*Experimental group*
$N_1 = 25$	$N_2 = 25$
$\bar{X}_1 = 619$	$\bar{X}_2 = 625$
$s_1^2 = 56$	$s_2^2 = 52$

a. Setting $\alpha = .05$, use the definitional Formula 8.5 for the independent t test and complete the four steps of the hypothesis-testing procedure. What can you conclude about the facilitative effect coaching has on LSAT scores?
b. Put your results in journal form.

2 Verbally describe what is meant by s^2_{pooled}.

3 You are interested in the effects of verbal praise on the ability of preschool children to copy geometric forms. You plan to do an experiment with "praise" and "nonpraise" conditions. Before conducting a large-scale experiment, you want to conduct a pilot study. You randomly select five preschoolers from a nursery school to participate. The children were asked to reproduce 20 equally complex forms. The overall accuracy of each child's reproductions were judged on a scale from 0 to 10, with 10 considered very accurate. You run the study by having the children reproduce ten of the figures without any encouragement on one day, and then reproduce the other ten figures one week later while you provide encouraging words such as "good," "excellent," "beautiful," and so on, spoken throughout the period. You obtain the following scores:

Subject	*Encouragement* X_1	*No encouragement* X_2
S_1	3	2
S_2	5	0
S_3	5	1
S_4	7	0
S_5	4	1

a. Setting $\alpha = .05$, use the definitional Formula 8.8 for the dependent t test and complete the steps of the hypothesis-testing procedure. What can you conclude about the effect of encouragement on the accuracy of reproductions?
b. Put your results in journal form.

4 Thinking about it further, you wonder whether it might be better to use independent groups to study the effect of verbal praise on preschool childrens' ability to copy geometric forms. Based on the results of your pilot study in Exercise 3, you want to know whether the encouragement significantly increases accuracy of reproductions. You obtain ten preschoolers and randomly assign five of them to the control group and the other five to the experimental group. The subjects in the experimental group were asked to reproduce ten forms while you utter encouraging words. The subjects in the control group were asked to reproduce the same ten forms, but without any encouragement. You obtain the following scores (judged on a scale from 0 to 10, with 10 being very accurate):

Encouragement *Subject*	X_1	*No encouragement* *Subject*	X_2
S_1	4	S_6	0
S_2	5	S_7	0
S_3	4	S_8	2
S_4	4	S_9	1
S_5	3	S_{10}	2

a. Setting $\alpha = .05$, use the computational Formula 8.9 for the independent t test and complete the hypothesis-testing procedure. What can you conclude about any facilitating effect of encouraging words on accuracy?
b. Put your results in journal form.

5 You believe a particular math instructor is an effective teacher. In an attempt to evaluate teacher effectiveness you decide to measure the "amount of knowledge gained" by students enrolled in this instructor's course. Your criterion for effectiveness is whether the amount of increase in their scores is significant at the .01 level of alpha. Thus, 20 students enrolled in the math course were given a difficult math test the first day of class (Group 1) and then were given an equally difficult test on the last day of class (Group 2). It was found that the average increase in the number correct was 8 points (thus $\bar{D} = -8$), and $s_D = 16$.
a. Setting $\alpha = .01$, use the definitional Formula 8.8 for the dependent t test and complete the hypothesis-testing procedure.
b. Put your results in journal form.

6 It has been suggested that assembly line workers who smoke cigarettes have impaired efficiency, and hence productivity declines. You decide to implement an antismoking campaign in a large factory and require randomly selected workers who smoke to attend a week of workshops and seminars aimed at providing information about the danger and economics of smoking. The average (mean) number of cigarettes smoked on the job was determined for each subject during the month before the scheduled workshops/seminars, and again for the month after attending the workshops/seminars. You predict the campaign will reduce the average number of cigarettes smoked at work per day. You obtain the following results:

Subject	*Before* X_1	*After* X_2
S_1	15	5
S_2	11	13
S_3	17	14
S_4	13	10
S_5	24	17
S_6	19	16
S_7	18	16
S_8	14	14

a. Which is the most appropriate hypothesis test to use in this study, the independent or dependent t test?
b. Set $\alpha = .05$ and complete the hypothesis-testing procedure using the appropriate computational formula for t. What can you conclude about the campaign's effectiveness in reducing cigarette smoking?
c. Put your results in journal form.

7 In the study conducted in Exercise 6, the passage of time is an uncontrolled extraneous variable that makes it difficult to interpret the results. That is, the workers may have reduced their cigarette smoking even if the antismoking

campaign had never been implemented. You need a control group, equivalent in smoking habits with the subjects in the previous study, who have not been subjected to and are not aware of the antismoking campaign. Thus, you select eight other workers who smoke on the job to see whether they will show a reduction in smoking over the same period even though they never see or hear the campaign. You obtain the following results.

Subject	*Before* X_1	*After* X_2
S_1	16	18
S_2	11	10
S_3	18	16
S_4	11	14
S_5	23	26
S_6	19	15
S_7	20	21
S_8	15	14

Setting $\alpha = .05$, use the hypothesis-testing procedure and examine the null hypothesis that the passage of time does not reduce cigarette smoking at work.

8 You want to determine whether "noise" affects college students' ability to memorize new information. You randomly divide 16 college sophomores into two groups of eight subjects each. The subjects in both groups are asked to memorize a list of 30 nonsense syllables, such as TSG, JMB, and so on, during a five-minute study period. The subjects in Group 1 study the nonsense syllables with a "noisy" background, while those in Group 2 study the syllables with a "no noise" background. After the study period all subjects receive a distractor task wherein they must add up lists of digits for 20 minutes. At the end of this 20 minutes, each subject is asked to recall as many of the original nonsense syllables as possible. The scores below are the number of nonsense syllables each subject correctly recalled.

Noisy subject	X_1	*No noise subject*	X_2
S_1	20	S_9	17
S_2	18	S_{10}	19
S_3	17	S_{11}	17
S_4	19	S_{12}	14
S_5	16	S_{13}	24
S_6	19	S_{14}	23
S_7	12	S_{15}	16
S_8	18	S_{16}	18

a. Which is the most appropriate hypothesis test to use in this study, the independent or dependent t test?

b. Set $\alpha = .05$ and complete the hypothesis-testing procedure using the appropriate definitional formula. What can you conclude about the effect of noise on recall?

c. Put your results in journal form.

9 A variety of studies have shown that environmental variables such as city size, amount of air pollution, availability of recreation facilities, and so on play an important role in job satisfaction. As head of personnel for a large tire factory employing people living in two nearby cities you are interested in the effect that a city's general unemployment rate may have on the job satisfaction of your workers. City 1 consistently has 2% higher unemployment than City 2. Ten pairs of your male workers were selected such that the members of each pair were matched on salaries, job status, marital status (married), and age. One member of each pair lived in City 1 and the other member of the pair lived in City 2. The following scores were obtained for the ten pairs of workers on a questionnaire designed to measure job satisfaction (high scores indicate high job satisfaction; 10 is the maximum score):

Pair number	*City 1*	*City 2*
1	1	4
2	2	1
3	5	9
4	0	5
5	7	8
6	1	4
7	7	9
8	6	4
9	4	4
10	6	10

a. Which is the most appropriate hypothesis test to use in this study, the independent or dependent t test?
b. Set $\alpha = .05$ and complete the hypothesis-testing procedure using the appropriate computational formula. What can you conclude about the effect of general unemployment on job satisfaction?
c. Put your results in journal form.

10 Tension headaches are related to increased muscle activity (electrical activity) in the frontalis muscle of the forehead. As a biofeedback researcher, you wish to test the assertion that relaxation plus auditory feedback about electrical activity in the frontalis is more effective in acquiring self-control of such activity than relaxation training alone. Twenty-three persons suffering from frequent tension headaches began the experiment and were divided by random assignment into one group of 12 and one group of 11. Prior to the biofeedback session all subjects were familiarized with progressive relaxation training (breathing and muscle exercises). Immediately afterward all subjects participated in a one-hour biofeedback session in which electrical activity was recorded (in millivolts) from the frontalis. All subjects were asked to relax, and then the subjects in the "auditory feedback" condition heard a tone decrease in loudness if their electrical activity decreased. The subjects in the control group were not provided any feedback while their electrical activity was being monitored. Each subject's average (mean) millivoltage was calculated during the one-hour period (low scores indicate better self-control). The results are:

No feedback subject	X_1	*Auditory feedback subject*	X_2
S_1	3.1	S_{13}	2.2
S_2	2.0	S_{14}	1.5
S_3	2.6	S_{15}	2.1
S_4	3.1	S_{16}	2.2
S_5	1.2	S_{17}	1.6
S_6	2.2	S_{18}	2.9
S_7	2.8	S_{19}	1.7
S_8	1.5	S_{20}	.9
S_9	4.1	S_{21}	3.1
S_{10}	1.9	S_{22}	1.9
S_{11}	3.1	S_{23}	3.4
S_{12}	1.8		

a. Which is the most appropriate hypothesis test to use in this study, the independent or dependent *t* test?

b. Set $\alpha = .01$ and complete the hypothesis-testing procedure using the appropriate computational formula. What can you conclude about the facilitative effect of auditory feedback on acquisition of self-control?

c. Put your results in journal form.

9

TWO-SAMPLE NONPARAMETRIC TESTS

CHAPTER GOALS

After studying this chapter, you should be able to:

1. Specify when to use the Mann-Whitney *U* test and the Wilcoxon test.
2. Assign ranks to scores when applying the Mann-Whitney *U* test and Wilcoxon test, and be able to deal with ties.
3. Perform the Mann-Whitney *U* test and Wilcoxon test.
4. State the assumptions for the Mann-Whitney *U* test and the Wilcoxon test.
5. Describe the difference in power between parametric and nonparametric tests, and identify the reason for this difference.

CHAPTER IN A NUTSHELL

In two-sample research situations, the independent or dependent *t* tests are often useful hypothesis-testing procedures. If the assumptions for the *t* tests are grossly violated, however, you should not use them. You can generally use nonparametric hypothesis-testing procedures instead, because nonparametric tests make less stringent assumptions about the population distribution. The Mann-Whitney *U* test and the Wilcoxon matched-pairs signed-ranks test are the nonparametric counterparts to the independent and dependent *t* tests, respectively, and are often appropriate when assumptions for the *t* tests cannot be met. These two nonparametric tests require ordinal or interval/ratio data and work by assigning ranks to scores. Since ranking ignores the magnitude of the distances between scores, these nonparametric tests are less powerful than their *t* test counterparts.

Introduction to nonparametric statistics

Once in a while you will encounter research situations in which you have two samples and you would like to apply a two-sample t test, but unfortunately the assumptions for the t tests are not satisfied. The assumptions are summarized again below:

1. For the independent and dependent t tests, the two population distributions are normal.
2. For the independent t test, the two populations have equal (homogeneous) variances.
3. The subjects in each sample are selected at random, and for the independent t test, the samples are independent; for the dependent t test, the samples are dependent.

A fourth assumption, which has not been mentioned until now, is an implicit assumption regarding the scale of measurement:

4. The data reflect an interval/ratio scale of measurement.

This assumption is implicit in the nature of the t tests because the formulas for t dictate that you calculate the variance or standard deviation. You will recall from Chapter 3 that you must have interval/ratio data to calculate a meaningful quantity for the variance and standard deviation.[1] For example, if you have ordinal data such as socioeconomic status or the rank orders of subjects completing a performance task, you should not apply a t test because the variance or standard deviation of these data is not very meaningful.

Furthermore, even if you do have interval/ratio data, you may still not want to apply a t test if any one of the first three assumptions mentioned above does not apply to your data. For example, if you have data from an extremely skewed population, such as with reaction time data, or with performance on a very difficult or very easy task, then your data do not conform with the first assumption of normality. Another example is if you are investigating a particular type of psychotherapy that has varying effects on different psychiatric patients; that is, it makes some patients better and some worse. You may produce a very spread-out distribution of scores on a measure of "mental adjustment." If you compare a group of patients who received this psychotherapy with a control group of patients who did not, you probably will have discrepant population variances and cannot meet the assumed requirement of homogeneity of variance.

Since the assumptions of normality and homogeneity of variance concern parameters of the populations from which your samples came,

[1] You should be aware, however, that this is not an entirely resolved issue. There are some research statisticians who argue that it is meaningful to calculate means and standard deviations for noninterval data and that it is appropriate to conduct parametric tests on such data.

parametric tests

the t tests are referred to as **parametric tests.** If we rely on the "extra" information about population parameters that is included in these assumptions, we find that the t tests are relatively powerful. That is, there is a relatively high probability of correctly rejecting a false H_0 (and hence detecting a real treatment effect or difference between population means). But there are other types of hypothesis tests that may be applied without having to conform to assumptions about population parameters. The hypothesis tests are appropriately referred to as **nonparametric tests.** Because nonparametric tests do not have the advantage of utilizing "extra" information about the population distributions, they are generally less powerful than parametric tests.

nonparametric tests

The two nonparametric tests discussed in this chapter are counterparts to the two-sample independent and dependent t tests. The nonparametric counterpart to the independent t test is the *Mann-Whitney U test,* or *U test* for short. Like the independent t test, it is not necessary to have the same number of subjects in each group to conduct the U test. The nonparametric counterpart to the dependent t test is the *Wilcoxon matched-pairs signed-ranks test,* or simply *Wilcoxon test.* Like the dependent t test, the subjects in each group must be the same subjects tested and retested, or different subjects paired up in some fashion. These nonparametric tests should be applied whenever one or more of the assumptions for the t tests are grossly violated. Nonparametric tests are frequently applied to ordinal data when parametric tests are inappropariate because of the fourth assumption mentioned above. Like most nonparametric tests, the Mann-Whitney U and Wilcoxon tests also have some assumptions that must apply, but these assumptions are more lenient than those required in parametric tests and can usually be satisfied.

The Mann-Whitney *U* test

assumptions

The Mann-Whitney U test is employed in research situations in which the two-sample independent t test might be appropriate but where all the assumptions for the t test do not apply.

The **assumptions** that should be satisfied before you apply the U test are:

1. The data reflect an ordinal or interval/ratio scale of measurement.
2. The data reflect a continuous underlying scale of measurement.
3. The subjects in each sample are selected at random, and the samples are independent.

The first assumption requires that data be at least ordinal, such as rankings. If your data reflect only a nominal scale of measurement, this assumption is not satisfied. The second assumption requires continuous data and requires a potentially infinite degree of precision in specifying scores. It is important to realize that whenever you have ordinal data in the form of ranks, if you want to apply the U test the ranks must be based on a continuous measurement scale. For example, suppose you rank the order (first place, second place, etc.) in which rats run in a straight alleyway. You will wind up with ordinal data in the form of ranks, which reflect the continuous variable of running speed. Hence you may consider the second assumption satisfied. On the other hand, if you are conducting a study in which the dependent variable is birth order (firstborn, second born, etc.), even though your data are ordinal, it is discrete. The U test is therefore not appropriate.[2] The third assumption requires the samples to be random, and requires the subjects in each group to be different subjects and not matched (paired up) in any way.

To illustrate the logic and application of the U test, suppose you are a statistics teacher interested in determining the effect of a new study aid on students' performance on a final exam. The study aid is a programmed workbook you devised. From past experience you know students tend to do well on your final, so the distribution of scores is strongly negatively skewed. You randomly divide the class into two groups or samples. Group 1 used the programmed workbook and Group 2 did not. After the final, you obtain the following scores:

Group 1: Uses Workbook		*Group 2: No Workbook*	
Subject	X_1	*Subject*	X_2
S_1	77	S_6	48
S_2	84	S_7	51
S_3	88	S_8	70
S_4	96	S_9	81
S_5	100	S_{10}	98
	$\bar{X}_1 = 89$		$\bar{X}_2 = 69.6$

It appears from the sample means that the sample using the programmed workbook performed better on the final. As usual, there are two possible explanations for this difference. One possibility is that the programmed workbook actually aids performance. Another way of saying this is that the two groups were drawn from differently distributed populations. That is, if you gave the programmed workbook to *everyone* in the population of students, these students would perform differently on the average than would a population of students without the workbook. The other explanation is that the difference between the two groups is due to the vagaries of random sampling—a sampling error. In other words, if you gave the programmed workbook to everyone in the population of students, they

[2]In studies with only nominal data, or with a discrete underlying scale of measurement, other hypothesis tests are available. For example, the chi-square tests described in Chapter 10 may be appropriate.

would show the same performance on the average as a population of students without the workbook. These two possible explanations of why the performance of the two groups differs are outlined in Figure 9-1.

To decide whether to attribute the difference in final exam performance to some facilitating effect of the workbook or to sampling error, you must perform some kind of hypothesis test. The two-sample independent t test

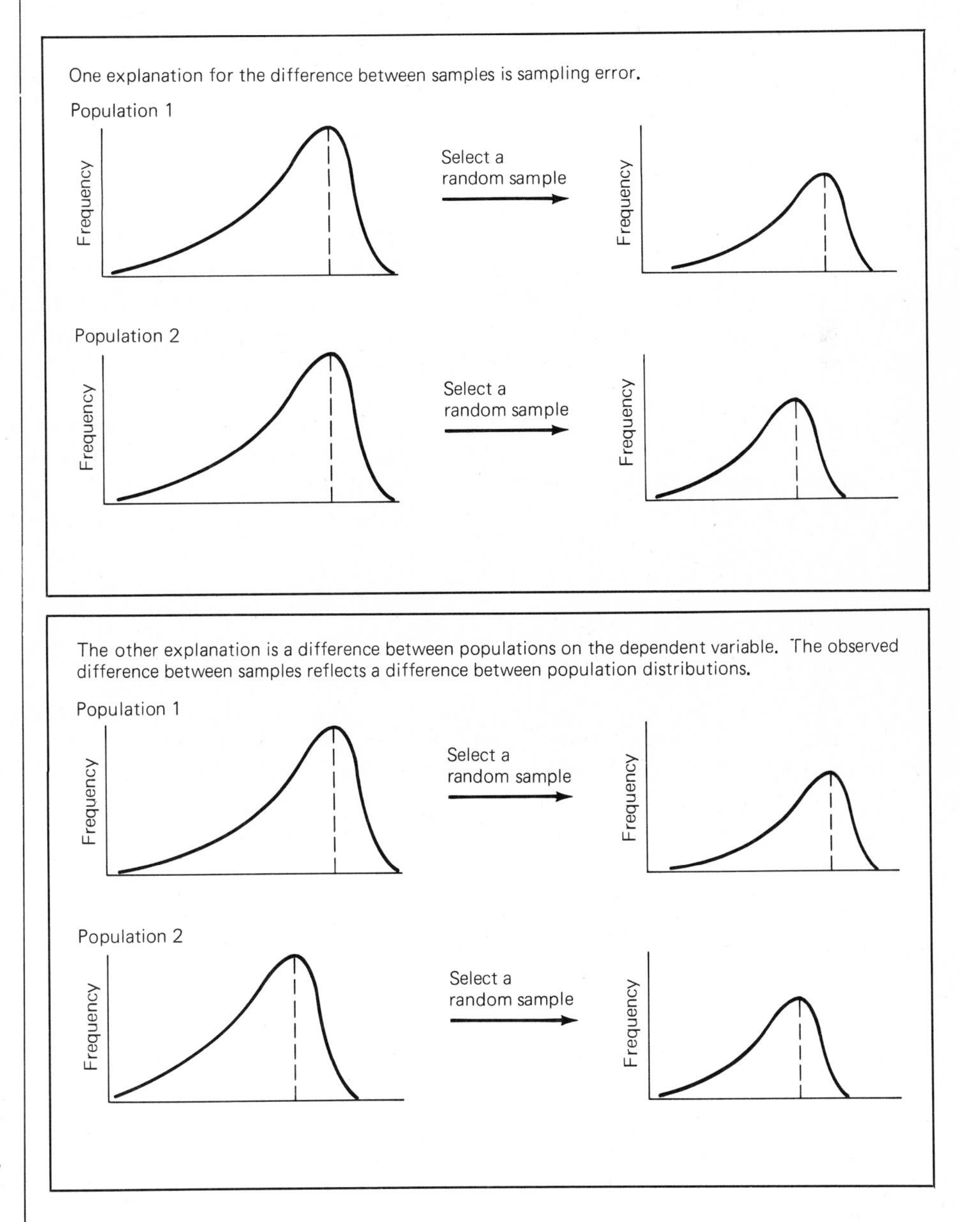

FIGURE 9-1

The two competing explanations for an observed difference between samples.

is not appropriate because the two population distributions are skewed and you therefore cannot satisfy the assumption of normality. When you look at the range of scores in the two samples you might also question whether homogeneity of variance is satisfied. Group 2 appears to have a larger variance than Group 1. Fortunately, the application of the Mann-Whitney U test does not require assumptions of normality and homogeneity of variance. The U test is appropriate here because performance on the final exam is measured with at least an ordinal scale (here we have an interval/ratio scale) and is based on what can be assumed to be continuous data. That is, it is theoretically possible to obtain any score between 0 and 100 on the final.

Whereas the independent t test compares the means of the scores in the two samples, the U test compares the *ranks* of the scores in the two samples. The logic here is to compare the two distributions of scores, and comparing the ranks of scores in the two samples is a relatively easy way to do this. Therefore, the first step in performing the U test is to combine all the raw scores from both samples into the same pool, so to speak, and assign ranks to each score in the pool. The smallest raw score should be given the rank of 1, the next smallest score the rank of 2, and so on up to the largest score. Although the scores from both groups are combined in the same pool for the moment, be sure to keep track of the group from which each score originated. The ranks need to be sorted out later according to their original group. A convenient way to organize the data for assigning ranks is shown below.

Given:

Group 1	*Group 2*
77	48
84	51
88	70
96	81
100	98

ranking scores

Then horizontally rank scores by groups as follows:

Assigned ranks:	1	2	3	4	5	6	7	8	9	10
Group 1:				77		84	88	96		100
Group 2:	48	51	70		81				98	

Sum of ranks for Group 1 $= R_1 = 4 + 6 + 7 + 8 + 10 = 35$

Sum of ranks for Group 2 $= R_2 = 1 + 2 + 3 + 5 + 9 = 20$

Once the raw scores are assigned ranks, the ranks are separated according to group. The U test compares the ranks between the two groups.

What does a comparison of the ranks tell you? If the scores in the population represented by sample Group 1 are much smaller on the average than the scores in the population represented by sample Group 2, then you expect the ranks of Group 1 to fall below the ranks of Group 2, as in the sample data in Figure 9-2(a). On the other hand, if the situation is reversed and the scores in the population represented by sample Group 1 are much larger on the average than the scores in the population represented by sample Group 2, then you expect the ranks in Group 1 to fall above the ranks of Group 2, as in the sample data in Figure 9-2(b). If the two populations have identical distributions, then you expect the ranks of Group 1 to be fairly evenly dispersed among the ranks of Group 2, as in the sample data in Figure 9-2(c). There will naturally be some deviation from perfect intermixing, but on the average you expect the ranks to be pretty well intermixed.

If the ranks of one group are fairly well above the ranks of the second group, this suggests that the populations are different because if the populations were identically distributed, your results would be unlikely to occur by chance. You can now use the general hypothesis-testing procedure. First assume that only chance is operating and that your results are due to sampling error. For a two-tailed test, H_0 is that the two populations are identically distributed. The next step is the measure the probability of obtaining differences in rankings as extreme as yours, assuming H_0 is true. How do you calculate this probability? You first need a measure of how much the ranks from the two samples are intermixed, and such a measure is called the ***U* statistic.**

U statistic

If there is absolutely no intermixing of the ranks, the sum of the ranks for the lower ranking group will be the smallest value possible. For example, look at the data in Figure 9-2(a) where all the scores in Group 1 lie below those in Group 2. Notice that the sum of the ranks for Group 1, $R_1 = 15$, is the smallest value possible for this sample size. Any intermixing of the ranks will increase R_1. When the sum of the ranks is at its minimum value, the ranks for the scores in the lower group all fall below the ranks in the upper group. The ranks in the lower group are the consecutive integers from 1 through the number of scores in that group, or 1, 2, 3, . . ., N_1. There is a formula to give you the value of this **minimum sum of ranks** for a group:

minimum sum of ranks

$$\text{minimum sum of ranks for Group 1} = 1 + 2 + 3 + \cdots + N_1 = \frac{(N_1)(N_1 + 1)}{2}$$

where N_1 is the number of scores in Group 1, which we are assuming is the lower ranked group. If H_0 is true and the populations are identically distributed, then the scores in your sample should be fairly well intermixed as in Figure 9-2(c). The sum of the ranks for Group 1 is increased in

(a)

An example of rankings to be expected when the population scores of Group 1 are much smaller than the population scores of Group 2.

Assigned ranks:			1	2	3	4	5	6	7	8	9	10	
Group 1:			15	23	31	34	43						
Group 2:								48	75	87	97	98	
Sum of ranks for Group 1	$= R_1$	=	1 +	2 +	3 +	4 +	5						= 15
Sum of ranks for Group 2	$= R_2$	=						6 +	7 +	8 +	9 +	10	= 40

(b)

An example of rankings to be expected when the population scores of Group 1 are much larger than the population scores of Group 2.

Assigned ranks:			1	2	3	4	5	6	7	8	9	10	
Group 1:								60	64	71	75	88	
Group 2:			22	23	32	37	47						
Sum of ranks for Group 1	$= R_1$	=						6 +	7 +	8 +	9 +	10	= 40
Sum of ranks for Group 2	$= R_2$	=	1 +	2 +	3 +	4 +	5						= 15

(c)

An example of rankings to be expected when the population scores of Group 1 and Group 2 are identically distributed.

Assigned ranks:			1	2	3	4	5	6	7	8	9	10	
Group 1:			28		41			63		78		91	
Group 2:				39		43	59		66		89		
Sum of ranks for Group 1	$= R_1$	=	1	+	3		+	6	+	8	+	10	= 28
Sum of ranks for Group 2	$= R_2$	=		2	+	4 +	5	+	7	+	9		= 27

FIGURE 9-2

A comparison of the assigned ranks under three different population conditions.

this example to 28. When all the scores in Group 1 lie above those in Group 2, again there is no intermixing of the ranks and the maximum possible value for the sum of the ranks in Group 1 occurs. This is shown in Figure 9-2(b), where $R_1 = 40$. Thus, as the rankings for the scores in Group 1 increase, the sum of the ranks, R_1, increases. Consequently, a measure of how much two groups of scores are intermixed can be obtained by calculating R_1 and determining how much this sum deviates from the minimum sum of ranks. If you find the difference between R_1 and the minimum sum of ranks by subtracting the minimum sum of ranks from R_1, the difference is called the U statistic. Thus the **definitional formula for the U test** is:

FORMULA 9.1
definitional formula for U

$$U = R_1 - \frac{(N_1)(N_1 + 1)}{2}$$

where R_1 is the sum of the ranks from Group 1 and N_1 is the number of scores in Group 1.

You can see that when all the scores in Group 1 are smaller than any of the scores in Group 2, U is at its extreme *minimum* value of 0. This is the case with the data in Figure 9-2(a):

$$U = 15 - \frac{(5)(5 + 1)}{2}$$

$$= 15 - \frac{30}{2} = 15 - 15 = 0$$

As the scores for Group 1 become intermixed with those of Group 2, the value of U increases and becomes *intermediate* in value as in Figure 9-2(c):

$$U = 28 - \frac{(5)(5 + 1)}{2}$$

$$= 28 - \frac{30}{2} = 28 - 15 = 13$$

The U statistic is at its extreme *maximum* value when all the scores in Group 1 are greater than any of the scores in Group 2, as in Figure 9-2(b):

$$U = 40 - \frac{(5)(5 + 1)}{2}$$

$$= 40 - \frac{30}{2} = 40 - 15 = 25$$

Therefore, assuming H_0 is true and the population distributions represented by your samples are identical, you expect to observe mostly *intermediate* values of U. However, extreme values of U will occasionally occur by sampling error. In typical hypothesis-testing fashion, if the probability of obtaining a U as extreme as yours is less than alpha, you can reject H_0 and conclude the population distributions are different. To determine the probability of a U value as extreme as yours, you need to compare your calculated U against a distribution of U values that occurs by chance. This distribution of U values is another theoretical sampling distribution whose characteristics can be mathematically inferred.

U^*

Critical values of U in the sampling distribution of U scores are abbreviated U^*, and are listed in Table U^*. Table U^* lists U^* values according to: alpha level; one- versus two-tailed tests; and as a function of the number of scores in each of your two groups. There are two critical values given for each combination of these three factors. Select the proper table (there are four pages of U^* tables) by looking at the top of the page for your value of alpha and whether you are conducting a one- or two-tailed test. Then find the pair of critical values by looking across the top of the table for N_1, the number of scores in Group 1, and look down the left side for N_2, the number of scores in Group 2. It is purely arbitrary which sample you label as Group 1. The pair of critical values lies at the intersection of the N_1 column and the N_2 row. If your calculated U is in the intermediate range and falls *between* (but not including) the two critical values, this indicates that the probability of obtaining a U value as extreme as yours is not very unlikely, and you should retain H_0. However, if your calculated U is greater than or equal to the larger critical value *or* is less than or equal to the smaller critical value, the probability of obtaining a U value as extreme as yours by chance is less than alpha and you should reject H_0. The population distributions are probably different.

The hypothesis-testing procedure using the Mann-Whitney U test may be summarized:

1. Determine the appropriate H_0 and H_A. For a two-tailed test, H_0 is that the populations from which the samples were drawn are identically distributed and H_A is that the populations are not identically distributed. For a one-tailed test, H_0 is that the Group 1 population lies above (below) or is distributed identically to the Group 2 population. H_A is that the Group 1 population lies below (above) the Group 2 population. Assume H_0 is true.

2. Set alpha and determine N_1, the size of Group 1, and N_2, the size of Group 2. Find the two critical values of U in Table U^* for your level of alpha, N_1 and N_2, and whether you are conducting a one- or two-tailed test.

Assign ranks to the combined scores in Groups 1 and 2 and then calculate R_1, the sum of the ranks in one group.

3. Calculate U with Formula 9-1:

$$U = R_1 - \frac{(N_1)(N_1 + 1)}{2}$$

If it is a two-tailed test proceed to Step 4a. If it is a one-tailed test proceed to Step 4b.

4a. For a two-tailed test, if U is greater than or equal to the larger U^*, *or* if U is less than or equal to the smaller U^*, reject H_0 and conclude that your samples were drawn from differently distributed populations.

If U falls between the two critical values, retain H_0 and conclude that there is not sufficient evidence to say your samples were selected from differently distributed populations.

4b. For a one-tailed test where you predict that the Group 1 population lies above the Group 2 population, reject H_0 only if your calculated U value is greater than or equal to the larger U^*. Otherwise retain H_0.

For a one-tailed test where you predict that the Group 1 population lies below the Group 2 population, reject H_0 only if your calculated U value is less than or equal to the smaller U^*. Otherwise retain H_0.

If you reject H_0, conclude that your samples were drawn from differently distributed populations as specified in H_A.

If you retain H_0, conclude that there is not sufficient evidence to say your samples were selected from differently distributed populations as specified in H_A.

Now apply this hypothesis-testing procedure to the study in which you want to determine whether the use of your programmed workbook aids performance on the final exam.

1. You have high hopes that the programmed workbook you devised will facilitate performance on the final. Therefore you conduct a one-tailed test.

   H_0: The workbook is not an effective aid. The distribution of final exam scores for the population of students using the workbook lies beneath or is identical to the distribution of scores for the population of students not using the workbook.

   H_A: The workbook is an effective aid. The distribution of final exam scores for the population of students using the workbook lies above the distribution of scores for the population of students not using the workbook.

2. Since you do not want to overlook any potentially facilitating effect of the workbook, you let $\alpha = .05$. Using Table U^*, find for a one-tailed test at $\alpha = .05$, with $N_1 = 5$ and $N_2 = 5$, that $U^* = 4$ and 21.

   You organize the data, assign ranks, and calculate $R_1 = 35$.

3. Now you calculate U with Formula 9.1:

$$U = R_1 - \frac{(N_1)(N_1 + 1)}{2}$$

$$= 35 - \frac{5(5 + 1)}{2} = 20$$

Since this is a one-tailed test proceed to Step 4*b*.

4*b*. You expect the scores in Group 1 to lie above those in Group 2. Since your calculated value of $U = 20$ is not greater than or equal to the larger U^* value of 21, retain H_0 and conclude that there is not sufficient evidence to say your programmed workbook aids performance.

Dealing with tied scores

A special problem arises when you attempt to assign ranks to tied raw scores, as in the data given below.

Group 1: Subject	X_1	*Group 2:* Subject	X_2
S_1	3	S_5	4
S_2	3	S_6	6
S_3	6	S_7	6
S_4	8	S_8	10

One way to deal with ties is to assign the average rank of the scores involved in the tie to each of the tied scores. For example, after combining the data, you can see in Figure 9-3 that the two scores of 3 occupy the ranks of 1 and 2. Therefore, calculate the average ranking of each by summing the ranks of the scores involved in the tie, and divide this sum by the number of scores in the tie. Assign the average ranking to each score involved in the tie. Thus the average ranking of 1.5 should be assigned to each score of 3:

$$\frac{(1 + 2)}{2} = 1.5$$

The score of 4 is ranked next and is assigned the rank of 3 because the 1 and 2 ranks have already been used. The three scores of 6 occupy the ranks of 4, 5, and 6. Therefore, assign the average of these rankings, 5, to each score of 6:

$$\frac{(4 + 5 + 6)}{3} = 5$$

Group 1: Subject	X_1	Group 2: Subject	X_2
S_1	3	S_5	4
S_2	3	S_6	6
S_3	6	S_7	6
S_4	8	S_8	10

Occupied ranks:	1	2	3	4	5	6	7	8
Group 1:	3	3				6	8	
	ties				ties			
Group 2:			4	6		6		10

Assign the average ranking to each score in the tie:

$$\frac{1 + 2}{2} = 1.5 \qquad \frac{4 + 5 + 6}{3} = 5$$

Then assign the average of these rankings as follows:

Assigned ranks:	1.5	1.5	3	5	5	5	7	8
Group 1:	3	3		6			8	
Group 2:			4		6	6		10

Sum of ranks for Group 1 $= R_1 = 1.5 + 1.5 + 5 + 7 = 15$

Sum of ranks for Group 2 $= R_2 = 3 + 5 + 5 + 8 = 21$

FIGURE 9-3
Dealing with ties.

The score of 7 occupies the rank of 7 so it is assigned the rank of 7, and the score of 8 is assigned the rank of 8.

The method of averaging ranks effectively deals with the problem of tied scores when the scores involved in a tie occur within one group. However, simply averaging ranks does not completely deal with the problem of tied scores when the scores involved in the tie occur in both groups (across groups). Therefore, you should not use the U test when there are very many tied scores across groups.

Dealing with large sample sizes

Notice that Table U^* can only be used for sample sizes between 5 and 20. Thus you cannot use Table U^* to find the critical values of U for larger samples. If both of your samples are larger than 20, you can still conduct a U test, however, because with sample sizes over 20:

sampling distribution of U

The **sampling distribution of U** is an approximately normal distribution with a mean of

$$\frac{(N_1)(N_2)}{2}$$

and a standard deviation equal to:

$$\sqrt{\frac{(N_1)(N_2)(N_1 + N_2 + 1)}{12}}$$

Since the sampling distribution of U is normal with a known mean and standard deviation, you can calculate a z for your observed U to see whether it is significant. In the z test Formula 7.1, where $z = \bar{X} - \mu/\sigma_{\bar{x}}$, simply make the relevant substitutions so that U replaces $\bar{X}$, $\frac{(N_1)(N_2)}{2}$ replaces μ, and

$$\sqrt{\frac{(N_1)(N_2)(N_1 + N_2 + 1)}{12}}$$

replaces $\sigma_{\bar{x}}$. Therefore, the z formula for finding the number of standard deviation units that your observed U score lies away from the mean of the sampling distribution of U is:

FORMULA 9.2

converting U to a z score

$$z = \frac{U - \frac{(N_1)(N_2)}{2}}{\sqrt{\frac{(N_1)(N_2)(N_1 + N_2 + 1)}{12}}}$$

Your value of U is significant if its corresponding z value is significant.

The following steps summarize the hypothesis-testing procedure using the Mann-Whitney U when both of the sample sizes are greater than 20:

1. State H_0 and H_A. Assume H_0 is true.
2. Set alpha. Determine z^* from Table z^* for your level of alpha and whether you are conducting a one- or two-tailed test.

   Assign ranks to the combined scores in Groups 1 and 2 and calculate R_1, the sum of the ranks in one group.
3. Calculate U with Formula 9.1:

$$U = R_1 - \frac{(N_1)(N_1 + 1)}{2}$$

Use Formula 9.2 to transform your calculated U to a z score:

$$z = \frac{U - \dfrac{(N_1)(N_2)}{2}}{\sqrt{\dfrac{(N_1)(N_2)(N_1 + N_2 + 1)}{12}}}$$

If it is a two-tailed test proceed to Step 4a. If it is a one-tailed test proceed to Step 4b.

4a. For a two-tailed test:

If $|z| > z^*$, reject H_0.
If $|z| \leq z^*$, retain H_0.

4b. For a one-tailed test where you predict that the population represented by Group 1 lies above the population of Group 2, then you predict a high value of U, so:

If z is positive and $|z| > z^*$, reject H_0; otherwise retain H_0.

For a one-tailed test where you predict that the population represented by Group 1 lies below the population of Group 2, then you predict a low value of U, so:

If z is negative and $|z| > z^*$, reject H_0; otherwise retain H_0.

To illustrate the use of this hypothesis-testing procedure, suppose you have large sample sizes with $N_1 = 25$ and $N_2 = 30$ and you want to show that there is a difference (two-tailed) between the two groups. Suppose $\alpha = .05$. The steps are:

1. H_0: The treatment has no effect. The distribution of scores for the Group 1 population is identical to the distribution of scores for the Group 2 population.

H_A: The treatment has some effect. The distribution of scores for the Group 1 population is different from the distribution of scores for the Group 2 population.

2. $\alpha = .05$. Using Table z^*, you find for a two-tailed test at $\alpha = .05$, $z^* = 1.96$. Assume that you assign ranks to the data (not shown here) and calculate $R_1 = 562.5$. Since this is a two-tailed test, you proceed to Step 3.

3. The calculated U is:

$$U = R_1 - \frac{(N_1)(N_1 + 1)}{2}$$

$$= 562.5 - \frac{(25)(26)}{2} = 237.5$$

Substitute known values into Formula 9.2 and convert your calculated U to a z score:

$$z = \frac{U - \dfrac{(N_1)(N_2)}{2}}{\sqrt{\dfrac{(N_1)(N_2)(N_1 + N_2 + 1)}{12}}}$$

$$= \frac{237.5 - \dfrac{(25)(30)}{2}}{\sqrt{\dfrac{(25)(30)(25 + 30 + 1)}{12}}}$$

$$= \frac{237.5 - \left(\dfrac{750}{2}\right)}{\sqrt{\dfrac{(25)(30)(56)}{12}}}$$

$$= \frac{237.5 - 375}{\sqrt{\dfrac{42000}{12}}}$$

$$= \frac{-137.5}{\sqrt{3500}} = \frac{-137.5}{59.161} = -2.324$$

4. Since $|-2.3241 > 1.96$, reject H_0 and conclude that your samples were drawn from differently distributed populations.

The Wilcoxon matched-pairs signed-ranks test

The Wilcoxon test is applicable in research situations in which the dependent t test might be appropriate except that one or more of the assumptions for the t test are grossly violated. As with the dependent t test, with the Wilcoxon test you have two samples of paired scores. The pairs of scores are provided either from the same subjects who have been tested and retested under two different conditions, or from different subjects who have been matched on the dependent variable or a related variable.

assumptions

The **assumptions** for the Wilcoxon test are:

1. The data reflect an ordinal or interval/ratio scale of measurement.
2. The differences between the pairs of scores (obtained by subtracting the

"after" scores from the "before" scores) may be ordered (ranked) in magnitude.

3. The subjects in each sample are selected on a random but dependent basis. That is, a pair of X_1 and X_2 scores either comes from the same randomly selected subject who is tested under two different conditions or from a randomly selected but matched pair of subjects.

To illustrate the logic and application of the Wilcoxon test, consider the reading speed scores (in words per minute) of 12 tenth-graders before and after the implementation of a new "reading skills" program:

Group 1:		*Group 2:*	
Before program		*After program*	
Subject	X_1	*Subject*	X_2
S_1	254	S_1	254
S_2	100	S_2	250
S_3	180	S_3	540
S_4	221	S_4	210
S_5	225	S_5	275
S_6	98	S_6	318
S_7	157	S_7	557
S_8	776	S_8	635
S_9	147	S_9	136
S_{10}	999	S_{10}	990
S_{11}	460	S_{11}	489
S_{12}	125	S_{12}	441
	$\bar{X}_1 = 311.8$		$\bar{X}_2 = 424.6$

The reading skills program was designed to increase students' reading speeds. Looking at the means of your samples provides some hope that your program is successful. It is possible, however, that the reading program has no effect on reading speed, or even has a detrimental effect. Your results might be due to sampling error—random fluctuations between the "before" and "after" scores. Alternatively, it is possible the reading program indeed facilitates reading speed. You might consider conducting a two-sample dependent t test to decide between these two possible explanations of your results. However, you know that the reading speed scores of the population of students you are interested in are very positively skewed. Given this consideration, what hypothesis test can you perform instead of a t test? The Wilcoxon test, of course, because it does not require normality of the population distributions.

Like the dependent t test, the Wilcoxon test compares the differences between "before" and "after" by subtracting "after" scores from "before" scores to create a set of difference scores (D scores). Assuming H_0 that your reading program is ineffective, you expect on the average that for each subject who slightly increases in reading speed, there is a subject who slightly decreases. Consequently, you expect individual chance fluctuations in reading speed to balance each other out such that: (a) about half of the subjects increase their scores and half decrease their scores; and (b) the magnitudes of the increases are about the same on the average as the magnitudes of the decreases. Now suppose you assign ranks to all the *D* scores regardless of whether they are increases (positive *D*'s) or decreases (negative *D*'s). That is, rank the absolute values of the *D* scores, giving a rank of 1 to the smallest *D*, and so on. Then separate the ranks according to whether they came from a positive *D* score or a negative *D* score, and get a sum of ranks for positive *D*'s and another sum of ranks for negative *D*'s. The Wilcoxon test compares the sum of ranks assigned to the positive *D*'s with the sum of ranks assigned to the negative *D*'s. Assuming H_0 that the two populations have identical distributions, you expect the sum of ranks for the positive *D*'s to be about equal to the sum of ranks for the negative *D*'s. This is because you expect the increases in the dependent variable to balance out the decreases. Conversely, with H_A that the population distributions are different, you expect the sum of the positive *D*'s to be different from the sum of the negative *D*'s.

A convenient way to organize your data for the Wilcoxon test is to set up columns as with the three sets of sample data in Figure 9-4. Columns 1 and 2 are the "before" and "after" scores, respectively. Column 3 is the difference (*D*) between the "before" and "after" scores (obtained by subtracting the "after" score from the "before" score). The absolute values of the *D* scores indicate the magnitude of each difference, and are given in Column 4. These magnitudes are then assigned ranks, which are listed in Column 5. In Columns 6 and 7 the assigned ranks are sorted according to whether they came from positive *D*'s or negative *D*'s.

If the scores in the "before" population are much smaller on the average than the scores in the "after" population, you expect the *D* scores to look something like those for the data in Figure 9-4(a). Notice in Column 3 that there are more negative *D*'s than positive *D*'s, and the magnitudes of the negative *D*'s are generally greater than the magnitudes of the positive *D*'s. Now if you look in Columns 6 and 7 at the ranks of the absolute magnitudes of these positive and negative *D*'s, you will see the correspondence. The sum of the ranks for the negative *D*'s is greater than the sum of the ranks for the positive *D*'s. The data in Figure 9-4(b) provide an example of the opposite situation, in which the scores in the "before" population are on the average considerably larger than the scores in the "after" population. Notice in Column 3 there are more positive *D*'s than negative *D*'s, and hence in Columns 6 and 7 you see that the sum of the ranks for the positive differences is greater than the sum of the ranks for the negative differences. However, if the "before" and "after" populations have identical distributions, you expect roughly equal numbers

Subject	Column 1 (Before) X_1	Column 2 (After) X_2	Column 3 D $(X_1 - X_2)$	Column 4 $\|D\|$	Column 5 Rank of $\|D\|$	Column 6 Ranks for positive D's	Column 7 Ranks for negative D's
(a) *An example of rankings to be expected when the scores in the "before" population are smaller on the average than the scores in the "after" population.*							
S_1	4	32	−28	28	7		7
S_2	2	68	−66	66	10		10
S_3	74	99	−25	25	6		6
S_4	70	77	− 7	7	2		2
S_5	32	45	−13	13	4		4
S_6	5	17	−12	12	3		3
S_7	55	40	15	15	5	5	
S_8	17	67	−50	50	8		8
S_9	35	99	−64	64	9		9
S_{10}	20	26	− 6	6	1		1
						Sum = 5	Sum = 50
(b) *An example of rankings to be expected when the scores in the "before" population are larger on the average than the scores in the "after population.*							
S_1	76	64	12	12	3	3	
S_2	19	9	10	10	2	2	
S_3	80	34	46	46	6	6	
S_4	45	2	43	43	5	5	
S_5	5	3	2	2	1	1	
S_6	14	39	−25	25	4		4
S_7	86	6	80	80	10	10	
S_8	87	17	70	70	9	9	
S_9	77	18	59	59	8	8	
S_{10}	66	14	52	52	7	7	
						Sum = 51	Sum = 4
(c) *An example of the rankings to be expected when the scores in the "before" population and the "after" population have essentially identical distributions.*							
S_1	37	10	27	27	5	5	
S_2	8	99	−91	91	10		10
S_3	12	66	−54	54	7		7
S_4	31	46	−15	15	4		4
S_5	63	73	−10	10	3		3
S_6	98	11	87	87	9	9	
S_7	88	83	5	5	1	1	
S_8	99	65	34	34	6	6	
S_9	74	80	−6	6	2		2
S_{10}	69	9	60	60	8	8	
						Sum = 29	Sum = 26

FIGURE 9-4 *A comparison of positive and negative D's under three different population conditions.*

of positive and negative D's, such as in the data in Figure 9-4(c). Of course, there are minor deviations due to chance differences in motivation and so on, but on the average, when the populations are identically distributed and you look at Columns 6 and 7, you find that the sum of the ranks for the positive D's is about equal to the sum of the ranks for the negative D's.

Although the sums of the ranks for the positive D's and negative D's are both calculated for the Wilcoxon test, the procedure in this Wilcoxon test focuses attention on only one of these two sums of ranks. Specifically, it focuses on the *smaller* sum of ranks, which is defined as **W**. If the two population distributions are identical, the smaller sum of ranks will be close to the larger sum. However, as the population distributions become increasingly different from each other, the smaller of the two sums of the ranks becomes relatively smaller. Therefore W, the smaller of the two sums of the ranks, provides a good indicator of how different your two groups are. The question is, what is the probability of observing a value of W as extreme as yours if H_0 is true and the two populations are identically distributed?

W

The theoretical sampling distribution of W values can be mathematically inferred and used to determine how unlikely your observed value of W is, assuming H_0 is true. If your W falls into the region of rejection, you can reject H_0. Critical values of W, symbolized W^*, are listed in Table W^*. If your observed W is *smaller* than the critical value listed in Table W^*, then the probability of getting your W is less than alpha. Be careful here because this is a different criterion for rejection than you are used to with other hypothesis tests which require larger values than the tabled critical values. You may reject H_0 if W is smaller than W^* and conclude that the two populations are differently distributed. The hypothesis-testing procedure for the Wilcoxon test is summarized as follows:

W^*

1. Determine the appropriate H_0 and H_A. For a two-tailed test, H_0 is that the population represented by your sample of "before" scores and the population represented by your sample of "after" scores are identically distributed. For a one-tailed test, H_0 is that the "before" population lies above (below) or is distributed identically to the Group 2 population. H_A is then that the "before" population lies below (above) the "after" population. Assume H_0 is true.
2. Set alpha and determine N, where N equals the number of untied pairs of raw scores. Determine the critical value of W from Table W^* for your level of alpha, N, and whether you are conducting a one- or two-tailed test.
3. Calculate W. First create a set of D scores by subtracting each subject's X_2 score ("after" score) from his or her X_1 score ("before" score). Assign ranks to the absolute values of these D scores from smallest to largest. Separately calculate the sum of the ranks for the positive D scores and the sum of the ranks for the negative D scores. The smaller of these two sums is W.

   If it is a two-tailed test, proceed to Step 4.

If it is a one-tailed test: if you predict the population of X_1 scores ("before" scores) to be larger than the population of X_2 scores ("after" scores), expect W to be the sum of the ranks for the negative D scores; if you predict the population of X_1 scores to be smaller than the population of X_2 scores, expect W to be the sum of the ranks for the positive D scores.

If W is associated with the expected sum of ranks, proceed to Step 4.

If W is not associated with the expected sum of ranks, retain H_0 and conclude that there is not sufficient evidence to say your sample was drawn from differently distributed populations, as specified in H_A.

4. If $W < W^*$, reject H_0 and conclude that your samples were drawn from differently distributed populations.

   If $W \geq W^*$, retain H_0 and conclude that there is not sufficient evidence to say your samples were drawn from differently distributed populations as specified in H_A.

There are two additional considerations worth mentioning that involve D scores. One consideration concerns ties in the D scores, and the other concerns zero D's. Ties in the D scores should be dealt with as they are in the Mann-Whitney U test. If two or more D scores have the same absolute value, assign the average of the ranks to each D score involved in the tie. D scores of zero will arise when a pair of X_1 and X_2 scores are tied, so $D = X_1 - X_2 = 0$. To deal with cases in which one or more pairs of X_1 and X_2 scores are tied, simply cross out the pairs of equal raw scores and ignore them (as though you never collected them). When you ignore pairs of equal scores, do not forget that the total number of pairs of scores is reduced, and so N must be reduced accordingly. Unlike the U test, it is not bad to have a very large number of equal scores in the Wilcoxon test.

To illustrate the hypothesis-testing procedure using the Wilcoxon test, consider the data from which you wanted to determine whether the reading skills program increases students' reading speeds. The steps in the Wilcoxon test are as follows:

1. The reading skills program is intended to increase reading speeds, so you conduct a one-tailed test.

   H_0: The distribution of reading speeds (in words per minute) for the population of tenth-graders who are tested "after" implementation of the reading skills program lies below or is identical to that of the population who are tested "before" the program. (The reading skills program does not increase reading speed.)

   H_A: The distribution of reading speeds for the population of tenth-graders who are tested "after" implementation of the reading skills program lies above that of the population tested "before" the program. (The reading skills program increases reading speed.)

2. Set $\alpha = .05$. Although you collect 12 pairs of scores, since one pair of X_1 and X_2 scores is equal, ignore this one pair; $N = 11$. Using Table W^*, for a one-tailed test at $\alpha = .05$ and $N = 11$, $W^* = 14$.

3. Create a set of D scores by subtracting X_2 scores ("after" scores) from X_1 scores ("before" scores). Since X_1 and X_2 in the first pair of scores are tied, this pair is ignored. After eliminating this pair, notice that two D scores are involved in a tie for the second and third ranks. Thus you assign the average ranking of 2.5 to both.

   Separate the ranks into those which came from positive D's, and those which came from negative D's, and then sum each. The sum of the ranks for the positive D scores is 12, and the sum of the ranks for the negative D scores is 54. Since the sum of the ranks for the positive D scores is smallest, $W = 12$.

Subject	*(Before)* X_1	*(After)* X_2	D $(X_1 - X_2)$	$\|D\|$	*Rank of* $\|D\|$	*Ranks for positive D's*	*Ranks for negative D's*
S_1	254	254	0				
S_2	100	250	−150	150	7		7
S_3	180	540	−360	360	10		10
S_4	221	210	11	11	2.5	2.5	
S_5	225	275	−50	50	5		5
S_6	98	318	−220	220	8		8
S_7	157	557	−400	400	11		11
S_8	776	635	141	141	6	6	
S_9	147	136	11	11	2.5	2.5	
S_{10}	999	990	9	9	1	1	
S_{11}	460	489	−29	29	4		4
S_{12}	125	441	−316	316	9		9
						Sum = 12 = W	Sum = 54

Since this is a one-tailed test and you predict that the "before" population lies below the "after" population, you expect W to be the sum of the positive D scores. Since W is associated with the expected sum of ranks, you proceed to Step 4.

4. Since $12 < 14$, reject H_0 and conclude that the new reading skills program increases reading speed in tenth-graders.

The power of two-sample parametric and nonparametric tests

Power is the probability of rejecting a false H_0, or the probability of detecting a real treatment effect or difference between population distributions. The Mann-Whitney U and Wilcoxon tests are fairly powerful compared with other nonparametric tests. However, since the Mann-Whitney U and Wilcoxon tests make fewer assumptions about the population distributions than do their parametric t test counterparts, they are not as powerful as the t tests. Thus, selecting a nonparametric test when a t test is possible would be a mistake because a researcher wants as much power as possible. Since the Mann-Whitney U and Wilcoxon tests are based on rankings, which do not give any information about relative distances between scores, the use of these tests with interval/ratio data discards useful information about the distribution of scores in the sample. As an example of how information is lost when you perform a nonparametric test on interval/ratio data for which a t test is appropriate, consider the two sets of interval/ratio data below. Each set of raw scores has been assigned ranks.

Set A: Raw score	Rank	*Set B:* Raw score	Rank
49	1	3	1
50	2	50	2
51	3	51	3
52	4	80	4
53	5	85	5
54	6	86	6
55	7	100	7
56	8	121	8
57	9	140	9
58	10	2000	10

It is clear that the raw scores transmit much more information than the ranks. You can see that the raw scores in Set A are actually much less variable than the raw scores in Set B. You would not be able to discern this if someone showed you only the ranks and not the raw data. Nor would you be able to discern that the extreme score of 2000 in Set B deviates by as much as 1997 from the lowest raw score. Ranking data only tells you the order of the scores and ignores the relative distance between scores. In general, researchers tend to use the t tests even when the assumptions for employing the t tests are violated because the t tests are very robust and relatively powerful.

Journal form

The journal form for the Mann-Whitney U and the Wilcoxon tests is very similar to that for the z test, because neither incorporates df. Since the U statistic and the W statistic are less common than z and t, it is a good idea to write out the name of these tests in research reports. For example, you might report, "The difference between the distributions was significant, Mann-Whitney $U = 26$, $p < .05$," or "The distribution of Group 1 was significantly greater than the distribution of Group 2, Wilcoxon $W = 10$, $p < .05$."

KEY CONCEPTS

1 The t tests belong to a general class of hypothesis tests called parametric tests because they require several assumptions about the parameters of

the population distributions. Nonparametric tests represent another general class of hypothesis tests that make less stringent assumptions about the population distributions than do the parametric tests.

2 The nonparametric counterpart to the two-sample independent t test is the Mann-Whitney U test. The nonparametric counterpart to the two-sample dependent t test is the Wilcoxon test.

3 The assumptions that should be met before applying the Mann-Whitney U test are: (a) the data are measured with an ordinal or interval/ratio scale; (b) there is a continuous underlying scale of measurement; and (c) the subjects in each sample are selected at random and the samples are independent.

4 The Mann-Whitney U test combines the scores from two independent samples into one set and assigns ranks to the scores. The ranks are then separated according to the group they represent. U is a measure of the similarity of the ranks in the two groups. Either high or low values of U indicate a difference in the rankings in the two groups. U is compared with two critical values.

5 When the sample size for each sample is 20 or less ($N_1 \leq 20$ and $N_2 \leq 20$), the hypothesis-testing procedure using the Mann-Whitney U test is as follows:

1. State H_0 and H_A. Assume H_0 is true.

2. Set alpha and find N_1 and N_2. Determine U^* from Table U^* for your level of alpha, N_1 and N_2, and whether you are conducting a one- or two-tailed test.

Assign ranks to the combined scores and calculate R_1, the sum of the ranks for one group.

3. Calculate U with the formula:

$$U = R_1 - \frac{(N_1)(N_1 + 1)}{2}$$

If it is a two-tailed test proceed to Step 4a. If it is a one-tailed test proceed to Step 4b.

4a. If U is greater than or equal to the larger U^*, or if U is less than or equal to the smaller U^*, reject H_0.

If U falls between the two critical values, retain H_0.

4b. If it is a one-tailed test and you predict that the Group 1 population distribution lies above that of the Group 2 population, reject H_0 only if your calculated U value is greater than or equal to the larger U^*. Otherwise retain H_0.

If it is a one-tailed test and you predict that the Group 1 population distribution lies below that of the Group 2 population, reject H_0 only if your calculated U value is less than or equal to the smaller U^*. Otherwise retain H_0.

6 For larger samples ($N_1 > 20$ and $N_2 > 20$), the sampling distribution of U will be approximately normal. U can then be converted to a z score to see whether it is less likely than alpha. The Mann-Whitney U test for larger samples proceeds as follows:

1. State H_0 and H_A. Assume H_0 is true.

2. Set alpha. Determine z^* from Table z^* for your level of alpha and whether you are conducting a one- or two-tailed test.

Assign ranks to the combined scores and calculate R_1, the sum of ranks for one group.

3. Calculate U with the formula:

$$U = R_1 - \frac{(N_1)(N_1 + 1)}{2}$$

Convert your U to a z score with the formula:

$$z = \frac{U - \dfrac{(N_1)(N_2)}{2}}{\sqrt{\dfrac{(N_1)(N_2)(N_1 + N_2 + 1)}{12}}}$$

If it is a two-tailed test proceed to Step 4a. If it is a one-tailed test proceed to Step 4b.

4a. For a two-tailed test:

If $|z| > z^*$, reject H_0. If $|z| \leq z^*$, retain H_0.

4b. For a one-tailed test where you predict that the population represented by Group 1 lies above that of Group 2:

If z is positive and $|z| > z^*$, reject H_0; otherwise retain H_0.

For a one-tailed test where you predict that the population represented by Group 1 lies below that of Group 2:

If z is negative and $|z| > z^*$, reject H_0; otherwise retain H_0.

7 The assumptions that should be met before applying the Wilcoxon test are: (a) the data are measured with an ordinal or interval/ratio scale; (b) the differences between pairs of scores can be ranked in magnitude; and (c) the subjects in each sample are selected at random and the samples are dependent.

8 The Wilcoxon test involves creating a set of D scores by subtracting each subject's X_2 score from the X_1 score and assigning ranks to the absolute values of the D scores. The ranks are separated according to those representing positive D's and those representing negative D's. W is a measure of the similarity of the raw scores in the two samples. Small values of W indicate that the scores in the two samples are different.

9 Using the Wilcoxon test, the hypothesis-testing procedure is as follows:

1. State H_0 and H_A. Assume H_0 is true.

2. Set alpha and calculate N, the number of untied pairs of raw scores. Determine the critical value of W from Table W^* for your level of alpha, N, and whether you are conducting a one- or two-tailed test.
3. Create a set of D scores by subtracting X_2 from X_1 for each untied pair of raw scores. Assign ranks to the absolute values of the D scores and calculate the sum of the ranks separately for the positive D's and the negative D's. W is the smaller of these two sums of ranks.

   If it is a two-tailed test, proceed to Step 4.

   If it is a one-tailed test: if H_A is that the scores in the Group 1 population distribution lie above those of the Group 2 population, expect W to be the sum of the ranks for the negative D's; if H_A is that the scores in the Group 1 population distribution lie below those of the Group 2 population, expect W to be the sum of the ranks of the positive D's. If W is associated with the expected sum, then proceed to Step 4. If it is not associated with the expected sum, retain H_0.
4. If $W < W^*$, reject H_0. If $W \geq W^*$, retain H_0.

10 When you have tied raw scores (in the Mann-Whitney U test) or tied D scores (in the Wilcoxon test), assign the mean of the ranks of the scores involved in the tie to each of them.

11 Since the nonparametric tests have less power than their parametric counterparts, the Mann-Whitney U and Wilcoxon test should not be used when the t tests are appropriate.

EXERCISES

1 You are interested in the newborn infant's contribution to the father-infant interaction process. You wish to determine whether male or female infants will be judged more "cuddly" by their fathers. Random samples of seven male infants and eight females were obtained. When the babies were ten days old, each father was given a neonatal assessment scale designed to measure the father's perception of his baby's cuddliness. The scale of cuddliness goes from 0 to 40, with high scores reflecting high degrees of cuddliness. The results are:

Male babies		*Female babies*	
Subject	X_1	*Subject*	X_2
S_1	9	S_8	34
S_2	20	S_9	29
S_3	13	S_{10}	30
S_4	33	S_{11}	37
S_5	22	S_{12}	35
S_6	28	S_{13}	38
S_7	31	S_{14}	26
		S_{15}	27

a. Which is the most appropriate two-sample nonparametric test to use in this study, the Mann-Whitney *U* test or the Wilcoxon test?
b. Why might a nonparametric test be preferred to a *t* test for these data?
c. Set $\alpha = .05$ and complete the hypothesis-testing procedure using the appropriate nonparametric test. What can you conclude about the effect of gender on father perceptions of cuddliness?
d. Put your results in journal form.

2 As the elementary school counselor for six elementary schools in the district, you are interested in developing strategies to increase parent-child interaction on school-related tasks, particularly reading achievement. A random sample of 15 third-graders reading below grade level were selected to participate in a pilot study. The parents are given instructions on how to engage their children in a "reading activities program" at home, and on how to make this reading activity pleasant and nonfrustrating. Parents were also encouraged to provide their child with undivided attention during this reading period for at least $\frac{1}{2}$ hour per evening. Each child's reading scores (in grade equivalences) were assessed with an achievement test the day before and one month after the reading activities program was initiated. The "before" and "after" reading grade equivalence scores are:

Subject	*Before* X_1	*After* X_2
S_1	2.5	3.0
S_2	2.4	3.2
S_3	2.3	2.3
S_4	2.2	2.7
S_5	2.3	3.0
S_6	1.9	2.8
S_7	1.1	2.9
S_8	2.4	3.1
S_9	2.1	2.8
S_{10}	2.0	2.9
S_{11}	1.8	3.0
S_{12}	2.3	2.0
S_{13}	1.8	1.8
S_{14}	2.2	3.1
S_{15}	1.2	2.3

a. Which is the appropriate two-sample nonparametric test to use in this study, the *U* test or the Wilcoxon test?
b. Why might a nonparametric test be preferred to a *t* test for these data?
c. Set $\alpha = .05$ and complete the hypothesis-testing procedure using the appropriate nonparametric test. What can you conclude about the benefit that the home reading program has on reading achievement?
d. Put your results in journal form.

3 You are interested in the development of gender stereotyping of adult occupations in children. Last year you selected a sample of 100 children ranging from 8 to 12

years of age and asked them to rate ten adult occupations on a scale from 1 to 5 concerning their beliefs about which sex has the ability to do the job. The children rated each occupation as follows:
1. only women had the ability
2. mostly women, but a few men had the ability
3. men and women are equivalent in their ability
4. mostly men, but a few women had the ability
5. only men had the ability

This year the same occupations were rated by the same children. The means of the childrens' ratings are presented below for each occupation:

Occupation	*Last year* X_1	*This year* X_2
Truck driver	4.48	4.47
Auto mechanic	4.57	4.55
Pilot	4.40	4.44
Farmer	4.72	4.75
Grocery clerk	2.26	2.25
Hair stylist	2.21	2.22
Nurse	1.56	1.56
Teacher	2.14	2.15
Secretary	1.90	1.88
Taxi driver	3.43	3.43

a. Setting $\alpha = .05$, employ the Wilcoxon test and complete the hypothesis-testing procedure to determine whether there has been any change in occupational stereotyping over the year.
b. Put your results in journal form.
c. Why might the Wilcoxon test be preferred to the dependent t test for this study?

4 You believe that new information may exist in one's memory first in a temporary state and, with time, become permanent or "consolidated" in long-term memory. You propose that increased neural activity should facilitate consolidation and hence improve memory. To test this hypothesis, you obtain 50 rats and randomly divide them into two groups of 25 each. All rats are then trained in a maze to turn either right or left at a choice point in order to obtain a food reward. A white square at the choice point indicates that only a right turn will be rewarded; a black square at the choice point indicates that only a left turn will be rewarded. Each rat receives 25 consecutive training trials (one minute apart). Amphetamine has been demonstrated to increase neural activity. Thus, in line with your proposal, injections of amphetamine prior to training should improve performance (reduce errors) by increasing neural activity during the consolidation period. The rats in the experimental group received injections of amphetamine 5 minutes before the first trial. The rats in the control group received a placebo 5 minutes before the first trial. The number of errors are given below:

Amphetamine				Placebo			
Subject	X_1	*Subject*	X_1	*Subject*	X_2	*Subject*	X_2
S_1	8	S_{14}	8	S_{26}	8	S_{39}	10
S_2	6	S_{15}	8	S_{27}	9	S_{40}	13
S_3	10	S_{16}	2	S_{28}	12	S_{41}	10
S_4	10	S_{17}	11	S_{29}	10	S_{42}	10
S_5	9	S_{18}	8	S_{30}	11	S_{43}	9
S_6	9	S_{19}	9	S_{31}	11	S_{44}	12
S_7	8	S_{20}	2	S_{32}	7	S_{45}	13
S_8	5	S_{21}	3	S_{33}	11	S_{46}	8
S_9	7	S_{22}	10	S_{34}	8	S_{47}	9
S_{10}	7	S_{23}	13	S_{35}	10	S_{48}	11
S_{11}	6	S_{24}	7	S_{36}	14	S_{49}	11
S_{12}	7	S_{25}	12	S_{37}	10	S_{50}	10
S_{13}	9			S_{38}	9		

a. Which is the appropriate two-sample nonparametric test to use in this study, the U test or the Wilcoxon test?
b. Why might a nonparametric test be preferred to a t test for these data?
c. Set $\alpha = .01$ and complete the hypothesis-testing procedure using the appropriate nonparametric test. What can you conclude about the facilitative effect of amphetamine on memory?
d. Put your results in journal form.

5 **a.** Employ the Mann-Whitney U test and complete the hypothesis-testing procedure to test the facilitating effects of auditory feedback on self-control of the frontalis muscle described in Exercise 10 of Chapter 8. Set $\alpha = .01$.
b. Put your results in journal form.

6 **a.** Employ the Wilcoxon test and complete the hypothesis-testing procedure to test the effectiveness of the antismoking campaign in reducing cigarette smoking as described in Exercise 6 of Chapter 8. Set $\alpha = .05$.
b. Put your results in journal form.

7 **a.** Employ the Wilcoxon test and complete the hypothesis-testing procedure to test the effect of general unemployment on job satisfaction for the study in Exercise 9 of Chapter 8. Set $\alpha = .05$
b. Put your results in journal form.

CHI-SQUARE

CHAPTER GOALS

After studying this chapter, you should be able to:

1 Perform the χ^2 goodness-of-fit test.

2 Perform the χ^2 test of independence.

3 Define the sampling distribution of χ^2.

4 Describe the principle involved in calculating the expected frequencies for the χ^2 test of independence.

5 State the assumptions for the χ^2 tests.

CHAPTER IN A NUTSHELL

A common research problem is to decide whether a sample of scores measured in discrete categories is a random sample from a population in which the relative frequency of scores in each category is known. The general hypothesis-testing procedure can be employed here. You assume the sample to be randomly drawn from the population and measure how much the observed frequencies deviate from the frequencies you expect to get by chance. The measure of this deviation is called chi-square, symbolized χ^2. The calculated value of χ^2 is compared with the sampling distribution of χ^2, a distribution of χ^2 values you would find by chance if samples are drawn at random from the population. There are two types of hypothesis tests using χ^2. In the χ^2 goodness-of-fit test the goal is to decide whether or not the pattern of frequencies observed on one variable deviates significantly from the expected pattern. The χ^2 test of independence aids you in deciding whether the pattern of frequencies observed along two variables indicates whether the variables are independent. This test is very similar to the χ^2 goodness-of-fit test except that the methods for calculating expected values and degrees of freedom are different.

Introduction to chi-square

So far we have discussed hypothesis testing in research situations in which the researcher is interested in determining whether or not there is a significant difference between two sample means. However, there are numerous occasions in behavioral science research when the researcher is interested in testing whether an entire frequency distribution of different objects/events follows a given pattern. It is possible to determine whether the frequencies do or do not follow the expected pattern with a hypothesis test known as the χ^2 test. The Greek letter χ is spelled chi and pronounced "kye" (rhymes with "rye"); squaring χ thus gives the *chi-square test* its name. There are two varieties of the χ^2 test. The first type is the χ^2 goodness-of-fit test. It is very similar to, and provides the foundation for, the second type, called the χ^2 test of independence.

χ^2

The χ^2 goodness-of-fit test: The one-variable case

Suppose you are trying to decide whether the pattern of the number of people in Beverly Hills, California, owning "subcompact," "compact," "medium," or "large" automobiles follows the distribution seen nationally. This is certainly a hypothesis-testing type of study, but it differs in two main ways from research situations we have previously discussed. First, this study deals only with frequencies in nominal categories of *one variable,* such as the number of car owners in each of the four automobile categories. The variable is the type of automobile. Notice that this study does not deal with continuous data such as a subject's numerical score on some ability/trait; it deals with discrete frequencies. Second, this type of problem can involve more than two categories of one variable. You are looking at how well the frequencies of car owners in the four categories fit the nationwide pattern of frequencies. Thus, in problems in which you are trying to see how well an observed pattern of frequencies along one variable conforms to a known, or *expected,* pattern of frequencies, you are concerned with testing the "goodness of fit" of the observed frequencies to the expected frequencies. It is possible to decide whether the observed frequencies do or do not fit the expected frequencies with the **χ^2 goodness-of-fit test.**

χ^2 goodness-of-fit test

To illustrate the hypothesis-testing procedure involving the χ^2 goodness-of-fit test, consider another example. Suppose you are interested in doing a survey of college students' attitudes about some current moral issue. A friend suggests a procedure for obtaining a random sample. However, you want to conduct a preliminary study to be sure the sampling method he suggests for randomly selecting 100 students does, indeed, give you a random sample of the students in the college. His suggestion is that you select every tenth student who enters the college bookstore on Monday morning, until you have selected 100 students in all. You know the college's entire population consists of 25% freshmen, 20% sophomores, 30% juniors, and 25% seniors. Using the suggested sampling method you find that of the 100 students in your sample, 35 are freshmen, 30 are sophomores, 20 are juniors, and 15 are seniors. Given the known percentages of freshmen, sophomores, juniors, and seniors in the population (of your college), you expected to obtain 25 freshmen (25% of 100), 20 sophomores, 30 juniors, and 25 seniors in your sample. These observed and expected frequencies are tabulated separately below, where you can see that the observed frequencies of 35-30-20-15 do not exactly match the expected frequencies of 25-20-30-25:

Observed frequencies

freshmen	*sophomores*	*juniors*	*seniors*
35	30	20	15

Expected frequencies

freshmen	*sophomores*	*juniors*	*seniors*
25	20	30	25

In each table the observed and expected frequencies are traditionally "boxed in." Because of this, each category (or level) of classification is referred to as a **cell.**

cell

In reality, due to sampling error, you probably would not expect the observed frequencies to be identical to the expected frequencies, but you also do not expect the observed frequencies to be too different from the expected frequencies either. If you have a truly random sampling method, you expect the observed frequencies to be close to the expected frequencies, and even be identical once in a while. Whenever you find that the observed frequencies differ from the expected frequencies, there are two competing explanations for how the difference came about. One explanation is that your sampling method is truly random, and your observed frequencies occurred by chance from a population whose pattern (or proportions) are represented by the expected frequencies. This says that the difference between the observed and expected frequencies is due to sampling error. The other explanation is that your sample is not a random sample of the entire school population after all because the sampling method introduces some bias into the selection process. Your observed frequencies came from a "biased" population whose pattern (or proportions) are not represented by the expected frequencies. In effect you have taken a sample from a population that is different from the entire school population. These two explanations are illustrated in Figure 10-1, and you want to decide between them. The explanation that your observed frequencies occur by chance from a population whose proportions are represented by the expected frequencies is H_0, because it allows you to calculate the probability of obtaining observed frequencies as deviant from the expected frequencies

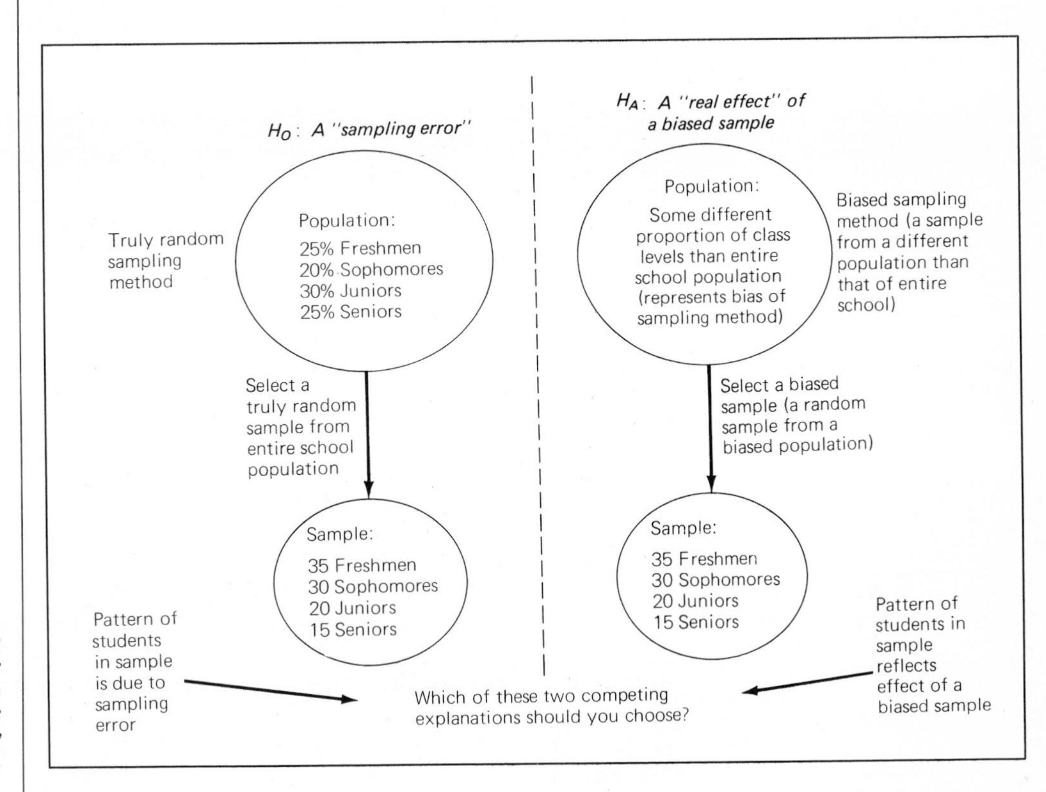

FIGURE 10-1

The two competing explanations of the observed distribution of college students obtained by a new sampling technique.

as yours. H_A is that your observed frequencies really come from some other population whose proportions are not represented by the expected frequencies.

In typical hypothesis-testing style, you begin by assuming H_0 is true and then calculate the probability of getting observed frequencies that deviate from the expected frequencies as much as yours do. If the difference between the observed and expected frequencies is too unlikely to attribute to sampling error, then you can reject H_0 and conclude that your sample is not a random sample from the entire school population. The question is, how do you calculate the probability of getting, by chance, observed frequencies as extreme as yours?

Before you can calculate the probability of getting observed frequencies as deviant from the expected frequencies as yours, you need to measure precisely how much your observed frequencies differ from the expected frequencies. By now you might realize that if you simply subtract the expected frequencies from the observed frequencies for each category and then add the differences, the positive and negative differences you get will cancel each other out (similar to the old zero-sum problem with the variance). For example, for freshmen, the observed frequency minus the expected frequency is $35 - 25 = 10$, but for juniors, it is $20 - 30$

= −10. To make these differences more usable, you can remove the signs of the differences between the expected and observed frequencies by squaring the differences. If you sum the squared differences you will have a pretty good measure of the deviation of the observed frequencies from the expected frequencies—except that the categories with high observed frequencies and/or high expected frequencies are emphasized more than the categories having small frequencies. This is because the categories with higher frequencies usually give considerably larger differences. Furthermore, when these larger differences are squared, their influence in the sum of the squared differences is magnified. For example, if you expect a frequency of 10 for a given category but actually observe 11, the difference of 1 is a 10% deviation and is not surprising. The square of the difference, 1^2, adds relatively little weight to the sum of the squared deviations. Suppose, though, that the expected value in another category is 100 and you actually observe 110. This difference of 10 is still only a 10% deviation and is also not surprising. Nevertheless, 10^2 adds much more weight to the sum of the squared difference than 1^2 does for the other category. You do not want the category with the higher frequency to overinfluence the measure of the deviation of the observed frequencies from the expected frequencies, since the observed frequency in both categories is the same percentage bigger than the expected frequency. Therefore, to de-emphasize the influence of the categories with higher frequencies, you need to have a measure of deviation that is akin to a percentage deviation. Such a measure is provided by dividing the squared difference of each category by the expected frequency of that category. You can see that this measure gives you the squared difference between your observed frequency and the expected frequency in terms of expected frequency units. The sum of these squared differences per unit of expected frequency is called χ^2, and is calculated by the following definitional formula:

FORMULA 10.1
definitional formula for χ^2

$$\chi^2 = \sum \frac{(O - E)^2}{E}$$

O
E

Here, O ("oh") stands for the observed frequency and E is the expected frequency for a particular cell. You can calculate a χ^2 of 16.3 for the sample of college students by starting off with tables of observed and expected frequencies and applying Formula 10.1 as follows:

Observed frequencies

freshmen	*sophomores*	*juniors*	*seniors*
$O = 35$	$O = 30$	$O = 20$	$O = 15$

Expected frequencies

$E = 25$	$E = 20$	$E = 30$	$E = 25$

$$\chi^2 = \sum \frac{(O - E)^2}{E} = \frac{(35 - 25)^2}{25} + \frac{(30 - 20)^2}{20} + \frac{(20 - 30)^2}{30} + \frac{(15 - 25)^2}{25}$$

$$= \frac{10^2}{25} + \frac{10^2}{20} + \frac{10^2}{30} + \frac{10^2}{25}$$

$$= \frac{100}{25} + \frac{100}{20} + \frac{100}{30} + \frac{100}{25}$$

$$= 4 + 5 + 3.3 + 4 = 16.3$$

For each cell the expected frequency is subtracted from the observed frequency to get the difference. The difference is then squared and divided by the expected frequency. The results from each cell are finally summed to give χ^2. The more the observed frequencies deviate from the expected frequencies, the larger χ^2 becomes. Whereas a χ^2 of 0 indicates that the observed and expected frequencies are identical for each category, what can you say about your calculated χ^2 of 16.3?

Assuming H_0 is true and your sample is truly a random sample, you need to determine the probability of getting a χ^2 as extreme as 16.3. To determine this probability, you can compare your calculated χ^2 with a distribution of χ^2 values that would occur by chance if H_0 is true. This distribution, called a **chi-square distribution,** is another type of sampling distribution whose characteristics can be mathematically inferred. Actually, there is a different chi-square distribution for every possible number of categories that might be examined. As the number of categories increases, you expect to get larger values of χ^2 by chance because there are more cells and hence more $(O - E)^2/E$ results to be added up.

chi-square distribution

Assuming your sample is random and H_0 is true, there should usually not be much deviation between the expected and observed frequencies, so most values of χ^2 will be relatively low. Large values of χ^2 will be relatively unlikely. Critical values of χ^2, symbolized χ^{2*} are listed in Table χ^{2*}. If your calculated χ^2 is greater than χ^{2*}, you know its probability of occurrence is less than alpha and you can reject H_0. Table χ^{2*} lists χ^{2*} values according to the level of alpha and degrees of freedom. For the χ^2 goodness-of-fit test, *df* is equal to $K_c - 1$, where K_c is the number of categories, or columns, in the row of cells ("column" is denoted with the subscript *c*).

χ^{2*}

K_c

Now apply the χ^2 goodness-of-fit test to your study using the sample of college students. Since there are four categories, $df = K_c - 1 = 4 - 1 = 3$. If $\alpha = .05$, you find $\chi^{2*} = 7.815$ from Table χ^{2*}. For your calculated χ^2 to be significant, it must be greater than χ^{2*}. Note that your calculated χ^2 will always be a positive number. If you get a negative value you made a computational error. Since your calculated χ^2 of 16.3 is greater than the critical value of 7.815, you reject H_0 and conclude that your sampling method did not produce a random sample of the population of students.

The general hypothesis-testing procedure using the chi-square goodness-of-fit test is as follows:

1. Determine the appropriate H_0 and H_A. H_0 is that your observed frequencies occurred by chance from a population whose proportions are represented by the expected frequencies. H_A is that your observed frequencies occurred from a population with proportions different from those represented by the expected frequencies. Assume H_0 is true.
2. Set alpha and calculate $df = K_c - 1$, where K_c is the number of categories of observations. Determine the critical value of χ^2 from Table χ^{2*} for your alpha and *df*.
3. If you have not done so already, create tables of the observed and expected frequencies. Calculate χ^2 for your data with Formula 10.1:

$$\chi^2 = \sum \frac{(O - E)^2}{E}$$

4. If $\chi^2 > \chi^{2*}$, reject H_0 and conclude that your observed frequencies came from a population whose proportions are different from those represented by the expected frequencies.

   If $\chi^2 \leq \chi^{2*}$, retain H_0 and conclude that there is not sufficient evidence to say your observed frequencies came from a population whose proportions are different from those represented by the expected frequencies.

Now apply this process to a new research problem concerning hyperactive children. Suppose you want to determine whether male and female elementary school children are equally likely to be diagnosed as hyperactive. You assume that there are equal numbers of males and females in the population of elementary school children. However, in a sample of 100 hyperactive children there were 65 males and 35 females. You expect to find 50 males and 50 females in such a sample. Are your observed frequencies significantly different from what you would expect by chance? The hypothesis-testing procedure you would use to answer this question is as follows:

1. H_0: Males and females are equally likely to be diagnosed as hyperactive. Your observed frequencies occurred by chance from a population with equal numbers of hyperactive males and females.

   H_A: Males and females are not equally likely to be diagnosed as hyperactive. Your observed frequencies arose from a population with unequal numbers of hyperactive males and females.

2. Let $\alpha = .01$. The .01 level is chosen because you do not want to claim "bias" without good cause and this more conservative level will reduce the probability of an alpha error.

   Since you have two categories of sex, $df = 2 - 1 = 1$. Using Table χ^{2*} you find that for $\alpha = .01$ and $df = 1$, $\chi^{2*} = 6.635$.
3. Create tables of the observed and expected frequencies.

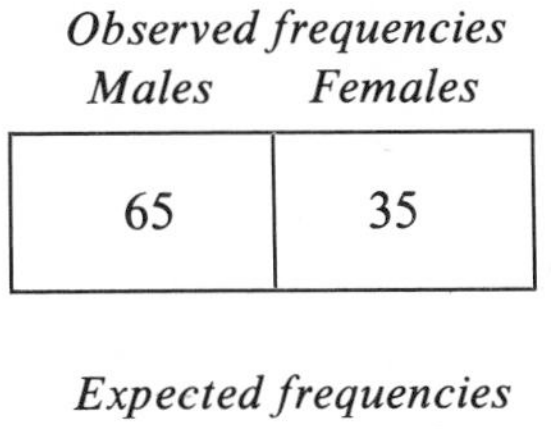

Observed frequencies

Males	*Females*
65	35

Expected frequencies

Males	*Females*
50	50

The calculated χ^2 for your data using Formula 10.1 is:

$$\chi^2 = \sum \frac{(O - E)^2}{E} = \frac{(65 - 50)^2}{50} + \frac{(35 - 50)^2}{50}$$

$$= \frac{15^2}{50} + \frac{15^2}{50}$$

$$= \frac{225}{50} + \frac{225}{50}$$

$$= 4.5 + 4.5 = 9.0$$

4. Since $9.0 > 6.635$, reject H_0 and conclude that males and females are not equally likely to be diagnosed as hyperactive.

The χ^2 test of independence: The two-variable case

χ^2 test of independence

There is another type of research situation in which Formula 10.1 for χ^2 can be applied. For example, you may wish to determine whether a voter's preference for a particular candidate depends on the voter's educational background. Are the preferences of college-educated and non-college-educated voters essentially the same, or does preference depend, at least in part, on whether the voter is college educated? A special application of χ^2 helps answer such questions concerning the "independence" of two variables measured using nominal categories. This application is called the **χ^2 test of independence** and is almost identical to the χ^2 goodness-of-fit test except for the way the expected frequencies are calculated.

To illustrate the χ^2 test of independence, suppose you are interested in determining whether or not peoples' level of aggressiveness at work is dependent on the amount of fear they experience in their working environment. The level of aggression is one variable, and the level of fear is the other variable. You go to a large company with diverse working conditions and randomly select a sample of 100 workers. On the basis of reports from supervisors and co-workers as to how aggressive each worker is, you classify the workers as either high or low in aggression. You also question the workers themselves to find out if they think most of their working time is spent under conditions of high, medium, or low fear. You tabulate your *observed frequencies* as follows:

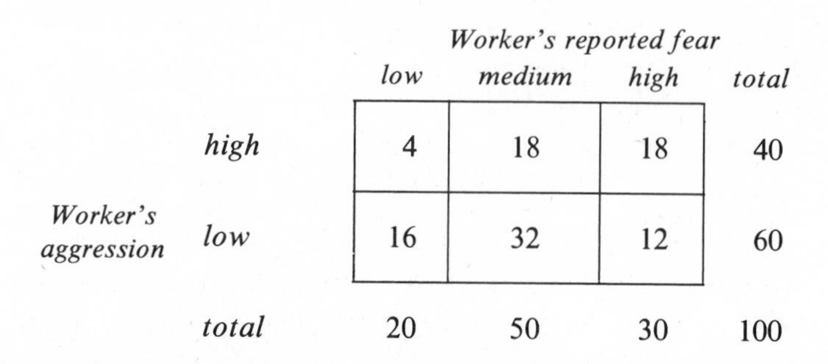

		Worker's reported fear			
		low	*medium*	*high*	*total*
Worker's aggression	*high*	4	18	18	40
	low	16	32	12	60
	total	20	50	30	100

The numbers in each cell represent how many workers fall into each category. Thus four workers were high in aggression and reported working under low fear, 18 workers were high in aggression and reported medium fear, and so on. Contrast this table with the type used for the χ^2 goodness-of-fit test. In the χ^2 goodness-of-fit test, the table of observed frequencies usually consists of only one row of cells. This is because usually one variable is examined. On the other hand, in the χ^2 test of independence, two variables are examined. This gives you at least two rows of observed frequencies and at least two columns of observed frequencies. A table of observed frequencies categorized along two variables is called a **contingency table.** The research question is whether the observed frequencies for one variable depend on the level of the other variable. For example, "Does reported fear at work depend on the worker's level of aggression?" If fear depends on level of aggression, you would find that workers high in aggression will have a different level of fear (either more or less) than workers low in aggression. If fear and aggression are *independent,* workers high in aggression will have the *same* level of fear, on the average, as workers low in aggression. In a hypothesis-testing situation, the word "depend" is often a cue that you should apply the χ^2 test of independence.

contingency table

Since you have the observed frequencies written in the contingency table, you want to apply Formula 10.1 and solve for χ^2 to see how much your observed frequencies deviate from the expected frequencies. However, you cannot do this until you determine the expected frequency for each cell. To determine the expected frequencies you need to make an assumption about how fear and aggression are related. There are two possibilities:

either level of aggression is dependent on the worker's reported fear (and vice versa), or level of aggression is independent of reported fear. To say level of aggression is independent of reported fear means that the same proportions of workers fall into high- and low-aggression categories, regardless of whether the workers report high, medium, or low fear. If you assume that fear and aggression are dependent, the question arises, "To what degree is aggression dependent on fear?" For example, fear might have a slight effect on aggression, a very strong effect, or an intermediate effect. Who knows? Assuming that fear and aggression are dependent leaves a number of questions unanswered. On the other hand, if you assume that fear and aggression are independent, you can avoid these questions and readily calculate the expected frequency for each cell. Thus, H_0 is that fear and aggression are independent.

Recall from Chapter 6 that when two events are independent, you can use the multiplicative rule to calculate the probability of both occurring together. The multiplicative rule can be used to find the expected frequencies you need. For example, suppose one event is that a worker is highly aggressive and experiences low fear. Simply multiply the probability of a worker being highly aggressive by the probability of a worker experiencing low fear:

$$\begin{matrix}\text{probability of high} \\ \text{aggression } \textit{and} \text{ low fear}\end{matrix} = \begin{pmatrix}\text{probability of} \\ \text{high aggression}\end{pmatrix}\begin{pmatrix}\text{probability of} \\ \text{low fear}\end{pmatrix}$$

This gives the probability of a worker fitting the description in the upper-left-hand cell of the contingency table. All you have to do now is find the probability of high aggression and the probability of low fear and plug these separate probabilities into the multiplicative rule.

You can determine the probability of a worker being highly aggressive if you know how many workers are highly aggressive and also the total number of workers. Adding the observed frequencies in the first row tells you how many workers are highly aggressive ($4 + 18 + 18 = 40$). Adding the observed frequencies in every cell gives you the total number of workers (100). Since probability is the number of possible ways a particular event occurs divided by the total number of possible outcomes, then

$$\begin{matrix}\text{probability of} \\ \text{high aggression}\end{matrix} = \frac{\text{number of highly aggressive subjects}}{\text{total number of subjects}} = \frac{40}{100}$$

Similarly you can calculate the probability of a worker experiencing low fear. Determine how many workers experience low fear by adding the number of workers in the first column ($4 + 16 = 20$). Since there are 100 total workers,

$$\begin{matrix}\text{probability of} \\ \text{low fear}\end{matrix} = \frac{\text{number of low-fear subjects}}{\text{total number of subjects}} = \frac{20}{100}$$

Now you can calculate the probability of being in the upper left-hand corner by substituting the separate probabilities you just calculated into the multiplicative rule:

$$\text{probability of high aggression } \textit{and} \text{ low fear} = \left(\frac{40}{100}\right)\left(\frac{20}{100}\right)$$

By assuming H_0 is true, you can follow this procedure to calculate the probability of a worker being in each cell in the contingency table. Once you know the probability of being in a given cell, you can calculate the expected frequency for that cell by multiplying the probability by the total number of outcomes. For example, if the probability of flipping a head on a fair coin is 1/2, or .5, and you flip the coin 100 times, how many heads do you expect to get? By multiplying (.5)(100), you calculate that you expect to get 50 heads. Therefore, if the probability of being in the upper left-hand corner of your contingency table is (40/100)(20/100), and the total number of workers is 100, the expected number of workers in that cell is:

$$\begin{aligned}\text{expected frequency of a particular cell} &= \left(\begin{matrix}\text{probability of}\\ \text{being in the}\\ \text{particular cell}\end{matrix}\right)\left(\begin{matrix}\text{total number}\\ \text{of subjects}\end{matrix}\right)\\ &= \left[\left(\frac{40}{100}\right)\left(\frac{20}{100}\right)\right](100)\end{aligned}$$

You can use this formula to calculate the expected frequency for each cell in the contingency table. However, there is an easier way to perform this calculation. To illustrate, notice that the probability of being highly aggressive is merely the number of subjects in the first row (40) divided by the total number of subjects (100). The probability of experiencing high fear is the number of subjects in the first column (20) divided by the total number of subjects (100). To calculate the expected frequency for the upper left-hand cell you multiply these probabilities together and then multiply by the total number of subjects. This last total number of subjects will cancel out with the total number of subjects in the denominator of the probability of experiencing high fear. This will be true regardless of the cell being considered. You are thus left with a row total times a column total, divided by the total number of subjects. Therefore, you can more easily calculate the expected frequency in any cell by using the following formula:

FORMULA 10.2

$$\text{expected frequency of a particular cell} = E = \frac{(R)(C)}{N}$$

R
C

R is the number of subjects in the row containing the particular observed frequency, ***C*** is the number of subjects in the column containing the particular observed frequency, and N is the total number of subjects.

The expected frequency for each cell in the contingency table of fear and aggression is calculated below by Formula 10.2.

Expected frequencies

	low fear	*medium fear*	*high fear*
high aggression	$E = \frac{(40)(20)}{100} = 8$	$E = \frac{(40)(50)}{100} = 20$	$E = \frac{(40)(30)}{100} = 12$
low aggression	$E = \frac{(60)(20)}{100} = 12$	$E = \frac{(60)(50)}{100} = 30$	$E = \frac{(60)(30)}{100} = 18$

As a partial check for mathematical errors, you can now verify that the sum of all expected values equals the sum of all observed values. This must be so if you performed all your calculations accurately. Since the sum of the E values in the above table is 100, everything checks out. You now have two tables by assuming that aggression and fear are independent (H_0): one of the observed frequencies, and one of expected frequencies. If H_0 is correct, you expect the observed frequencies to fit the expected frequencies fairly well. But if the observed frequencies are very different from the expected frequencies, you should suspect that your assumption of independence is not correct. The hypothesis-testing procedure for the χ^2 test of independence is performed like the χ^2 goodness-of-fit procedure except for a minor difference in how degrees of freedom are calculated. The degrees of freedom are calculated by multiplying one less than the number of rows times one less than the number of columns. Symbolically, $df = (K_r - 1)(K_c - 1)$, where $\boldsymbol{K_r}$ is the number of categories, or *number of rows,* in the row variable (hence subscript "r"), and K_c is the number of categories, or *number of columns,* in the column variable (hence subscript c).

K_r

The hypothesis-testing procedure with the χ^2 test of independence enables you to decide whether aggression and fear are independent. The steps for this hypothesis test are as follows:

1. H_0 Aggression and fear are independent. Your observed frequencies occurred by chance from a population whose composition is represented by the expected frequencies.

H_A: Aggression and fear are dependent. Your observed frequencies came from a population whose composition is different from that represented by the expected frequencies.

2. Let $\alpha = .05$. Since you have two categories of aggression (two rows) and three categories of fear (three columns),

$$df = (K_r - 1)(K_c - 1) = (2 - 1)(3 - 1) = (1)(2) = 2$$

Using Table χ^2* you find, for $\alpha = .05$ and $df = 2$, $\chi^2 = 5.991$.

3. Create tables of the observed and expected frequencies. The table of expected frequencies is derived using Formula 10.2:

$$\text{expected frequency of a particular cell} = E = \frac{(R)(C)}{N}$$

Observed frequencies

	low fear	*medium fear*	*high fear*
high aggression	4	18	18
low aggression	16	32	12

Expected frequencies

	low fear	*medium fear*	*high fear*
high aggression	8	20	12
low aggression	12	30	18

The calculated χ^2 for your data using Formula 10.1 is as follows:

$$\chi^2 = \sum \frac{(O - E)^2}{E} = \frac{(4 - 8)^2}{8} + \frac{(18 - 20)^2}{20} + \frac{(18 - 12)^2}{12} + \frac{(16 - 12)^2}{12} + \frac{(32 - 30)^2}{30} + \frac{(12 - 18)^2}{18}$$

$$= \frac{(4)^2}{8} + \frac{(2)^2}{20} + \frac{(6)^2}{12} + \frac{(4)^2}{12} + \frac{(2)^2}{30} + \frac{(6)^2}{18}$$

$$= \frac{16}{8} + \frac{4}{20} + \frac{36}{12} + \frac{16}{12} + \frac{4}{30} + \frac{36}{18}$$

$$= 2 + .2 + 3 + 1.33 + .33 + 2$$

$$= 8.67$$

4. Since $8.67 > 5.991$, reject H_0 and conclude that level of aggression depends on the worker's level of reported fear.

Assumptions of the χ^2 tests

Before you perform either the χ^2 goodness-of-fit test or the χ^2 test of independence, several assumptions should be satisfied:[1]

1. The data are frequency data only.
2. The observed frequencies in each cell are independent.
3. The subjects (or objects, events, processes, etc.) comprising the sample are selected on a random basis.
4. The sample is fairly large.

You must meet the first assumption because χ^2 is based on frequencies of observed and expected values, and not on measures of some ability/trait on ordinal or interval/ratio scales. If you have a measurement of an ability/trait on an ordinal or interval/ratio scale, it must be reduced to frequency data before applying χ^2. For example, the χ^2 test could be applied to IQ data if it is reduced to frequencies of people with high, medium, and low IQ. The second assumption that the observed frequencies in each cell are independent indicates that each subject can provide only one tally. For example, if you are comparing the distribution of various automobile sizes for car owners living in Beverly Hills with the nationwide distribution, you cannot ask the same car owner twice (if the person owns two cars) because you would then have two tallies representing the same owner, nor could you sample the same single-car owner at different times or under different conditions. Each subject must provide one and only one tally, which appears in one and only one cell. This second assumption is *very important* and is often violated even by researchers who should know better. Be careful not to violate this assumption when you employ χ^2. The third assumption requires that subjects are selected randomly so as to eliminate any potential systematic bias and to allow inferences from your sample to the population.

The fourth assumption, that the sample is fairly large, is considered met when the *expected values* for each cell are at least five. The reason why the sample must be fairly large is because you are using the theoretical sampling distribution of χ^2 to approximate the empirical distribution, and this is not a good approximation with small samples.

Journal form

Journal form for reporting the results of both the χ^2 goodness-of-fit test and the χ^2 test of independence are alike. First give the name of the hypothesis test, χ^2, and then give the degrees of freedom in parentheses after χ^2. Next give the calculated value of χ^2, followed by the probability that your observed frequencies arose by chance from a population repre-

[1]Sophisticated readers might notice that the Yates correction for continuity which is usually applied when $df = 1$ is missing from this chapter. Recent research indicates that the Yates correction is not to be recommended. See Camilli and Hopkins, *Psychological Bulletin,* 1978, *85,* 163–167.

sented by the expected frequencies. For example, you may read in a journal article, "The incidence of alcoholism among the first, second, and third child in three-sibling families was not randomly distributed, $\chi^2(2) = 10.21$, $p < .05$." Or, briefly, you may read, "These frequencies are random, $\chi^2(6) = 9.3$, N.S."

KEY CONCEPTS

1. χ^2 is a measure of the deviation of observed frequencies from expected frequencies.

2. The χ^2 goodness-of-fit test is used to determine how well the observed frequencies in two or more categories of one variable correspond with the expected frequencies. H_0 is that the observed frequencies occurred by chance from a population represented by the expected frequencies.

3. The χ^2 test of independence is actually a goodness-of-fit test applied to determine how well the observed frequencies on each of two variables correspond with the expected frequencies. The expected frequencies are determined under H_0 that the two variables are independent. Assuming that the variables are independent allows you to use the multiplicative rule to calculate the probability of a given combination of categories of the two variables. Then you can multiply this probability by the total number of subjects and get the expected frequency. This procedure can be simplified to:

$$E = \frac{(R)(C)}{N}$$

4. The general hypothesis-testing procedure for both the χ^2 goodness-of-fit test and the χ^2 test of independence is:
   a. State H_0 and H_A. Assume H_0 is true.
   b. Set alpha and calculate the degrees of freedom.
   If it is the χ^2 goodness-of-fit test, $df = K_c - 1$. If it is the χ^2 test of independence, $df = (K_r - 1)(K_c - 1)$. K_c refers to the number of categories of observation for the χ^2 goodness-of-fit test and the number of columns for the χ^2 test of independence, and K_r is the number of rows for the χ^2 test of independence.
   Determine the critical value of χ^2 from Table χ^{2*} for your alpha and *df*.
   c. Calculate χ^2 with the formula

$$\chi^2 = \sum \frac{(O - E)^2}{E}$$

   d. If $\chi^2 > \chi^{2*}$, reject H_0. If $\chi^2 \leq \chi^{2*}$, retain H_0.

5 The assumptions that should be met before applying either the χ^2 goodness-of-fit test or the χ^2 test of independence are: (a) the data are frequency data; (b) the observed frequencies in each cell are independent; (c) the subjects comprising the sample are selected at random; and (d) the expected frequency for each cell is at least 5.

EXERCISES

1 You want to decide whether a new minting process produces "fair" coins. You take 100 new coins from the mint and toss each one once. You get 66 heads. Setting $\alpha = .01$, use the χ^2 goodness-of-fit test to test the null hypothesis that the coins are fair. Use the four steps of the hypothesis-testing procedure.

2 You are interested in determining whether the incidence of various types of crimes varied from one precinct of a city to another. The city was divided into three precincts, and the crimes were classified into five categories: willful homicide, forcible rape, aggravated assault, robbery, and larceny. Records kept over a three-month period revealed the following frequencies:

	Type of crime				
	willful homicide	*forcible rape*	*aggravated assault*	*robbery*	*larceny*
Precinct 1:	10	35	22	18	33
Precinct 2:	13	20	8	2	8
Precinct 3:	15	5	14	5	12

a. You want to determine whether the frequency of the type of crime depends on the particular precinct. You employ the χ^2 test of independence and calculate $\chi^2 = 28.656$. Setting $\alpha = .05$, is your result significant? Does the type of crime depend on the specific precinct?

b. What if a larger study of types of crime and precinct is made with nine categories of crimes and six precincts, and your calculated $\chi^2 = 49.148$. With $\alpha = .05$, does the type of crime depend on the given precinct?

3 Assume you work with a "hot-line" type of community crisis intervention and suicide prevention center. You are interested in whether there are any seasonal influences affecting the frequency of suicide-related phone calls the center receives. You hypothesize that incidence of suicide-related calls is not equally distributed across the four seasons. You search through the past year's records of phone calls and tally the following frequencies of suicide-related calls for each season:

	Season			
	fall	*winter*	*spring*	*summer*
Observed frequencies:	27	58	71	44

a. Setting $\alpha = .05$, use the χ^2 goodness-of-fit test and complete the hypothesis-testing procedure. What can you conclude about the hypothesis that suicide-related calls are not equally likely for each season?
b. Put your results in journal form.
c. What assumption might you have violated in conducting this study?

4 You are interested in the occupational status of women holding office in state legislatures. A sample of 100 female state senators are compared in terms of their occupations with those of state senators throughout the United States. Measurements were taken and you found women senators fell into one of five main occupational categories:

	Occupation				
	teachers, social workers	*lawyers*	*executives*	*real estate sales*	*major government employee*
Observed frequencies:	14	45	30	8	3

a. Set $\alpha = .01$ and test your observed distribution of occupations to see if it differs significantly from the expected distribution based on the occupations of all state senators throughout the United States: teachers, social workers = 18; lawyers = 61; executives = 10; real estate sales = 5; major government employees = 6. Use the hypothesis-testing procedure.
b. Put your results in journal form.
c. What assumption would be violated if you change two frequencies in the expected distribution so that real estate sales = 4 instead of 6, and executives = 12 instead of 10?

5 You are interested in people's knowledge about two common pollutants of water: mercury and nitrates/phosphates. You talk with 25 randomly selected college students and ask, "Which water pollutant, mercury or nitrate/phosphate, do you know more about?" Seventeen people said they knew more about mercury pollution, and eight said they knew more about nitrate/phosphate pollution.
a. Setting $\alpha = .05$, can you conclude that college students are not equally knowledgeable about these water pollutants? Use the hypothesis-testing procedure.
b. Put your results in journal form.
c. What assumption of χ^2 would be violated if each person was asked to rate his knowledge of the two pollutants on a scale from 1 to 5 as follows:
   (1) I know only of mercury as a water pollutant.
   (2) I know of both, but know more about mercury.
   (3) I know of both about equally as water pollutants.
   (4) I know of both, but know more about nitrates/phosphates.
   (5) I know only of nitrate/phosphate as a water pollutant.

6 You have observed that total strangers such as seatmates on a plane or train sometimes provide you with quite intimate information about themselves. This phenomenon is referred to as "self-disclosure." You are interested in examining

the impact of sex differences on self-disclosure to strangers. The aid of two male and two female friends to serve as "approachers" is obtained. Each "approacher" approaches 25 adult males and 25 adult females at random in the lobby of Los Angeles International Airport and explains to the travelers that they are collecting "self-descriptions" as part of a class project. You assume that the travelers who provide self-descriptions are also providing a measure of self-disclosure, and hypothesize that the sex of the self-disclosing traveler and the approacher are independent. Of the 200 travelers approached, 118 refused to participate. Of the 82 who provided self-descriptions, the following contingency table was obtained:

	Self-disclosing travelers	
	Male	*Female*
Male approachers	5	9
Female approachers	41	27

a. Which is the appropriate χ^2 test to use in this study, the χ^2 goodness-of-fit test or the χ^2 test of independence?

b. Set $\alpha = .05$ and complete the hypothesis-testing procedure using the appropriate χ^2 test. What can you conclude about your hypothesis that the sex of the self-disclosing traveler and the approacher are independent?

c. Put your results in journal form.

7 You are interested in the pattern of drug use among senior high school students living in the Northeast, South, and West. In a pilot study you survey seniors and ask, "Which one of the following drugs would you be most likely to use: alcohol; amphetamines; barbiturates; cocaine; hallucinogens; or marijuana?" You show the frequencies of answers in the table below.

	Type of drug					
Region	*Alcohol*	*Amphetamines*	*Barbiturates*	*Cocaine*	*Hallucinogens*	*Marijuana*
Northeast	25	13	12	6	6	13
South	29	5	18	2	4	10
West	21	12	6	10	12	16

a. Setting $\alpha = .01$, test the null hypothesis that there is no relationship between drug use and the region in which the high school seniors live. That is, that drug use and region are independent. Use the hypothesis-testing procedure.

b. Put your results in journal form.

11

CHAPTER GOALS

After studying this chapter, you should be able to:

1 Construct a scatter plot.
2 Identify the types of relationships observable in a scatter plot.
3 Interpret correlation coefficients of various values.
4 Calculate the correlation coefficient by the definitional and computational formulas.
5 Describe the relationship between correlation and causality.
6 Describe the effect on the correlation coefficient if the range of scores along one or both variables is restricted.
7 Describe the sampling distribution of the correlation.
8 Describe the difference between a one-tailed test and a two-tailed test of the correlation coefficient, and verbally and symbolically state H_0 and H_A.
9 Perform a significance test for a sample correlation coefficient.

CHAPTER IN A NUTSHELL

It is often useful to describe the relationships between two variables. For example, as one variable increases, does the other also increase, does it decrease, or does it remain the same? You can graphically describe the relationship between two variables by constructing a scatter plot. Each point in a scatter plot depicts one subject's pair of scores along the two variables. You can numerically describe the relationship between two variables by calculating the Pearson product-moment correlation coefficient, abbreviated *r*. The correlation coefficient is a number between −1 and +1 which tells you the type and strength of the relationship. It is possible to observe a chance relationship between two variables in a sample when actually no relationship exists in the population. The general hypothesis-testing procedure can be used to decide whether the correlation coefficient observed in the sample is due to sampling error or to a true relationship in the population. The procedure involves comparing the sample correlation coefficient with a distribution of correlation coefficients that would occur by chance in random samples drawn from a population in which there is actually no relationship between the two variables. This distribution of correlation coefficients is called the sampling distribution of the correlation.

Introduction to correlation

There are numerous research situations in which you want to describe the relationship between two variables. For example, "How is LSD use related to the frequency of paranoid symptoms?" or "How is the number of years of formal education related to annual income?" or "How is height related to weight?" or "How is swimming ability related to mathematical ability?" There are two common methods for describing such relationships. One method is to describe the relationship graphically with a *scatter plot.* The other method summarizes the type and strength of the relationship in a single number called the *correlation coefficient.*

Scatter plots

scatter plot

Graphic displays of data are useful because they allow you to summarize large amounts of data and to visualize trends and relationships. You can visually depict the relation between two variables with a special kind of graph called a **scatter plot.** For example, suppose you are interested in determining the relationship between the heights and weights of women in your statistics class. You ask the 20 women in your class to write down their heights (in inches) and their weights (in pounds). You obtain the following pairs of scores from each subject:

Subject	*Height (inches)* *X*	*Weight (pounds)* *Y*
S_1	63	110
S_2	63	118
S_3	63	103
S_4	68	125
S_5	66	120
S_6	64	125
S_7	64	120
S_8	69	120
S_9	67	115
S_{10}	68	145
S_{11}	66	110
S_{12}	70	140
S_{13}	61	103
S_{14}	67	128
S_{15}	62	115
S_{16}	62	100
S_{17}	66	123
S_{18}	60	105
S_{19}	66	130
S_{20}	60	100

To construct a scatter plot, you do not have to make a frequency distribution, but it is helpful to tabulate the scores in pairs for each subject, as shown above. The actual construction of a scatter plot proceeds very much like the graphic displays discussed in Chapter 2. The values in the X column of your tabulation are represented on the abscissa, and the values in the Y column are represented on the ordinate. Both axes must have tick marks and be labeled. The units of measurement (inches, pounds, seconds, etc.) should also be designated. Dots are placed in the scatter plot to represent each subject's pair of X and Y scores. The height of the dot corresponds to the subject's score on the Y variable, and the left-to-right (horizontal) position corresponds to the subject's score on the X variable. Each dot is called a data point, and represents one woman's height and weight. A scatter plot of the womens' height and weight scores is presented in Figure 11-1.

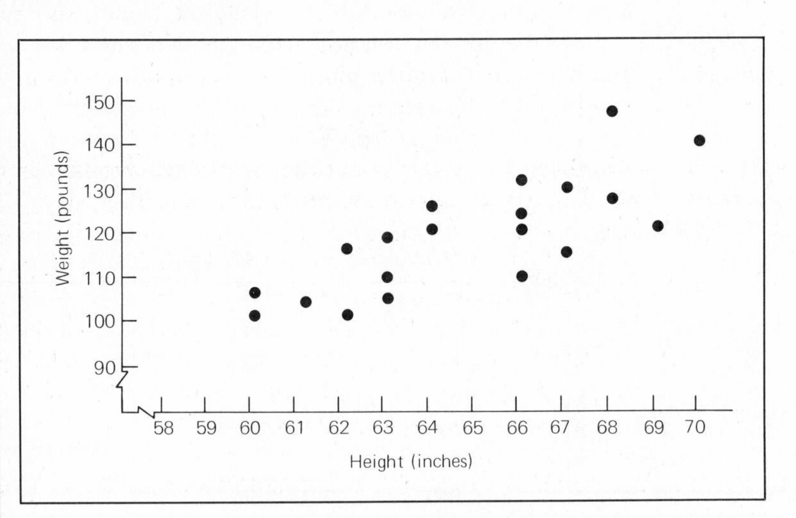

FIGURE 11-1
Scatter plot for women's height and weight data.

If you look at the scatter plot in Figure 11-1, you can see that there is a relationship between height and weight, just as you might expect. Low values of X tend to go with low values of Y, and high values of X tend to go with high values of Y. In other words, shorter women tend to be lighter, and taller women tend to be heavier. Another way to visually summarize the relationship between two variables in a scatter plot is to draw a straight line through the middle of the data points that approximates the trend in the data. Such a line is drawn in the scatter plot in Figure 11-2. The simplest way to approximate this **line of best fit** is to "eyeball" it through the data points so that it is as close as possible to the greatest number of points. Actually, there is a mathematical technique that allows you to precisely determine a line of best fit, which will be discussed in Chapter 12. For the present, however, it will suffice to simply sketch in an appropriate line through the middle of the data points on the scatter plot.

line of best fit

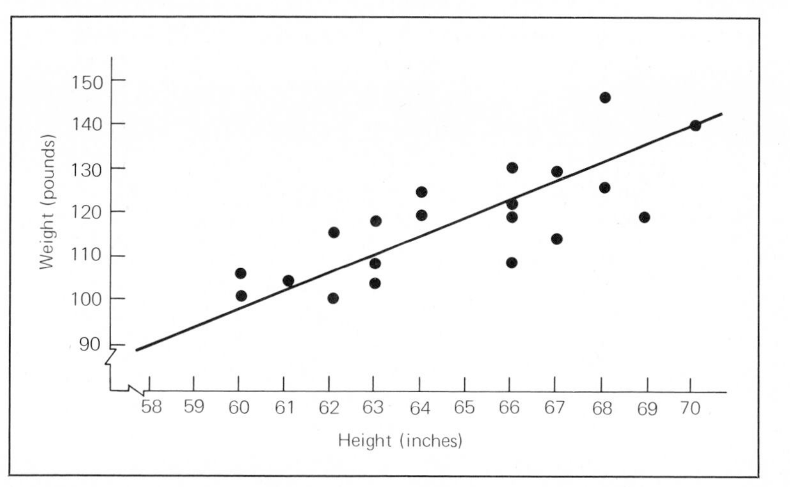

FIGURE 11-2
Scatter plot for height and weight data with the line of best fit sketched in. The line of best fit goes uphill from left to right, indicating a direct (or positive) relationship.

Types of relationships

direct relationship

There are many types of relationships that can exist between two variables. These relationships are classified according to the nature of the scatter plot and the associated line of best fit. One type of relationship is called a **direct relationship.** The relationship shown in Figures 11-1 and 11-2 is a direct relationship because, on the average, as the value of one variable increases, the value of the other variable also increases. Conversely, as the value of one variable decreases, the value of the other variable decreases. For example, as height increases, weight increases, and as height decreases, weight decreases. This is reflected by the fact that the line of best fit goes uphill from left to right. If you remember your geometry you will recognize that this uphill line has a positive slope. Consequently, a direct relationship is also called a **positive relationship.**

positive relationship

inverse relationship

A second type of relationship is called an **inverse relationship.** An inverse relationship can be illustrated if you take a number of elementary school children and determine their age (in years) and the number of errors they make on a standardized achievement test in spelling. You will discover that, on the average, older children are better spellers and hence make fewer errors than younger children. A scatter plot of these data is given in Figure 11-3. Notice that as the value of one variable increases, the value of the second variable decreases. Inverse relationships are often referred to as **negative relationships** because the line of best fit has a negative slope.

negative relationships

no relationship

A third type of relationship is when there is **no relationship** between two variables. This can be illustrated if you record the last digit in a person's phone number and that person's waist size in a scatter plot. The result will generally reveal that there is *no* relationship between these two variables. As the last digit in the phone number increases, there is no consistent change, on the average, in waist size. As shown in Figure 11-4,

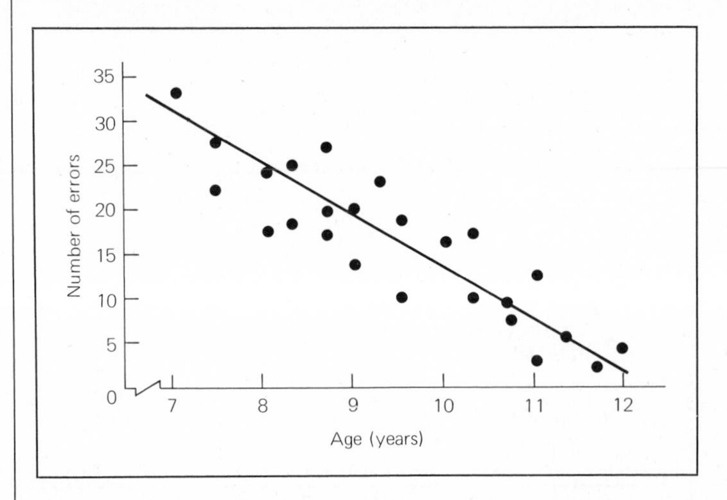

FIGURE 11-3

Scatter plot of age of elementary school children and number of errors on a standard spelling test. The line of best fit goes downhill from left to right, indicating an inverse (or negative) relationship.

the line of best fit will be horizontal. You might argue that the line of best fit in Figure 11-4 ought to be vertical (or even some other orientation), but there is a reason which will be discussed in Chapter 12 for why the line should be horizontal.

perfect direct relationship

perfect inverse relationship

Stronger, more clear-cut relationships are indicated when the points in the scatter plot are relatively close to the line of best fit. For example, consider the scatter plot in Figure 11-5, which might be obtained if you measure people's height in inches and also in centimeters. The dots lie perfectly along the line of best fit. This is an example of **perfect direct relationship.** A **perfect inverse relationship** would be illustrated if you measure how long it takes (in minutes) for people to drive a 60-mile stretch of open highway, and also their speed in miles per hour. The scatter plot

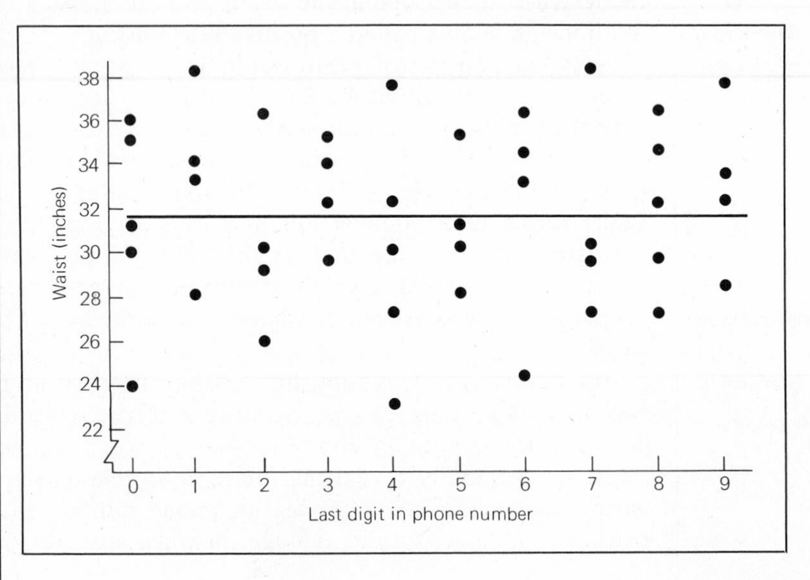

FIGURE 11-4

Scatter plot of the last digit in phone number and waist size. The line of best fit runs horizontally, indicating no relationship.

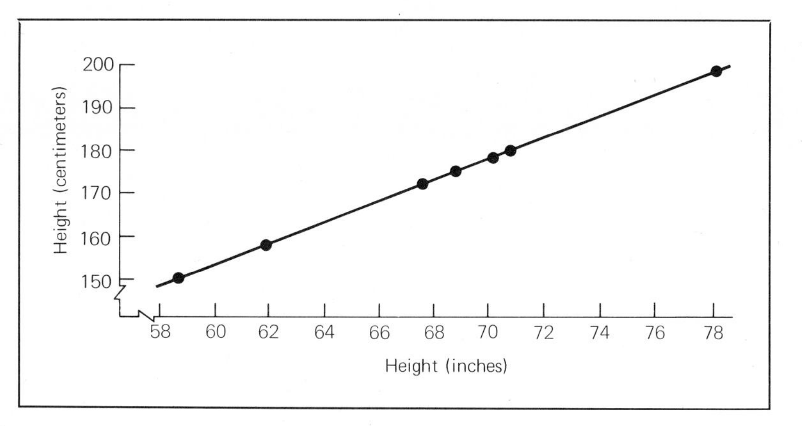

FIGURE 11-5
Scatter plot for height in inches and height in centimeters of seven people, showing a perfect direct relationship.

in Figure 11-6 depicts this relationship, which is perfectly described by the line of best fit. Strong relationships are indicated by scatter plots in which the dots lie fairly close to the line of best fit. The farther away

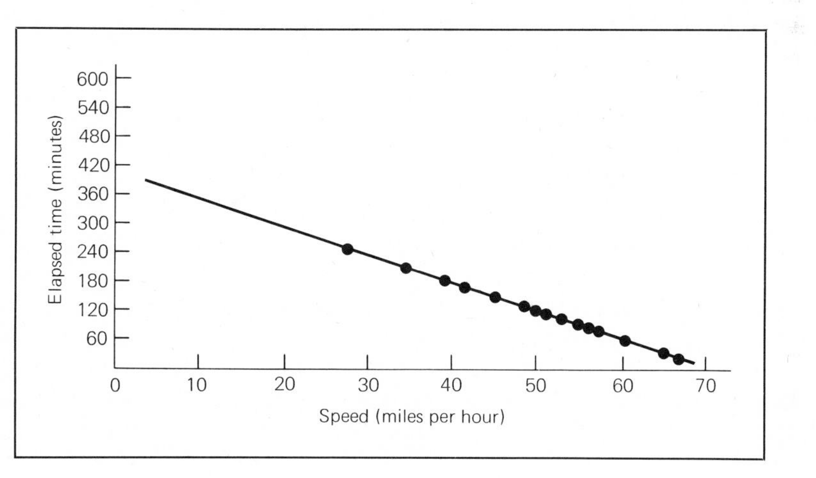

FIGURE 11-6
Scatter plot for driving speed in miles per hour and elapsed time in minutes for 15 people who drove a distance of 60 miles. This illustrates a perfect inverse relationship.

the dots are from the line of best fit, the weaker the relationship becomes. Notice the poor fit of the data in Figure 11-4, in which there is no relationship.

The Pearson product-moment correlation coefficient: *r*

Whereas a scatter plot visually describes the approximate relationship between two variables, the *Pearson product-moment correlation coefficient,* or simply **correlation coefficient,** is abbreviated ***r*** and is a single number that precisely describes the type and strength of the relationship between two variables. Numerical descriptors of data are useful because you can condense information about a large amount of data into a single number. To derive a single number describing a relationship between two variables, you need to compare the scores on the two variables for each subject.

In a direct relationship, scores on the two variables tend to increase and decrease together. When scores are high on one variable, they also tend to be high on the other variable, and when scores are low on one variable, they also tend to be low on the other variable. Therefore, the scores on the two variables tend to be close together. In an inverse relationship, when scores on one variable are high, scores on the other variable tend to be low, and vice versa. Therefore, the scores on inversely related variables tend to be far apart. Unfortunately, however, raw scores on the two variables are usually measured in different units and cannot be compared without first transforming them to equivalent units of measurement. For example, a score of 55 (inches) in height cannot be directly compared to a score of 65 (pounds) in weight. Is this person heavier than tall, or taller than heavy? Recall from Chapter 4 that z scores are useful in comparing the scores of an individual on two different variables. Therefore, the first step in deriving the correlation coefficient is to transform each pair of raw scores to z scores so that they can be meaningfully compared. Suppose you take your original data on the 20 women's heights (variable X) and weights (variable Y) and convert them to z scores using Formula 4.3:

$$z_{\text{height}} = \frac{X - \mu_x}{\sigma_x} \quad \text{and} \quad z_{\text{weight}} = \frac{Y - \mu_y}{\sigma_y}$$

To apply Formula 4.3, you need to find the mean and standard deviation for height, denoted μ_x and σ_x respectively, and the mean and standard deviation for weight, μ_y and σ_y. These values are calculated on page 241. Because you want to describe a relationship in your group of 20 women and you are not now interested in making inferences outside this, you treat them as a population and consider the mean to be μ instead of $\bar{X}$, and the standard deviation to be σ instead of s. This is a general convention when calculating r, as seen in the data on page 241.

You can calculate the z scores in a systematic fashion by setting up a table like that in Figure 11-7. Set up a separate column for each variable in which the mean is subtracted from each score. Then set up a third column in which you divide the result in the second column by the standard deviation to get the z scores. By comparing the two columns of z scores, you can detect the relationship between height and weight. There is a tendency for a subject's z score on one variable to be close to the z score on the other variable. When a subject's z score is high on one variable, it tends to be high on the second variable as well, and when the z score is low on one variable, the corresponding z score tends to be low on the other variable also. There are a few exceptions in Figure 11-7, but the z scores on the variables *tend* to vary together. A measure of the strength of this tendency can be obtained by finding the difference between the z scores in each pair. When the z scores in each pair are

Height		Weight	
X	X^2	Y	Y^2
63	3969	110	12,100
63	3969	118	13,924
63	3969	103	10,609
68	4624	125	15,625
66	4356	120	14,400
64	4096	125	15,625
64	4096	120	14,400
69	4761	120	14,400
67	4489	115	13,225
68	4624	145	21,025
66	4356	110	12,100
70	4900	140	19,600
61	3721	103	10,609
67	4489	128	16,384
62	3844	115	13,225
62	3844	100	10,000
66	4356	123	15,129
60	3600	105	11,025
66	4356	130	16,900
60	3600	100	10,000
$\Sigma X = 1295$	$\Sigma X^2 = 84{,}019$	$\Sigma Y = 2355$	$\Sigma Y^2 = 280{,}305$

$$\mu_x = \frac{\Sigma X}{N} = \frac{1295}{20} = 64.75 \qquad \mu_y = \frac{\Sigma Y}{N} = \frac{2355}{20} = 117.75$$

$$\sigma_x = \sqrt{\frac{\Sigma X^2 - \dfrac{(\Sigma X)^2}{N}}{N}} \qquad \sigma_y = \sqrt{\frac{\Sigma Y^2 - \dfrac{(\Sigma Y)^2}{N}}{N}}$$

$$= \sqrt{\frac{84019 - \dfrac{(1295)^2}{20}}{20}} \qquad = \sqrt{\frac{280305 - \dfrac{(2355)^2}{20}}{20}}$$

$$= 2.90 \qquad = 12.26$$

closer together, you have a stronger direct relationship. An example of a perfect direct relationship is given in Figure 11-8(a), where the z scores in each pair are identical. With weaker direct relationships, the z scores in each pair are farther apart. When there is no relationship between the two variables, as in Figure 11-8(b), the z scores in each pair are fairly

	Height			Weight	
X	$X - \mu_x$	$z_x = \dfrac{X - \mu_x}{\sigma_x}$	Y	$Y - \mu_y$	$z_y = \dfrac{Y - \mu_y}{\sigma_y}$
63	−1.75	−0.60	110	−7.75	−0.63
63	−1.75	−0.60	118	0.25	0.02
63	−1.75	−0.60	103	−14.75	−1.20
68	3.25	1.12	125	7.25	0.59
66	1.25	0.43	120	2.25	0.18
64	−0.75	−0.26	125	7.25	0.59
64	−0.75	−0.26	120	2.25	0.18
69	4.25	1.46	120	2.25	0.18
67	2.25	0.78	115	−2.75	−0.22
68	3.25	1.12	145	27.25	2.22
66	1.25	0.43	110	−7.75	−0.63
70	5.25	1.81	140	22.25	1.81
61	−3.75	−1.29	103	−14.75	−1.20
67	2.25	0.78	128	10.25	0.84
62	−2.75	−0.95	115	−2.75	−0.22
62	−2.75	−0.95	100	−17.25	−1.45
66	1.25	0.43	123	5.25	0.43
60	−4.75	−1.64	105	−12.75	−1.04
66	1.25	0.43	130	12.25	1.00
60	−4.75	−1.64	100	−17.75	−1.04

subtract 64.75 from each score

find z_x by dividing $X - \mu_x$ by 2.90

subtract 117.75 from each score

find z_y by dividing $Y - \mu_y$ by 12.26

FIGURE 11-7

Calculating the z scores for women's height and weight data.

dissimilar. With an inverse relationship the z scores in each pair are even more dissimilar. When you have a perfect inverse relationship, as in Figure 11-8(c), the z scores in each pair are, on the average, as far apart as possible. The z scores are identical but opposite in sign. You can thus derive a measure of the strength of a relationship by finding the average distance between the z scores in each pair of scores. Small differences in the z scores indicate that you have a strong direct relationship. As the average distance between the z scores increases, the strength of the relationship decreases up to a point until you have no relationship. Beyond this point, further increases in the average differences between pairs of z scores indicate a stronger and stronger inverse relationship.

A reasonable measure of the type and strength of a relationship therefore is the average difference between each pair of z scores. To find the average difference you first need to calculate the difference between each pair

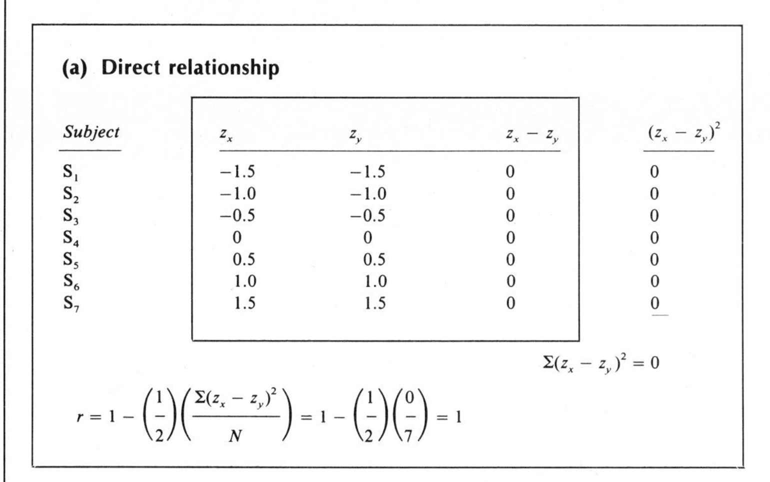

(a) Direct relationship

Subject	z_x	z_y	$z_x - z_y$	$(z_x - z_y)^2$
S_1	−1.5	−1.5	0	0
S_2	−1.0	−1.0	0	0
S_3	−0.5	−0.5	0	0
S_4	0	0	0	0
S_5	0.5	0.5	0	0
S_6	1.0	1.0	0	0
S_7	1.5	1.5	0	0

$$\Sigma(z_x - z_y)^2 = 0$$

$$r = 1 - \left(\frac{1}{2}\right)\left(\frac{\Sigma(z_x - z_y)^2}{N}\right) = 1 - \left(\frac{1}{2}\right)\left(\frac{0}{7}\right) = 1$$

(b) No relationship

Subject	z_x	z_y	$z_x - z_y$	$(z_x - z_y)^2$
S_1	−1.5	0	−1.5	2.25
S_2	−1.0	0.5	−1.5	2.25
S_3	−0.5	−0.5	0	0
S_4	0	1.0	−1.0	1
S_5	0.5	−1.0	1.5	2.25
S_6	1.0	−1.5	2.5	6.25
S_7	1.5	1.5	0	0

$$\Sigma(z_x - z_y)^2 = 14$$

$$r = 1 - \left(\frac{1}{2}\right)\left(\frac{\Sigma(z_x - z_y)^2}{N}\right) = 1 - \left(\frac{1}{2}\right)\left(\frac{14}{7}\right) = 0$$

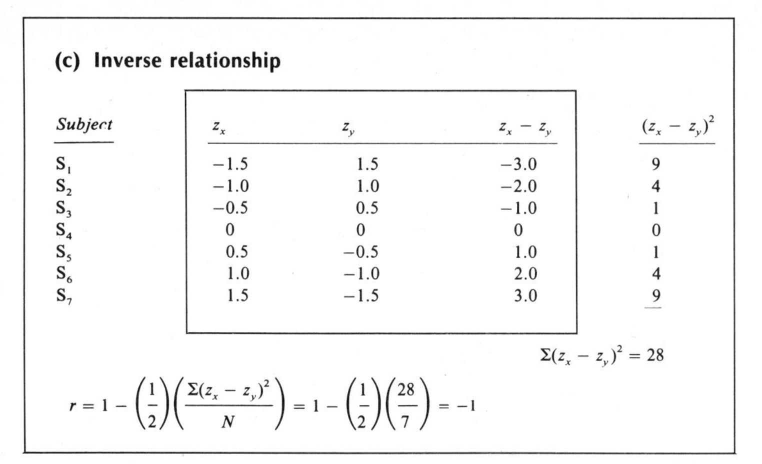

(c) Inverse relationship

Subject	z_x	z_y	$z_x - z_y$	$(z_x - z_y)^2$
S_1	−1.5	1.5	−3.0	9
S_2	−1.0	1.0	−2.0	4
S_3	−0.5	0.5	−1.0	1
S_4	0	0	0	0
S_5	0.5	−0.5	1.0	1
S_6	1.0	−1.0	2.0	4
S_7	1.5	−1.5	3.0	9

$$\Sigma(z_x - z_y)^2 = 28$$

$$r = 1 - \left(\frac{1}{2}\right)\left(\frac{\Sigma(z_x - z_y)^2}{N}\right) = 1 - \left(\frac{1}{2}\right)\left(\frac{28}{7}\right) = -1$$

FIGURE 11-8
Three different types of relationships between two variables are illustrated with the pairs of z scores in the small boxes.

of z scores by subtracting one z score from the other, and then sum these differences before dividing by the total number of pairs. However, finding the sum of the differences between z scores presents problems because some of the differences are positive and some are negative. The positive and negative differences will tend to cancel each other out. This is the zero-sum problem again. To obtain a measure of the average difference between z scores in each pair, you have to eliminate the signs of the differences before summing them. This is done by squaring the differences before adding. Therefore, if you find the difference between the z scores in each pair, square the difference, sum the squares, and then divide by the number of pairs of scores, you will have a useful measure of the strength and type of a relationship. This measure is called the **mean squared difference between z scores.** The mean squared difference between z scores for the women's height-weight scores is calculated on page 245.

mean squared difference between z scores

$$\text{mean squared difference between } z \text{ scores} = \frac{\Sigma(z_x - z_y)^2}{N} = \frac{8.79}{20} = .44$$

where N is the number of *pairs* of z scores. It can be proven mathematically that the mean squared difference between z scores will always be a number between 0 and 4. When it equals 0 there is a perfect direct relationship; when it equals 2 there is no relationship; and when it equals 4 there is a perfect inverse relationship. Intermediate values between 0 and 2 represent intermediate degrees of direct relationships. Intermediate values between 2 and 4 represent intermediate degrees of inverse relationships.

For several reasons researchers have decided it would be more convenient if their measure of relationship varied between -1 and $+1$ instead of 0 and 4. Fortunately, to arrive at a measure that varies from -1 to $+1$, all you have to do is subtract one-half of the mean squared difference between z scores from 1. This measure defines the Pearson product-moment correlation coefficient, abbreviated r. The **definitional formula for r** is:

FORMULA 11.1
definitional formula for r

$$r = 1 - \left(\frac{1}{2}\right)\left(\frac{\Sigma(z_x - z_y)^2}{N}\right)$$

When you have a perfect direct relationship, the mean squared difference between z scores is 0, so $r = 1$. When you have no relationship, the mean squared difference between z scores is 2, so $r = 0$. When you have a perfect inverse relationship, the mean squared difference between z scores is equal to 4, so $r = -1$. Intermediate values between 0 and $+1$ represent intermediate degrees of direct relationships, and intermediate values between 0 and −1 represent intermediate degrees of inverse relationships. This arrangement makes the interpretation of r very simple. All you have to do is look at the sign and magnitude of the number representing r. If the sign is positive the relationship is direct, and if the sign is negative the relationship is inverse. The closer r is to 0, the weaker the relationship is. Conversely, the closer r is to ±1, the stronger the relationship is. Using Formula 11.1, values of r have been calculated in Figure 11-8 for each of the three types of relationships depicted there.

Now you can calculate the correlation coefficient for the women's height and weight data:

Height z_x	*Weight* z_y	$z_x - z_y$	$(z_x - z_y)^2$
−0.60	−0.63	0.03	0.00
−0.60	0.02	−0.62	0.38
−0.60	−1.20	0.60	0.36
1.12	0.59	0.53	0.28
0.43	0.18	0.25	0.06
−0.26	0.59	−0.85	0.72
−0.26	0.18	−0.44	0.19
1.46	0.18	1.28	1.64
0.78	−0.22	1.00	1.00
1.12	2.22	−1.10	1.21
0.43	−0.63	1.06	1.12
1.81	1.81	0.00	0.00
−1.29	−1.20	−0.09	0.01
0.78	0.84	−0.06	0.00
−0.95	−0.22	−0.73	0.53
−0.95	−1.45	0.50	0.25
0.43	0.43	0.00	0.00
−1.64	−1.04	−0.60	0.36
0.43	1.00	−0.57	0.32
−1.64	−1.04	−0.60	0.36
			$\Sigma(z_x - z_y)^2 = 8.79$

$$r = 1 - \left(\frac{1}{2}\right)\left(\frac{\Sigma(z_x - z_y)^2}{N}\right) = 1 - \left(\frac{1}{2}\right)\left(\frac{8.79}{20}\right)$$

$$= 1 - (.5)(.44)$$

$$= 1 - .22 = .78$$

Your calculated $r = .78$ indicates that there is a moderate direct relationship between height and weight, because r is positive and intermediate between 0 and +1. As height increases there is a moderate tendency for weight to increase also. If your calculated r had been .9 instead, you would have had an even stronger direct relationship than the one you actually observed. On the other hand, had r been .2, you would have had a weaker direct relationship than the one you observed. Similarly, a correlation coefficient of −.8 represents a stronger inverse relationship than $r = -.5$, and a correlation coefficient of −.3 indicates a weaker inverse relationship than $r = -.5$.

Computational formulas for *r*

There are two computational formulas for r that make calculation substantially easier. One formula is directly derived from the definitional formula through a little mathematical manipulation. The derivation will not be given here, but the mathematically oriented student might find it a challenging task. This computational formula is referred to as the **z score computational formula:**

FORMULA 11.2
z score computational formula for r

$$r = \frac{\Sigma z_x z_y}{N}$$

Formula 11.2 still uses z scores, but it is much easier to use than the definitional Formula 11.1 because it does not require you to repeatedly subtract pairs of z scores. The use of the z score computational formula is illustrated below for the women's height and weight data.

Height z_x	*Weight* z_y	$z_x z_y$
−.060	−0.63	0.38
−0.60	0.02	−0.01
−0.60	−1.20	0.72
1.12	0.59	0.66
0.43	0.18	0.08
−0.26	0.59	−0.15
−0.26	0.18	−0.05
1.46	0.18	0.26
0.78	−0.22	−0.17
1.12	2.22	2.49
0.43	−0.63	−0.27
1.81	1.81	3.28
−1.29	−1.20	1.55
0.78	0.84	0.66

Height z_x	*Weight* z_y	$z_x z_y$
−0.95	−0.22	0.21
−0.95	−1.45	1.38
0.43	0.43	0.18
−1.64	−1.04	1.71
0.43	1.00	0.43
−1.64	−1.04	1.71
	$\Sigma z_x z_y =$	15.05

$$r = \frac{\Sigma z_x z_y}{N} = \frac{15.05}{20} = .75$$

You first have to create a new column in which you multiply each pair of *z* scores together to obtain $z_x z_y$ for each pair. You then sum this column and divide by *N*, the number of pairs of scores, to find *r*. Mathematically speaking, the value of *r* obtained using the computational Formula 11.2 should be exactly the same as that calculated using the definitional Formula 11.1. However, in practice, the values obtained by these two formulas might differ slightly because of accumulated rounding errors.

Formulas 11.1 and 11.2 are most appropriate when the data are already given in *z* scores. However, most often you obtain data in raw-score form, and it is a cumbersome task to compute *r* from raw scores using computational Formula 11.2 because you must first transform all the raw scores to *z* scores.

There is a second computational formula that can be used when your data are in the form of raw scores. This computational formula does not require you to transform the raw scores to *z* scores. It is derived by substituting the formula for *z* scores (Formula 4.3 modified so the calculation of the mean and standard deviation are incorporated in it) and then applying some mathematical manipulations. This yields a computational formula for *r* in terms of raw scores. Again, the derivation will not be given here, but will be left as an exercise for interested students. This computational formula is called the **raw-score computational formula:**

FORMULA 11.3
raw-score computational formula for *r*

$$r = \frac{N\Sigma XY - \Sigma X \Sigma Y}{\sqrt{[N\Sigma X^2 - (\Sigma X)^2]\,[N\Sigma Y^2 - (\Sigma Y)^2]}}$$

The use of the raw-score computational formula is demonstrated below for the women's height and weight data.

Height		Weight		
X	X^2	Y	Y^2	XY
63	3969	110	12,100	6930
63	3969	118	13,924	7434
63	3969	103	10,609	6489
68	4624	125	15,625	8500
66	4356	120	14,400	7920
64	4096	125	15,625	8000
64	4096	120	14,400	7680
69	4761	120	14,400	8280
67	4489	115	13,225	7705
68	4624	145	21,025	9860
66	4356	110	12,100	7260
61	3721	103	10,609	6283
70	4900	140	19,600	9800
67	4489	128	16,384	8576
62	3844	115	13,225	7130
62	3844	100	10,000	6200
66	4356	123	15,129	8118
60	3600	105	11,025	6300
66	4356	130	16,900	8580
60	3600	100	10,000	6000
$\Sigma X = 1295$	$\Sigma X^2 = 84{,}019$	$\Sigma Y = 2355$	$\Sigma Y^2 = 280{,}305$	$\Sigma XY = 153{,}045$

$$(\Sigma X)^2 = (1295)^2 = 1{,}677{,}025 \qquad (\Sigma Y)^2 = (2355)^2 = 5{,}546{,}025$$

$$r = \frac{N\Sigma XY - \Sigma X \Sigma Y}{\sqrt{[N\Sigma X^2 - (\Sigma X)^2][N\Sigma Y^2 - (\Sigma Y)^2]}}$$

$$= \frac{(20)(153{,}045) - (1295)(2355)}{\sqrt{[(20)(84{,}019) - 1{,}677{,}025][(20)(280{,}305) - 5{,}546{,}025]}}$$

$$= \frac{3{,}060{,}900 - 3{,}049{,}725}{\sqrt{(1{,}680{,}380 - 1{,}677{,}025)(5{,}606{,}100 - 5{,}546{,}025)}}$$

$$= \frac{11{,}175}{\sqrt{(335)(60{,}075)}}$$

$$= \frac{11{,}175}{\sqrt{201{,}551{,}625}} = \frac{11{,}175}{14{,}196.89} = .787$$

When using the raw-score formula you must add some extra columns. First square each raw score for both variables to obtain the values listed under the columns labeled X^2 and Y^2. Sum these columns to get ΣX^2 and ΣY^2, respectively. Then multiply each pair of raw scores together to obtain the values listed under the XY column. Sum this column to get ΣXY. Substitute the obtained values into Formula 11.3 to find r = .787. The value of r obtained with Formula 11.3 is the most accurate because of the relative absence of rounding errors.

Limitations of the correlation coefficient

correlations are not proportions

There are four limitations of the correlation coefficient. First, **correlation coefficients are not proportions** and hence do not represent the proportionate strengths of different relationships. For example, $r = .4$ does not represent twice as strong a relationship as $r = .2$. While it is true that as the correlation coefficient becomes larger in absolute value, the relationship it describes also becomes stronger, the strength of the relationship is not in direct proportion to the absolute magnitude of r. In the next chapter this matter is discussed more fully.

linear relationships

A second limitation of the Pearson product-moment correlation coefficient is that it is a measure only of **linear relationships.** A linear relationship is one such that the points in a scatter plot lie approximately along a straight line. This is important because it is possible to have an extremely strong, systematic relationship that cannot be summarized with a straight line. For example, consider the scatter plot in Figure 11-9. This scatter

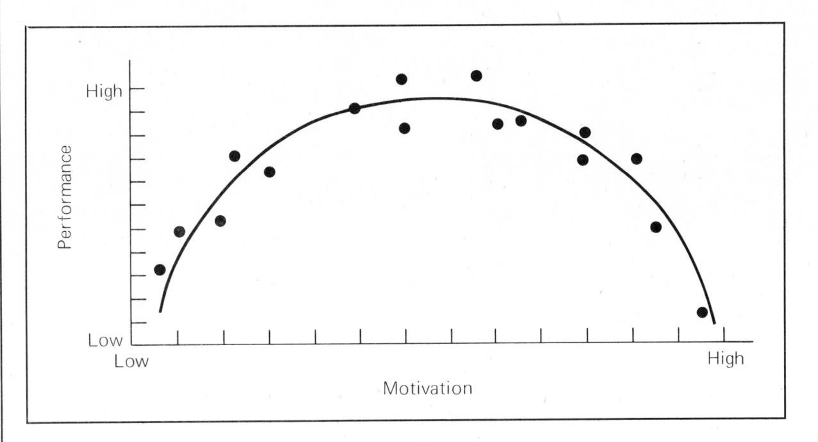

FIGURE 11-9
Scatter plot of motivation and performance, revealing a nonlinear relationship.

plot represents the nonlinear relationship, which might be obtained between motivation and performance. In numerous tasks, as the level of motivation increases, so does the performance—but only up to a point. As motivation increases beyond this point, there is a tendency for additional motivation to result in a decrease in performance. If you actually test subjects in an experiment designed to determine the relationship between motivation and performance, you will probably get a value of r close to 0 because **r is an accurate measure only of linear relationships.** The application of Pearson's correlation coefficient will invariably underestimate the actual strength of nonlinear relationships. As a precaution, before calculating r, you should always construct a scatter plot of the data to see whether it is reasonable to assume that the relation is approximately linear. If the relation is clearly nonlinear, there are other types of correlation coefficients that can measure its strength, but these methods are beyond the scope of this book.

r measures only linear relationships

The third limitation has to do with the fact that the **correlation coefficient is sensitive to the range of the scores** observed for each variable. As a general rule, restricting the range of scores along one or both variables leads to a decrease in the absolute value of r, so r moves closer to 0. Thus, relationships observed between variables with restricted ranges are generally weaker than relationships observed between variables over more extended ranges. For example, consider the relationship depicted in Figure 11-10 between IQ and achievement in science (as measured by the number

r is sensitive to the range of scores

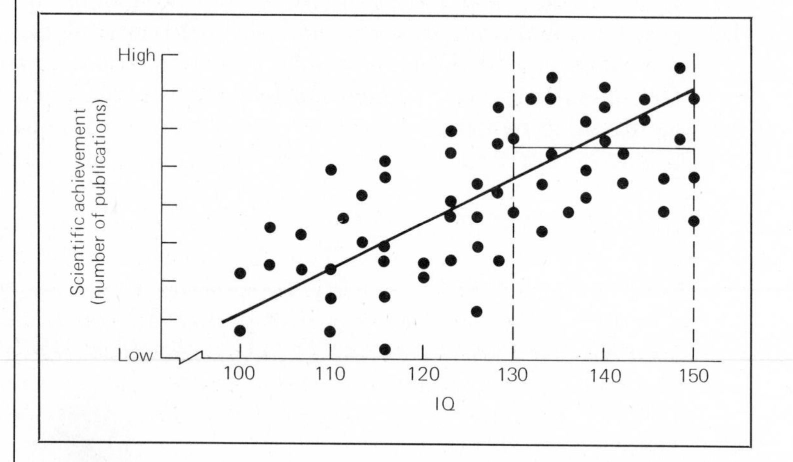

FIGURE 11-10
Scatter plot of IQ and scientific achievement measured in terms of the number of publications. The line of best fit for the IQs between 100 and 150 shows a positive relationship. However, by restricting the range of IQs to those between 130 and 150, the horizontal line of best fit indicates no relationship.

of publications). If you look at a group of scientists having a wide range of IQs (say, between 100 and 150), there is a small direct relationship between IQ and scientific achievement. However, if you restrict the range of IQ to only those having the highest IQs (say, between 130 and 150), you find very little, if any, relationship between IQ and scientific achievement. Those scientists with IQs around 150 have, on the average, the same number of publications as scientists with IQs around 130.

A fourth limitation of the correlation coefficient is that r is a descriptive statistic which only describes observed relationships. The **correlation coefficient alone cannot be used to infer causality.** A correlation might exist between two variables, X and Y, because changes in X cause changes in Y, or

correlation alone does not imply causality

because changes in Y cause changes in X, or because changes in a third variable, Q, cause changes in both X and Y. It is also possible that the correlation observed in a sample of subjects is **spurious**, which means that it is a result of random sampling error and no correlation truly exists in the population. The subjects in a sample may have been randomly selected in such a way as to produce a strong correlation in the sample when in fact no correlation between the two variables exists in the population. Consider the relationship commonly observed between smoking and heart disease. It has been found that people who smoke a lot have a high incidence of heart disease. However, from this correlation alone, you cannot conclude that smoking causes heart disease. While that remains a possibility, it is also possible that those people are predisposed to heart disease find smoking very rewarding. Or it could be there is a third factor, perhaps stress or anxiety, that leads to the tendencies to smoke *and* to develop heart disease. Or it is possible that no relationship between heart disease and smoking exists in the general population. By chance, your sample might have contained some subjects who have heart disease and who smoke a lot, and other subjects who are free of heart disease and who do not smoke. Actually, enough studies on smoking and heart disease have been conducted to rule out this last interpretation. But sampling error remains a possible interpretation in other research areas. It is wise to remember this limitation because people often mistakenly infer causality from correlation alone.

spurious *r*

Testing *r* for significance

It is possible for a correlation to be observed in a sample when actually no relationship exists between the variables in the population. Spurious strong correlations are particularly likely when you sample only a few subjects. For example, imagine you select three subjects and measure each subject's waist size (in inches) and also ask for the last digit in their phone number. It is entirely possible that the person with the largest waist size also has the largest phone digit; the person with the middle waist size has the intermediate phone digit; and the person with the smallest waist has the smallest phone digit. This chance combination of waist sizes and phone digits in your sample leads you to calculate a spurious correlation. If you draw a much larger sample, say of 100 subjects, it is very unlikely that you will observe by chance a strong correlation between waist size and phone digit.

A question that therefore naturally arises once you have calculated a sample r is, "Does your observed r reflect a true relationship between the variables in the population, or is it the result of chance sampling error and there is actually no relationship between the variables in the population?" How would you decide between these two possibilities?

Inferential statistics enters into the decision process again by offering a type of hypothesis-testing procedure to help you decide between the two possible explanations. You can assume an H_0, set a level of alpha, and calculate the probability of getting your observed sample r by chance. If this probability is less than alpha you can reject H_0. Suppose you want to decide whether the sample r of .787 you observed between women's height and weight is due to chance. H_0 is that there is no relationship between height and weight in the population, and hence your observed r is due to chance sampling error. Your H_0 can be symbolically written as $\rho = 0$, where the Greek letter ρ, spelled "rho" and pronounced "roe," designates the population correlation coefficient. H_A, the logical compliment of H_0, is that height and weight are related in the population, or symbolically, $\rho \neq 0$. Assuming H_0 is true, you now need to find the probability of observing by chance a sample r as strong as yours. You can do this by comparing your $r = .787$ with a distribution of r values that would occur by chance in samples the same size as yours if H_0 is true. This distribution of sample r's is called the **sampling distribution of the correlation.** The characteristics of the sampling distribution of the correlation can be inferred mathematically. The mean of the sampling distribution is 0, because if you take many random samples from a population in which $\rho = 0$, you expect that the sample correlation coefficients will be 0, on the average. There will be some variability around this value, but the average correlation coefficient should be 0. There is actually a different sampling distribution of the correlation for every possible sample size because it is easier to observe large spurious correlations in small samples than in larger samples. Figure 11-11 shows how the sampling distribution of the correlation varies

sampling distribution of the correlation

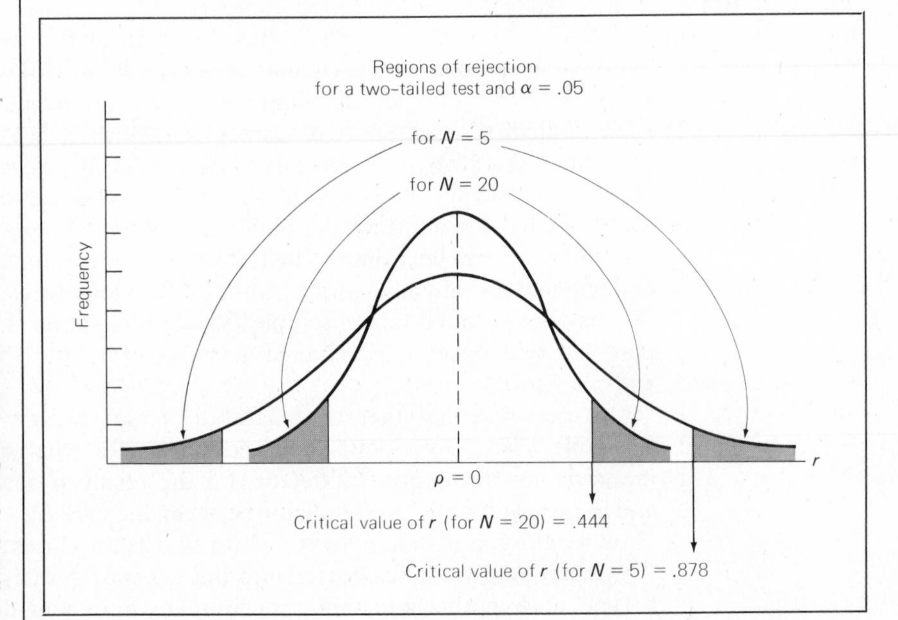

FIGURE 11-11
The sampling distribution of the correlation varies for different sample sizes. Comparing the regions of rejection in the sampling distribution of the correlation for samples of N = 5 (the broader distribution) and N = 20 shows that large spurious r's are more likely to occur by chance in small samples than in large samples.

for different sample sizes. Note that the sampling distribution is broader for the smaller sample size of 5 than for the larger sample size of 20. This reflects the fact that large, chance values of r are more frequently observed in small samples than in larger samples.

r^*

Critical values of r, symbolized r^*, are listed in Table r^* of the Appendix. Any sample r that is greater in absolute value than r^* will have a probability of occurrence that is less than alpha. Table r^* lists r^* values according to the level of alpha, N, and whether you are conducting a one- or a two-tailed test.[1] N is the number of subjects, or pairs of scores, in your sample. Notice in Table r^* that as the sample size increases, the r^* values become smaller. For example, with a sample size of 5, r^* for a two-tailed test at $\alpha = .05$ is .878, but with a sample size of 20, r^* for a two-tailed test at $\alpha = .05$ is .444. This is also shown in the regions of rejection in Figure 11-11. Because you are less likely to get large spurious r's with larger sample sizes, it is easier to reject H_0 with larger samples than with smaller ones.

Now use the critical values of r found in Table r^* to determine whether your $r = .787$ is less likely to occur by chance than alpha. For $\alpha = .05$ and $N = 20$, you find $r^* = .444$. Since your sample $r = .787$ is greater than r^*, reject H_0 and conclude $\rho \neq 0$. That is, height and weight are related in the population. Had your sample size been smaller, say $N = 5$, then for $\alpha = .05$, r^* from Table r^* would have been .878. With the smaller sample size you would have had to retain H_0. This illustrates that it is easier to reject H_0 with larger sample sizes than with smaller ones. It is an interesting fact that with a sample size of 20 and a two-tailed test at $\alpha = .05$, since $r^* = .444$, 5% ($\alpha = .05$) of the sample r's that would occur by chance are greater than only .444. For the same situation, but with only five subjects, $r^* = .878$, so 5% ($\alpha = .05$) of the sample r's are at least as large as .878!

correct conclusion when rejecting H_0

A note of caution is necessary for making the **correct conclusion in situations in which H_0 is rejected.** Rejecting H_0 tells you only that the population correlation coefficient is not 0. You may therefore conclude that there is some relationship between the two variables in the population. However, rejecting H_0 does not tell you how strong the relationship in the population is. For example, since your sample $r = .787$ between height and weight, you rejected H_0. This does not tell you that the population correlation coefficient is .787. You know only that it is not 0. There are many possible values for ρ that could give rise to a sample r of .787, and each of these values of ρ would also lead to rejecting H_0. For example, the correlation coefficient between height and weight in the population could be $\rho = .9$, and you obtained by chance an $r = .787$ in your sample. It is also possible that the same sample $r = .787$ could have arisen by

[1]Some tables of r^* are indexed in the left column in terms of df instead of N. For r, $df = N - 2$.

chance from a population in which $\rho = .8$, or $\rho = .7$, or many other values. You therefore cannot determine an exact value for ρ. You only know it is not 0.

two-tailed significance tests of *r*

Although **two-tailed tests of *r*** are much more common, there are certain research situations in which you may want to establish that the correlation coefficient of the population is in one particular direction, either positive or negative. This will occur if you know beforehand that the relationship is probably direct (or inverse), or if you are interested in a relationship only if it is in one particular direction. For example:

Many previous studies have found a strong direct relationship between verbal IQ and performance IQ (as measured with the WISC-R intelligence test) in the average 6- to 16-year-old child born in the United States. You are trying to replicate these findings, in part, using a sample of 6- to 16-year-old Vietnamese children who have lived in the United States for the past five years.

You are trying to establish that the percentage of alcohol in the bloodstream is inversely related to the percentage of time spent dreaming while asleep. That is, the more you drink, the less you dream.

If you want to establish that the correlation coefficient of the population is positive, then H_A is $\rho > 0$ and H_0 is $\rho \leq 0$. Conversely, if you want to establish a negative correlation, then H_A is $\rho < 0$ and H_0 is $\rho \geq 0$.

The general hypothesis-testing procedure using critical values of *r* can be summarized as follows:

1. Determine the appropriate H_0 and H_A. For a two-tailed test, H_0 is $\rho = 0$ and H_A is $\rho \neq 0$. For a one-tailed test, H_0 is $\rho \leq 0$ or $\rho \geq 0$, and H_A is $\rho > 0$ or $\rho < 0$, respectively. Assume H_0 is true.
2. Set alpha and determine the critical value of *r* from Table r^* for your alpha, *N*, and whether you are conducting a one- or two-tailed test.
3. If you have not done so already, calculate *r*.
   If you are conducting a two-tailed test, proceed to Step 4.
   If you are conducting a one-tailed test, examine the sign of your calculated *r* to see whether it is in the direction expected with H_A.
   If it is in the expected direction, then proceed to Step 4. If it is in the opposite direction, retain H_0 and conclude that there is not sufficient evidence to say the correlation in the population from which your sample was drawn is different from that specified in H_0.
4. If $|r| > r^*$, reject H_0 and conclude that there is a relationship between the two variables in the population as specified in H_A.
   If $|r| \leq r^*$, retain H_0 and conclude that there is not sufficient evidence to say the population correlation coefficient is different from that specified in H_0.

Now, take this hypothesis-testing procedure and apply it to a new research study investigating the relationship between blood pressure and the amount of time spent viewing TV. An informal pilot study you have conducted suggests that people who watch a lot of TV tend to have higher blood pressure than those who watch little TV. You now want to do a more formal study to replicate this finding of a direct relationship. You take a random sample of 25 subjects and determine their blood pressure and the number of hours spent watching TV per week. You calculate r and find $r = .295$. Does this sample $r = .295$ reflect a true relationship in the population between blood pressure and TV viewing habits, or is there no correlation in the population and your results are due to sampling error?

1. Since previous pilot research suggests that blood pressure is positively correlated with TV viewing, you therefore expect a positive correlation coefficient and conduct a one-tailed test. H_0 and H_A are:

H_0: $\rho \leq 0$ (Blood pressure and TV viewing are not directly related)
H_A: $\rho > 0$ (Blood pressure and TV viewing are directly related)

2. Set $\alpha = .05$ since this is exploratory research and you want to maximize power.
Using Table r^*, you find that for a one-tailed test at $\alpha = .05$ and $N = 25$, $r^* = .337$.

3. $r = .295$. Since this is a one-tailed test, you verify that the sign of your calculated r is positive, the direction specified in H_A.

4. Since $|.295| < .337$, retain H_0 and conclude that there is not sufficient evidence to say blood pressure and TV viewing are directly related.

Journal form

The journal form for reporting the significance of your sample r is very similar to that of the z test. You might report, "The correlation between IQ and academic achievement is .61, $p < .05$." Or for the same analysis you might say, "The variables of IQ and academic achievement are directly related with each other, $r = .61$, $p < .05$." As another example, you might report, "The length of time on the job does not significantly correlate with the frequency of on-the-job injuries, $r = -.49$, N.S." Note that the sign of r is reported in journal form.

KEY CONCEPTS

1 A scatter plot visually describes the relationship between two variables. The scores along one variable are represented on the abscissa, and the scores along the other variable are represented on the ordinate. Each data point in the graph corresponds to an individual subject's pair of scores on the two variables.

2 The Pearson product-moment correlation coefficient is abbreviated r and provides a numerical measure of the type and strength of a relationship.

It has values between −1 and +1. The sign indicates the type of relationship. A plus sign denotes a direct (positive) relationship, in which scores on the two variables increase and decrease together; and a minus sign denotes an inverse (negative) relationship, in which scores on the two variables increase and decrease inversely. An r of 0 indicates no relationship. The closer the absolute value of r is to 1, the stronger is the relationship. The Pearson correlation coefficient measures only linear (straight-line) relationships.

3 After each subject's pair of scores along the two variables are transformed to z scores, the definitional formula for r can be used:

$$r = 1 - \left(\frac{1}{2}\right)\left(\frac{\Sigma(z_x - z_y)^2}{N}\right)$$

Assuming the raw scores have been transformed to z scores, an easier formula for computing r is the z score computational formula:

$$r = \frac{\Sigma z_x z_y}{N}$$

Usually, each subject's scores are given in raw scores. Then the easiest formula for computing r is the raw-score computational formula:

$$r = \frac{N\Sigma XY - \Sigma X \Sigma Y}{\sqrt{[N\Sigma X^2 - (\Sigma X)^2]\,[N\Sigma Y^2 - (\Sigma Y)^2]}}$$

In each of these formulas, N is the number of pairs of scores.

4 The correlation coefficient by itself does not indicate causality. If X is correlated with Y, changes in X could cause changes in Y, changes in Y could cause changes in X, or changes in a third variable Q could cause changes in both X and Y. Also a correlation in a sample can be spurious; that is, it could occur by chance from a population in which there is no relation between the two variables.

5 Restricting the range of scores along one or both variables generally produces a decrease in the magnitude of r.

6 The sampling distribution of the correlation is a distribution of correlation coefficients of equal-sized samples that would occur by chance assuming there is no correlation in the population. The population correlation is abbreviated ρ.

7 You can test whether your observed r could have occurred by chance from a population in which there is zero correlation between the two variables. Critical values of r, symbolized r^*, are in Table r^* such that in the sampling distribution of the correlation any sample r that is greater in absolute value than r^* is less likely to occur than alpha. If your r is

greater than r^* you can reject H_0 and conclude that the correlation in the population is not 0.

8 The general hypothesis-testing procedure for determining whether your sample r could have occurred by chance is as follows:

a. State H_0 and H_A. Assume H_0 is true.

b. Set alpha and determine r^* from Table r^* for your level of alpha, N, and whether you are conducting a one- or two-tailed test.

c. If you have not done so already, calculate r.

If it is a two-tailed test, proceed to Step d. If it is a one-tailed test, examine the sign of your calculated r to see whether it is in the direction expected with H_A.

If r is in the expected direction, proceed to Step d. If r is in the opposite direction, retain H_0.

d. If $|r| > r^*$, reject H_0. If $|r| \leq r^*$, retain H_0.

EXERCISES

1 Construct a scatter plot for each of the data sets below.

(a)		(b)		(c)		(d)		(e)	
X	Y	X	Y	X	Y	X	Y	X	Y
5	4	1.0	−2.0	12	10	90	1.4	0	9
8	5	0.0	.5	17	12	83	3.0	12	9
7	7	−.5	1.0	18	4	89	1.6	4	4
1	2	1.5	−1.0	13	9	89	3.4	1	7
5	6	.5	0.0	13	6	87	2.5	8	4
4	6	−.5	0.0	12	11	90	3.5	6	2
1	3	.5	−1.0	13	10	84	1.4	6	3
9	8	−1.0	1.5	14	6	77	1.3	9	8
6	5			14	7	74	3.5	8	3
4	3			19	10	86	1.5	0	11
				16	5	73	3.0	2	6
				15	7	83	2.7	2	8
						85	1.7	11	12
						89	2.5	5	1
						79	2.3		
						76	3.2		
						70	1.4		

2 Identify the type of relationship observed in each of the scatter plots constructed in Exercise 1.

3 Compute the correlation coefficient for each of the data sets in Exercise 1 using the raw-score computational Formula 11.3.

4 Employ the hypothesis-testing procedure using critical values of r to test the significance of each r computed for data sets (a), (b), (c), and (d) in Exercise 1. Set $\alpha = .05$ and use a two-tailed test.

5 You conducted a study and found that the correlation coefficient between two variables based on a set of 30 pairs of scores is .46.

a. Setting $\alpha = .01$ and using a two-tailed test of significance, what can you conclude about the relationship?

b. Using the same alpha level ($\alpha = .01$) and a two-tailed test as in (a), what would be the effect of having a larger sample size than 30 on the significance of an $r = .46$? Describe the general effect of larger sample sizes on the significance of r.

6 A friend of yours describing the results of a study tells you that the correlation coefficient between the two variables of interest was 1.62. Comment on your friend's statement.

7 You have heard it said that scores on standardized college entrance examinations are positively related to subsequent achievement in college. To test this for yourself, you gather 40 college students at random who had taken a college entrance examination before college and obtain their grade point averages (GPA) for their freshman year. The following values are available:

$$N = 40$$

$$\Sigma(z_x - z_y)^2 = 31.2$$

$$\Sigma z_x z_y = 24.4$$

a. Compute the correlation coefficient using the definitional Formula 11.1.

b. Compute the correlation coefficient using the z score computational Formula 11.2.

c. Setting $\alpha = .05$, complete the hypothesis-testing procedure using critical values of r. What can you conclude about the hypothesis that college entrance examination scores and GPA are positively related?

d. Put your results in journal form.

8 You have begun developing a multiple-choice test to assess "leadership potential" in persons enlisting in the Coast Guard. To determine whether the test provides a consistent (reliable) measure of this potential, you randomly select a sample of 14 new Coast Guard enlistees and give them half of the test items chosen at random. You then transform their raw scores into z scores. One month later you give the same enlistees the other half of the test items and transform their scores into z scores. You expect the scores on the two halves of the test to be directly related to each other. Below are each enlistee's z scores on both halves of the test.

Subject	*First half* Z_x	*Second half* Z_y
S_1	−1.51	−1.12
S_2	−.92	−.77

Subject	*First half* Z_x	*Second half* Z_y
S_3	.25	.65
S_4	−1.51	−1.48
S_5	.25	.30
S_6	.84	−.76
S_7	1.43	1.38
S_8	−1.22	−1.12
S_9	−.33	−.05
S_{10}	.25	−.06
S_{11}	1.42	2.09
S_{12}	1.15	1.02
S_{13}	.55	.31
S_{14}	−.64	−.40

a. Compute the correlation coefficient using the definitional Formula 11.1.
b. Compute the correlation coefficient using the z score computational Formula 11.2.
c. Setting $\alpha = .01$, complete the hypothesis-testing procedure using critical values of r. What can you conclude about the hypothesis that the two halves of the test are directly related?
d. Put your results in journal form.

9 You predict that the average number of hours spent viewing TV each evening after 6:00 P.M. is inversely related to educational level. You select 12 adults at random and determine their years of education and the mean number of hours spent watching TV per evening. You obtain the following scores:

Subject	*Years of education* X	*Hours watching TV* Y
S_1	16	1.0
S_2	12	2.9
S_3	12	2.0
S_4	15	1.5
S_5	12	1.4
S_6	10	2.6
S_7	12	1.6
S_8	16	1.2
S_9	13	1.5
S_{10}	11	2.5
S_{11}	9	2.4
S_{12}	15	.8

a. Construct a scatter plot of the data to determine whether it is reasonable to assume that the relationship between the two variables is approximately linear.
b. Compute the correlation coefficient using the raw-score computational Formula 11.3.

c. Setting $\alpha = .05$, complete the hypothesis-testing procedure using critical values of r. What can you conclude about the prediction that TV viewing is inversely related to educational level?

d. Put your results in journal form.

10 Describe the effect on the correlation coefficient of restricting the range of scores sampled along one or both variables.

11 **a.** Assume that the amount of cigarettes smoked per day and the incidence of respiratory diseases such as emphysema are found to be significantly related, with $r = .8$. Is this significant correlation coefficient to be interpreted that cigarette smoking causes respiratory disease? Comment.

b. Assume that the amount of cigarettes smoked per day and "assertiveness" are found to be significantly related, with $r = .4$. Is the $r = .8$ between cigarette smoking and respiratory disease twice as strong a relationship as the $r = .4$ between cigarette smoking and assertiveness? Comment.

12

REGRESSION

CHAPTER GOALS

After studying this chapter, you should be able to:

1. Use the regression line to make predictions on one variable when given a particular value of the other variable.
2. Use the regression equation to find the regression line in a scatter plot.
3. Use the regression equation to make predictions on one variable when given a particular value on the other variable.
4. Describe the meaning of "regression toward the mean."
5. Define the standard error of the estimate and describe its usefulness in measuring the accuracy of prediction.
6. Define the coefficient of determination and describe its relation to the usefulness of a correlation coefficient in making predictions.

CHAPTER IN A NUTSHELL

A scatter plot can be used to make predictions about unknown values on one variable (*Y*) from known values on another variable (*X*) by finding the "line of best fit" through the data points. This line is also called the regression line, and can be objectively determined through a formula called the regression equation. The regression line is the line that minimizes the average squared distance that the data points lie away from it, thereby minimizing the errors in prediction. Both the regression line and the regression equation may be used to make predictions about the value of *Y* given a particular value of *X*. The accuracy in predicting *Y* from *X* is measured by the standard error of the estimate, which is the standard deviation of the distribution of errors in prediction. Accuracy in predicting *Y* from *X* is best when $r = \pm 1$, and worst when $r = 0$. The coefficient of determination, r^2, is a measure of how much it helps to know the value of *X* when you are trying to predict a value of *Y*.

Introduction to regression

One of the most useful aspects of the correlation coefficient is that once you have a significant correlation coefficient from a sample of subjects, you can generalize to new subjects and make predictions from one variable to another when you have a score on only one of the variables. It might be impractical or impossible to obtain the score on the second variable directly, but knowing the correlation coefficient from a previous sample allows you to predict the score on the second variable from the known score on the first variable. For example, assume that a previous study has established that two variables are strongly directly related. Some time later, an individual comes along whose score on only one of the variables is available. You want to know the score on the other variable, but you cannot measure it at this time. Fortunately, knowing that a strong direct relationship exists, you can use the one available score you do have to predict the unknown score. If the available score is high, you can predict that the unknown score will probably be high. If the available score is low, you can predict that the unknown score will probably be low.

By using a scatter plot it is possible to make predictions which are more accurate than merely "high" or "low." For example, the data plotted in Figure 12-1 represents the midterm and final exam scores for students

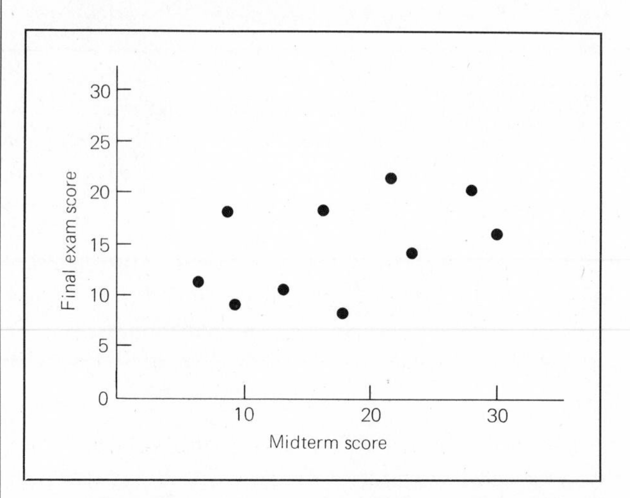

FIGURE 12-1 *Scatter plot for midterm and final exam scores.*

in a statistics class. The correlation coefficient for these scores is $r = .52$. This indicates a moderately strong direct relationship: students scoring high on the midterm tend to score high on the final. Suppose a student becomes very ill and cannot take the final. On the basis of the available midterm score, you can predict how well this student would do on the final by first constructing a scatter plot and then sketching in a "line of best fit." This is a straight line that, on the average, passes as closely as possible to all the data points. Later in this chapter you will see that there is a precise method for determining this line of best fit. For now, though, you can approximate it just by "eyeballing" (visualizing) a straight line through the data points. This line of best fit represents the trend in the data and can be used to make specific predictions about an unknown score on one variable from a known score on the other variable. For example, imagine a "good-fitting" straight line running through the data points as in Figure 12-2. Using this line you can predict that if the ill student's midterm score is 20, the score on the final should be 16. All you have to do is locate the known midterm score of 20 on the abscissa and project a straight line upward until it intersects the line of best fit. Then find the predicted final score by projecting another straight line over to the ordinate. Where it crosses the ordinate is the predicted final score, 16, which uniquely corresponds to a midterm score of 20.

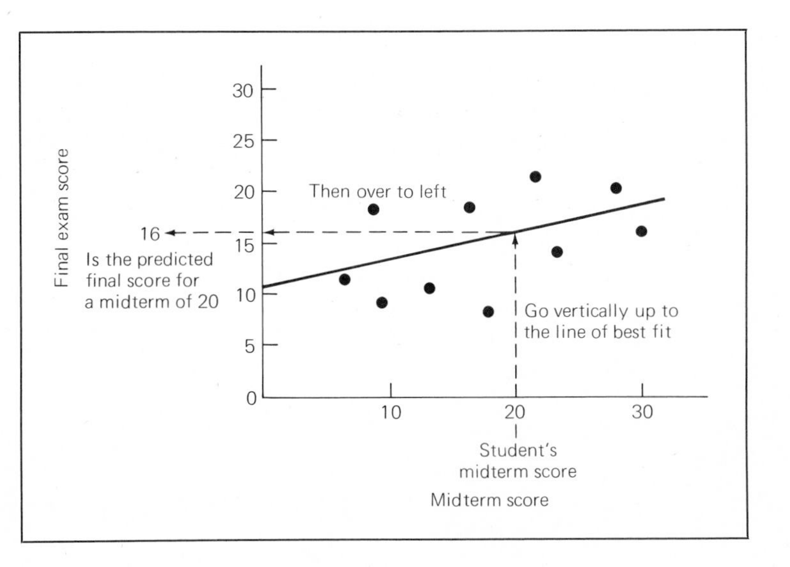

FIGURE 12-2
Scatter plot for midterm and final exam scores and "line of best fit" sketched in. The dashed lines show how a final score of 16 is predicted from the student's midterm grade of 20.

A problem arises when using the scatter plot procedure described above because different people may differ in their opinions about the straight line that offers the "best fit." There needs to be some way of being more objective about determining the best straight line to use in making predictions. You can see in Figure 12-2 that the predicted score is frequently not the same as that which may actually have been scored by a particular subject. For example, for a midterm score of 30, the straight line predicts a final exam score of 19. However, the student who actually received a score of 30 on the midterm actually scored a 16 on the final. The vertical distance between the straight line and the actual final score is a measure of an **error in prediction.** The error in prediction for a midterm score of 30 is thus −3, derived by subtracting the predicted final score of 19

error in prediction

from the observed final score of 16. Figure 12-3 identifies this error of prediction (and others) in terms of an approximate line of best fit. Thus, the problem of determining the best straight line to use for making predictions is to find the one that, on the average, makes the smallest errors in prediction.

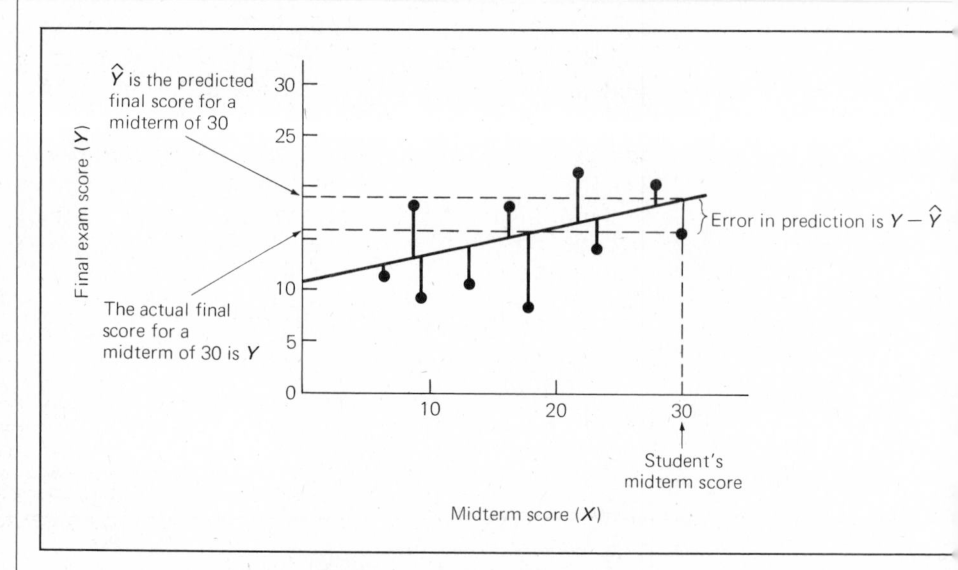

FIGURE 12-3
Scatter plot for midterm and final exam scores showing the errors in prediction and how they are calculated. Vertical lines between points and the regression line represent errors in prediction.

predictor variable
predicted variable
$\hat{Y}$

The formal process of determining the best straight line can be illustrated with a general case, in which the two variables are denoted X and Y. Consider the scores along X as known scores, and the scores on Y as those you want to predict. In other words, X is the **predictor variable** and Y is the **predicted variable.** Using the data in Figure 12-3, $\hat{Y}$ (pronounced "Y hat") denotes the value predicted from an actual X score. You might think that if you measure the error in prediction by calculating $Y - \hat{Y}$ for each data point, and then sum all the errors, you will get a measure of how well the line fits the actual data. Unfortunately, however, $Y - \hat{Y}$ will be positive for all the data points lying above the straight line, and $Y - \hat{Y}$ will be negative for all those lying below it. Consequently, the positive and negative errors might cancel each other out when you add them up. The solution to this familiar zero-sum problem is to square each $Y - \hat{Y}$, to make all the error scores positive. When the errors of prediction are squared and then added up, you have an excellent measure of how good an approximation your chosen line is to the actual data points. The farther away your line lies from the data points, the larger the errors in prediction will be, and hence the larger the sum of the squared errors in prediction. Conversely, the closer the line lies to the data points, the better an approximation it is, and the smaller the sum of the squared errors. Your measure of the errors in prediction is written:

measure of errors in prediction = $\Sigma(Y - \hat{Y})^2$

The idea behind calculating the error in prediction is that you can use it to precisely identify the line of best fit as the line having the smallest errors in prediction. However, it would be a very laborious, time-consuming task to sketch in the many straight-line "candidates" vying for line of best fit, and then measure the errors in prediction for each. Fortunately, you can locate the line of best fit by applying some math. You may recall from your geometry days that the general **equation for a straight line** is:

equation for straight line

$$\hat{Y} = mX + c$$

slope

***Y* intercept**

where $\hat{Y}$ is the predicted variable, m is the **slope,** X is the predictor variable, and c is the ***Y* intercept.** $\hat{Y}$ is used in this case instead of Y to indicate that the equation is used here for predicting scores on the Y variable from given values of X. You may also recall thât the slope is the distance the straight line rises divided by the distance the line extends to the right, relative to any 90° triangle, as shown in Figure 12-4. The Y intercept

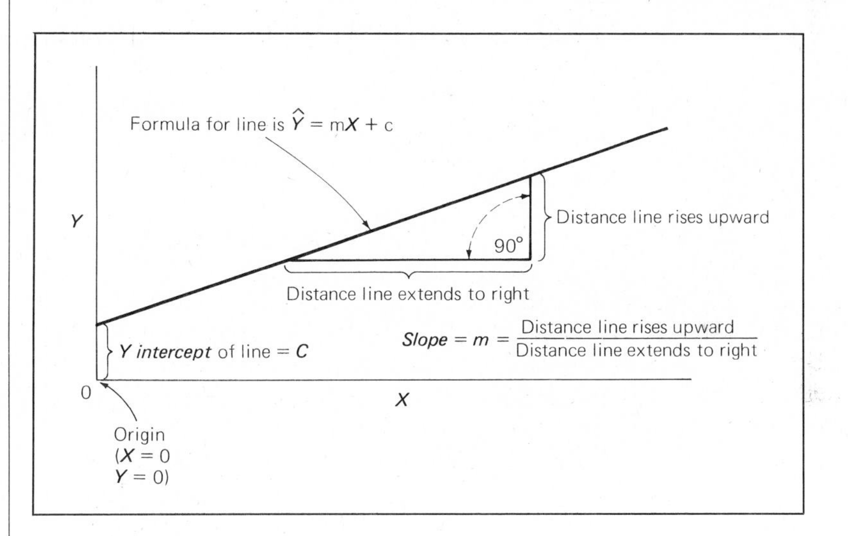

FIGURE 12-4

The slope m and Y intercept c of the straight line $\hat{Y} = mX + c$.

is the distance from the origin[1] the straight line intersects the ordinate. If you used a little calculus to determine the values of the slope and Y intercept that minimize the errors in prediction, you would discover that the actual line of best fit has a slope m, and Y intercept c, as follows:

$$m = r\frac{\sigma_y}{\sigma_x} \quad \text{and} \quad c = \bar{Y} - r\frac{\sigma_y}{\sigma_x}\bar{X}$$

[1]The origin is the point on a graph where $X = 0$ and $Y = 0$. This is the point where the axes cross each other.

where r is the Pearson product-moment correlation coefficient, σ_x and σ_y are the standard deviations of the X and Y variables respectively, and $\bar{X}$ and $\bar{Y}$ are the means of the X and Y variables respectively. If you substitute these formulas for the slope and Y intercept into the general equation for a straight line, you will get:

$$\hat{Y} = r\frac{\sigma_y}{\sigma_x}X + \bar{Y} - r\frac{\sigma_y}{\sigma_x}\bar{X}$$

Further mathematical simplification renders the **formula for the line of best fit:**

FORMULA 12.1
formula for line of best fit, regression equation

$$\hat{Y} = r\frac{\sigma_y}{\sigma_x}(X - \bar{X}) + \bar{Y}$$

Your reward for this mathematical manipulation is that you can now precisely predict the ill student's final exam score given a midterm score of 20. All you have to do is make the necessary substitutions in Formula 12.1. From the data shown in Figure 12-1 you obtain

Midterm (X)	*Final (Y)*	*Correlation*
$\bar{X} = 17.3$	$\bar{Y} = 15.6$	$r = .52$
$\sigma_x = 7.92$	$\sigma_y = 3.95$	

Substituting $X = 20$ for the student's midterm score, the predicted final exam score works out to 16.3; that is,

$$\begin{aligned}
\hat{Y} &= r\frac{\sigma_y}{\sigma_x}(X - \bar{X}) + \bar{Y} \\
&= .52\left(\frac{3.95}{7.92}\right)(20 - 17.3) + 15.6 \\
&= .52\,(.5)(2.7) + 15.6 \\
&= .7 + 15.6 \\
&= 16.3
\end{aligned}$$

You can see that it is not absolutely necessary to actually make a scatter plot and draw in the line of best fit to find predicted scores along Y. You can use Formula 12.1 directly. However, a scatter plot with the line of best fit has the advantage that it visually captures the predictions of Y over the entire range of X. It also allows you to determine whether a straight line is appropriate for describing the trend in your data, and whether the Pearson product-moment correlation coefficient can be appropriately applied. Do not use Formula 12.1 (or compute an r) if you do not have an approximately linear relationship. If you do have an approximately linear relationship, the easiest way to draw the line of best fit in your scatter plot is to pick out two values of X at opposite ends of the abscissa, and solve for the values of $\hat{Y}$ corresponding to each. For example, the $\hat{Y}$ predictions for the X scores of 8 and 30 are 13.2 and 18.9, respectively:

For X = 8	*For X = 30*
$\hat{Y} = r\frac{\sigma_y}{\sigma_x}(X + \bar{X}) + \bar{Y}$	$\hat{Y} = r\frac{\sigma_y}{\sigma_x}(X - \bar{X}) + \bar{Y}$
$= .52\left(\frac{3.95}{7.95}\right)(8 - 17.3) + 15.6$	$= .52\left(\frac{3.95}{7.95}\right)(30 - 17.3) + 15.6$
$= 13.2$	$= 18.9$

After you do this, you can plot the points corresponding to the two pairs of scores ($X = 8$ and $\hat{Y} = 13.2$; $\bar{X} = 30$ and $\hat{Y} = 18.9$) and connect the two points[2] with a straight line as in Figure 12-5. Technically speaking,

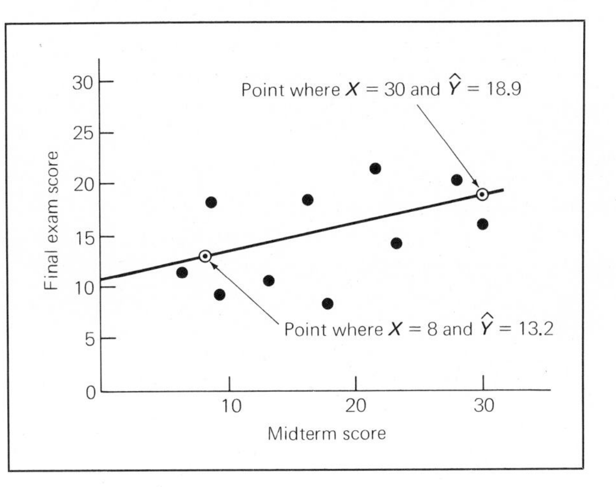

FIGURE 12-5
Draw the line of best fit by using values of X at opposite extremes on the abscissa. Use Formula 12.1 and solve for $\hat{Y}$. Then draw a straight line through these two points.

the line of best fit drawn accurately by using Formula 12.1 is called the **regression line.** Consequently, Formula 12.1 is often referred to as the **regression equation.**

regression line
regression equation

[2]You might notice here that if one of the values you pick for X is $\bar{X}$, when you plug this into Formula 12.1 you get $\hat{Y} = \bar{Y}$. Thus the regression line always goes through the point corresponding to the mean of X and the mean of Y. Thus you really need only calculate one other point.

If you want to make a prediction the other way around, that is, predict a score along X given a score on Y, all you have to do is switch the X and Y labels for the two variables and then use Formula 12.1. When you use Formula 12.1 you must predict the variable you label Y from the variable you label X because the regression line is set up to minimize the error in prediction when predicting $\hat{Y}$ from X. For example, if a student scored 19 on the final exam but missed the midterm, and you would like to estimate the missing midterm score, you can simply label the final exam scores as X and the midterm scores as Y:

Final (X)	*Midterm (Y)*	*Correlation*
$\bar{X} = 15.6$	$\bar{Y} = 17.3$	$r = .52$
$\sigma_x = 3.95$	$\sigma_y = 7.92$	

Substituting $X = 19$ for the final score, the predicted midterm exam works out to 20.8:

$$\begin{aligned}\hat{Y} &= r\frac{\sigma_y}{\sigma_x}(X - \bar{X}) + \bar{Y} \\ &= .52\left(\frac{7.92}{3.95}\right)(19 - 15.6) + 17.3 \\ &= 20.8\end{aligned}$$

Formula 12.1 is a very useful equation. It can be used to predict missing scores (such as grades), to predict subsequent school performance on the basis of "entrance" examination scores (such as the Graduate Record Examination, or GRE; the Law School Admission Test, or LSAT; the Medical College Admissions Test, or MCAT; the Dental Admission Test, or DAT; and so on), to predict performance on a specific trait by a correlated performance on another trait (such as predicting IQ on the basis of locomotor activity and coordination in infancy, predicting annual income on the basis of education level, predicting an automobile driver's likelihood of being involved in an accident on the basis of age, or for predicting children's heights from the heights of their parents). Whenever two variables are linearly related, you can use the regression equation to predict an unknown score on one variable given a known score on the other variable, provided you know the correlation coefficient, the means, and the standard deviations of the two variables. Notice we do not give a computational formula counterpart to Formula 12.1 (which would include a computational formula for r) that might save you time in calculating $\hat{Y}$. We assume that the user of Formula 12.1 has already calculated the correlation coefficient by whatever is the most useful method in a particular situation. The researcher should always know the value of the correlation coefficient for descriptive purposes before attempting to make precise predictions.

The regression equation has several interesting characteristics relevant to predictions. To help illustrate these characteristics, take Formula 12.1 and change it around as follows:

1. Begin with Formula 12.1:

$$\hat{Y} = r\frac{\sigma_y}{\sigma_x}(X - \bar{X}) + \bar{Y}$$

2. Subtract $\bar{Y}$ from both sides of the equation to give

$$\hat{Y} - \bar{Y} = r\frac{\sigma_y}{\sigma_x}(X - \bar{X})$$

3. Then divide both sides by σ_y to give

$$\frac{\hat{Y} - \bar{Y}}{\sigma_y} = r\frac{X - \bar{X}}{\sigma_x}$$

4. Notice that since $\frac{(X - \bar{X})}{\sigma_x}$ is the z score formula for X, and $\frac{(\hat{Y} - \bar{Y})}{\sigma_Y}$ is the z score formula for $\hat{Y}$, then

FORMULA 12.2

$$z_{\hat{y}} = rz_x$$

Where $z_{\hat{y}}$ designates the z score for the predicted $\hat{Y}$ value, and z_x designates the z score for a predictor value of X. Formula 12.2 reveals the interesting characteristics of using the regression equation for prediction. First, if you are trying to predict $\hat{Y}$ from X and $r = 0$, then the z score for $\hat{Y}$ will always be 0:

if: $r = 0$

then: $z_{\hat{y}} = 0(z_x) = 0$

predict the mean when $r = 0$

Remember that 0 is the z score that always corresponds to the mean. Consequently, when there is no correlation, regardless of what the given predictor value of X is, you always predict that the corresponding $\hat{Y}$ score is the mean of the Y scores. The mean of Y is actually the most reasonable prediction under the circumstances. Because a correlation of 0 indicates that the two variables are unrelated, knowing the predictor value of a score along X does not tell you anything at all about the subject's score along Y. The best thing to do when there is no correlation (and, in fact, whenever you want to make a prediction on a variable and you have no predictor) is to guess the mean of the Y variable. By guessing the mean of Y you will be the least far off on the average, since the average distance between the mean and all the scores in a distribution is 0. For example, suppose you are trying to guess a person's IQ and you know nothing about that person. You should guess 100, the mean IQ for the population. If you find out the last digit in the person's phone number is 8 and you still want to guess IQ, then you should again guess 100 because there is no correlation between the last phone digit and IQ. The last digit in a person's phone number tells you nothing about IQ. The regression line for variables having no correlation is a straight horizontal line extending from the mean of Y, as in Figure 12-6. This indicates that you will predict the mean of Y regardless of the value of X.

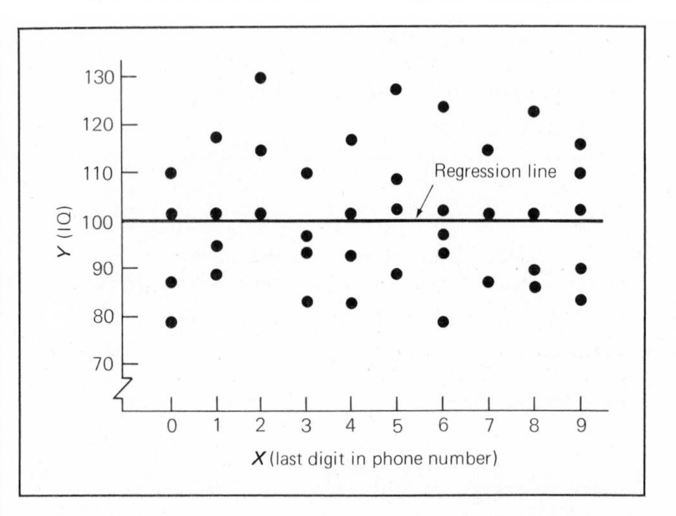

FIGURE 12-6

Scatter plot and regression line for no correlation ($r = 0$) between IQ and the last digit in phone number.

with perfect correlation, predict same z score as predictor

The second interesting characteristic about prediction revealed by Formula 12.2 is that if $r = \pm 1$, then the absolute value of the z score for $\hat{Y}$ will be the same as the z score for X:

if: $r = \pm 1$

then: $z_{\hat{y}} = +1(z_x)$ or $z_{\hat{y}} = -1(z_x)$

so: $|z_{\hat{y}}| = |z_x|$

For example, suppose you are predicting a person's height in centimeters from the person's height in inches. Since $r = +1$, you predict a z score for the person's height in centimeters that is equal to the z score for the person's height in inches. Consequently, if a person is one standard deviation below the mean height in inches, you predict that person will be one standard deviation below the mean height in centimeters as well. With $r = \pm 1$, you have perfect predictive power because all the data points in the scatter plot will lie exactly on the regression line, as in Figure 12-7. Either a perfect direct relationship or a perfect inverse relationship will allow you to make predictions from one variable to the other without error.

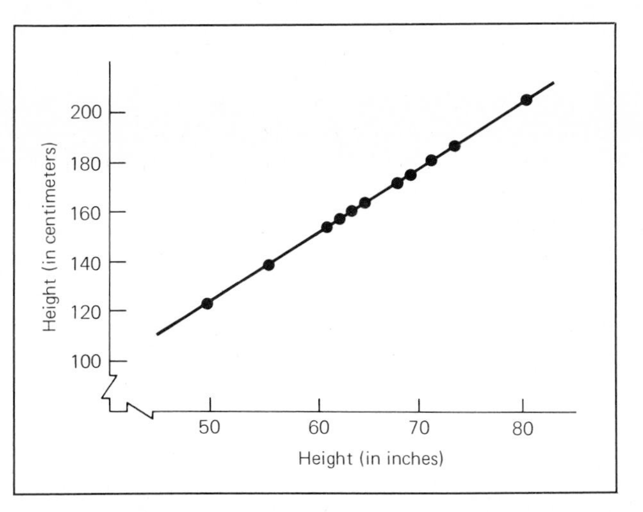

FIGURE 12-7
Scatter plot and regression line for perfect correlation ($r = 1$) between height measured in centimeters and height measured in inches.

The third and most interesting thing Formula 12.2 reveals about prediction is when r is a nonzero value somewhere between -1 and $+1$. In this case the correlation coefficient is a fraction, which means that in Formula 12.2 the z score for the given X score is multiplied by a fraction to get the predicted z score for $\hat{Y}$:

if: $\quad 0 < |r| < 1$

then: $\quad |z_{\hat{y}}| < |z_x|$

for imperfect correlation, prediction is closer to mean than predictor

This indicates that the predicted z score for $\hat{Y}$ will always be less in absolute value than the z score for the predictor X. When a z score decreases in absolute value, it approaches 0. Therefore, since 0 is the z score corresponding to the mean of a variable, the predicted z score for $\hat{Y}$ will be closer to the mean than the z score for the predictor X. In raw-score terms, this means that the predicted $\hat{Y}$ score is closer to the mean of the Y distribution than the given X score is to the mean of the X distribution, and your predicted $\hat{Y}$ score is said to *regress* toward the mean [of the Y variable]. This is where the term **regression** comes from. Sir Francis Galton (1822–1911) noted this when he tried predicting the heights of children from the heights of their parents, for which the correlation coefficient is somewhere between -1 and $+1$ (actually, it is about $+.50$). He found that children tend to have heights closer to the mean height of the population than do their parents. In other words, on the average, very tall parents tend to have children shorter than themselves, and very short parents tend to have children somewhat taller than themselves. Many people erroneously conclude that since there is a **regression toward the mean,** after a few generations everyone should be the same height. This does not occur because you are dealing with averages, and we must not forget

regression

regression toward the mean

that there is variability in the heights of the offspring. Although, on the average, children of very tall parents tend to be shorter than their parents, there is enough variability in the childrens' heights that some of the children are taller than their parents and some are shorter. This variability keeps the distributions of successive generations of childrens' heights from converging to the mean.

The nature of regression towards the mean becomes clearer when you consider that an r somewhere between -1 and $+1$ is imperfect and conveys a varying degree of knowledge. For example, as r approaches ± 1, and hence the relationship between X and Y becomes stronger, knowledge of the given X score is providing more and more knowledge about the value of Y. This is reflected in the fact that $|z_{\hat{y}}|$ approaches $|z_x|$. That is, in terms of absolute values, the z score for your predicted $\hat{Y}$ value approaches the z score for your predictor X value. On the other hand, as r approaches 0, and hence the relationship between X and Y becomes weaker, knowledge of X provides less and less information about Y. This is reflected by the fact that $z_{\hat{y}}$ approaches 0. That is, your predicted $\hat{Y}$ value approaches the mean of the distribution. Thus, with an imperfect correlation, the z score for $\hat{Y}$ should be somewhere between the mean of the distribution and the z score for X. Thus, in general, as the strength of the relation between two variables increases, the given X score is taken more into account to predict $\hat{Y}$. As the correlation coefficient weakens, the given X score is taken less into account and more reliance has to be placed on the mean of the Y distribution. For example, suppose you want to predict a random person's IQ from your knowledge of that person's height. Since the correlation between height and IQ is very weak, you will predict an IQ value close to the population mean IQ of 100. Thus your prediction relies more on your knowledge that the person is an "average" (random) person than on your knowledge of his height. Now suppose you want to predict a person's IQ from your knowledge of his GPA. Since IQ and GPA are strongly related, you will predict an IQ that has a z score close to that person's z score on GPA. Thus your prediction will rely more on your knowledge of that person's GPA than on your knowledge that the person is an "average" (random) person. Another way of saying this is that GPA tells you more about IQ than does height.

Regression towards the mean leads to some interesting situations. For example, suppose your instructor informs you that those who miss the final will have their final exam scores predicted from their midterm scores. You can expect that regression toward the mean will exert an effect. Using the regression equation, if you scored very high on the midterm, your instructor will predict a final score that is somewhat lower than your midterm score (relative to how the rest of the class scored). This is because the correlation between the midterm and final scores is probably between 0 and $+1$, and you scored way above the mean of the class. On the other hand, if you did very poorly on the midterm, your instructor will predict that you would score relatively higher on the final than the midterm. This regression toward the mean is a general phenomenon that occurs whenever you have an imperfect correlation.

Measuring accuracy of prediction

You can more or less tell how accurate your predictions will be if you look at the scatter plot to see how closely the data points lie to the regression line. However, there are two numerical descriptors that enable you to more precisely summarize errors in prediction. One is the *standard error of the estimate,* and the other is the *coefficient of determination.*

The standard error of the estimate. To illustrate the meaning of the first of these two numerical descriptors of errors in prediction, suppose you make a frequency distribution of the errors in prediction, that is, of the $Y - \hat{Y}$ values. To construct such a distribution, take each point in the scatter plot (or you may work from a distribution of X and Y scores) and subtract the predicted $\hat{Y}$ value from the actual Y value. This yields the vertical distance each point lies from the regression line, or the *error* in prediction for each point in the scatter plot. Now you can make a frequency polygon for these errors in prediction. Frequency polygons of the errors in prediction for the data in Figure 12-8(a) and (c) are shown in Figure 12-8(b) and (d) respectively. Because of the precise mathematical manner in which the regression line is formulated, it can be shown that

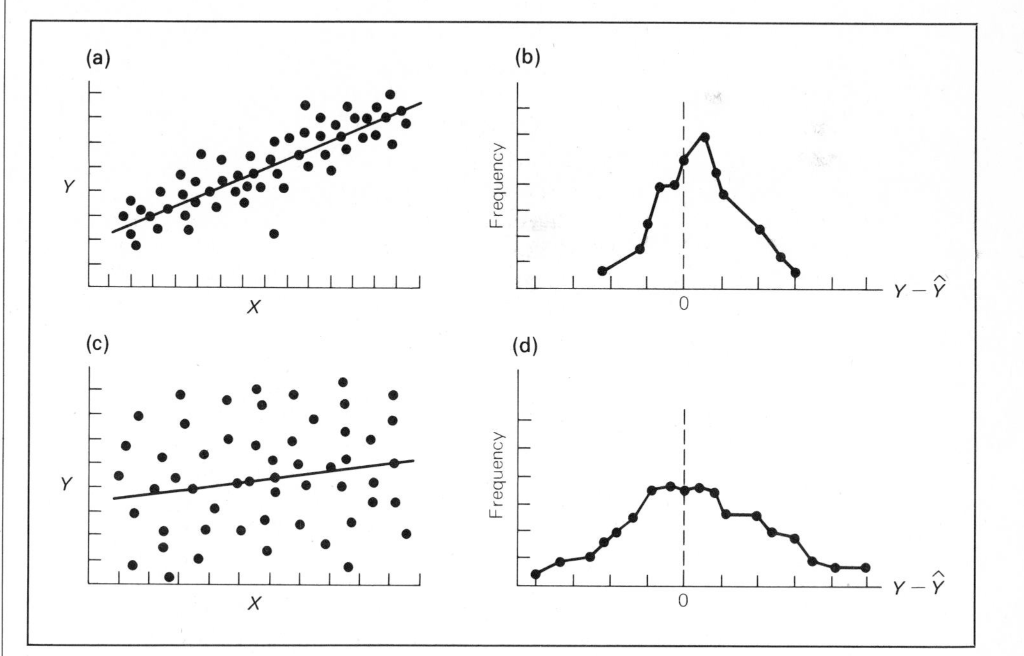

FIGURE 12-8

Scatter plots and associated frequency polygons of errors in prediction when r is close to ±1 and when r is close to 0.

(a) An example of a scatter plot for predicting $\hat{Y}$ from X when r is close to ±1. The points lie relatively close to the regression line. (A positive r is illustrated. The regression line would have a negative slope for negative r's.)

(b) Frequency polygon of errors in prediction for data in Figure 12-8(a).

(c) An example of a scatter plot for predicting $\hat{Y}$ from X when r is close to 0.

(d) Frequency polygon of errors in prediction for data in Figure 12-8(c).

the mean of the distribution of errors in prediction will always be zero. In other words, the positive errors ($Y - \hat{Y}$ for points above the regression line) will always exactly cancel out the negative errors ($Y - \hat{Y}$ for points below the regression line). You can examine the frequency distribution of errors to determine how accurate your predictions will be. If the frequency distribution is tightly bunched around the mean of zero, as in Figure 12-8(b), then your predictions will be very accurate. The more the distribution spreads out from the mean, as in Figure 12-8(d), the larger the errors will be and hence the less accurate your predictions will be. You can see that the variability of the distribution of errors provides a good measure of the accuracy of prediction. The more variability there is in the distribution of errors, the larger the errors will be on the average, and the worse the predictions.

standard error of the estimate $\sigma_{\hat{y}}$

A measure of the variability of the distribution of errors is the standard deviation of that distribution. The standard deviation of the distribution of errors is called the **standard error of the estimate,** and is abbreviated $\sigma_{\hat{y}}$. The formula for $\sigma_{\hat{y}}$ is derived by taking definitional Formula 3.7 for the standard deviation, σ, and making some substitutions. Because you have a distribution of error scores that has a mean of zero, you substitute $Y - \hat{Y}$ (your error score) for X and you substitute 0 for μ:

Formula 3.7: $$\sigma = \sqrt{\frac{\Sigma(X - \mu)^2}{N}}$$

Formula for standard deviation of distribution of error scores: $$\sigma_{\hat{y}} = \sqrt{\frac{\Sigma[(Y - \hat{Y}) - 0]^2}{N}}$$

Since it is not necessary to write the zero in the formula above, **the definitional formula for the standard error of the estimate** becomes simply:

FORMULA 12.3
definitional formula for $\sigma_{\hat{y}}$

$$\sigma_{\hat{y}} = \sqrt{\frac{\Sigma(Y - \hat{Y})^2}{N}}$$

The closer the points in the scatter plot are to the regression line, the smaller the errors in prediction will be. In other words, the errors will be closer to the mean error of zero. Consequently, the standard error of the estimate will be smaller. On the other hand, the farther away the points are from the regression line, the larger the errors will be, and hence the errors will be farther away from their mean of zero. Therefore, the standard error of the estimate will be larger.

It is fairly cumbersome to calculate $\sigma_{\hat{y}}$ from the definitional Formula

12.3 because you must make a prediction ($\hat{Y}$) from the regression line (or regression equation) for each Y value in the data, then subtract the predicted values of $\hat{Y}$ from the actual values of Y, then square these differences, and so on. There is another formula for $\sigma_{\hat{y}}$ that is much easier to compute. It also has the advantage that it sheds some light on the relationship between the value of the correlation coefficient and the accuracy of prediction. This new formula is called the computational formula, and it can be derived by mathematically manipulating the definitional Formula 12.3. The derivation will not be given here, but will be left as a challenging exercise for the interested, mathematically oriented student. The computational formula is:

FORMULA 12.4
computational formula for $\sigma_{\hat{y}}$

$$\sigma_{\hat{y}} = \sigma_y \sqrt{1 - r^2}$$

In Formula 12.4, σ_y is the standard deviation of the Y scores. Formula 12.4 allows you to calculate $\sigma_{\hat{y}}$ directly from the value of the correlation coefficient r and from the value of the standard deviation of the Y scores. This formula also directly illustrates an interesting and useful relationship between the value of r and the accuracy of prediction as measured by $\sigma_{\hat{y}}$. For example, when $r = \pm 1$, as in Figure 12-7, all of the data points lie along the regression line and there are no errors in prediction. If you substitute either $r = 1$ or $r = -1$ into Formula 12.4, you can calculate that $\sigma_{\hat{y}} = 0$:

$$\begin{aligned} \sigma_{\hat{y}} &= \sigma_y \sqrt{1 - r^2} \\ &= \sigma_y \sqrt{1 - (1)^2} = 0 \end{aligned}$$

Since there are no errors in prediction when $r = \pm 1$, you can make a perfect prediction. The standard deviation of the distribution of errors will be zero because all of the errors are 0. There is no variability about the mean error of zero, which represents *no error*. On the other hand, when $r = 0$, you will always predict the mean of Y, no matter what value of X you have. Your errors in prediction will be determined by how much the actual Y scores vary about their mean. If you substitute $r = 0$ into Formula 12.4, you will find that $\sigma_{\hat{y}} = \sigma_y$:

$$\begin{aligned} \sigma_{\hat{y}} &= \sigma_y \sqrt{1 - r^2} \\ &= \sigma_y \sqrt{1 - (0)^2} \\ &= \sigma_y \end{aligned}$$

This indicates that the standard error of the estimate is equal to the standard deviation of the Y scores. Thus the more variable the Y scores are, the larger σ_y will be and consequently the larger $\sigma_{\hat{y}}$ will be, so the accuracy of your prediction will be poorer. This is illustrated in Figure 12-9(a). However, if all the Y scores are tightly clustered about their mean, as in Figure 12-9(b), then σ_y is small and consequently $\sigma_{\hat{y}}$ is small, so your prediction will not be too far off.

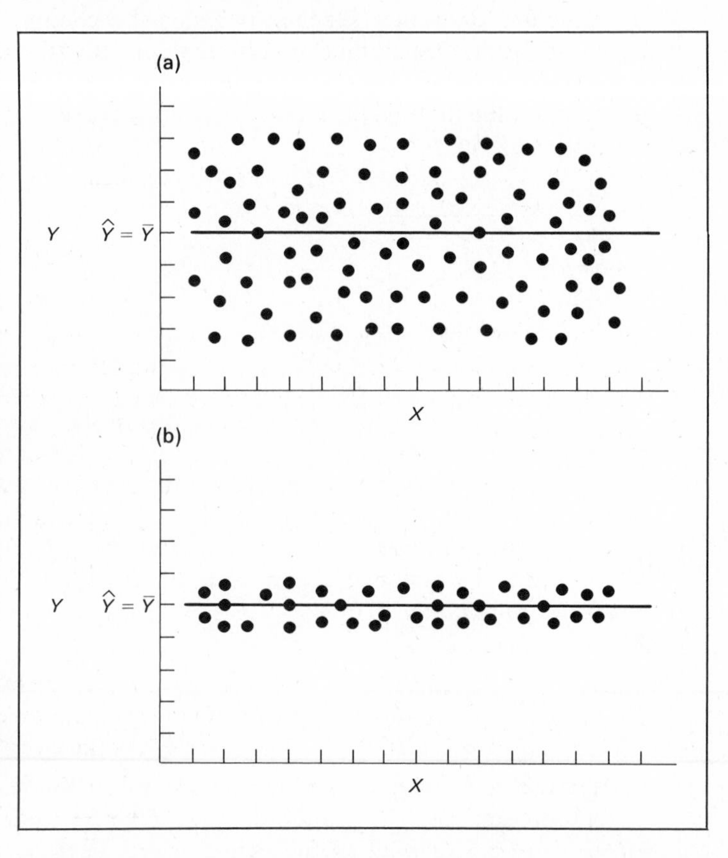

FIGURE 12-9

When $r = 0$, the accuracy in prediction depends on the variability of the Y scores about their mean.

(a) When $r = 0$ and there is considerable variability in the Y scores, you predict the mean of Y, and the errors in prediction are relatively large.

(b) When $r = 0$ and there is little variability in the Y scores, you predict the mean of Y, and the errors in prediction are relatively small.

Regardless of the exact value of σ_y, you have the least accuracy in prediction when $r = 0$. This is because knowing the value of X does not provide any information about Y. It is as if you have no X value from which to make a prediction, so you predict the mean of Y. For values of r between ± 1 and 0, the standard error of the estimate will be somewhere between 0 and σ_y, and there will thus be intermediate degrees of accuracy in prediction. When $r = \pm 1$, the standard error of the estimate is 0 and you have perfect prediction with no errors.

The coefficient of determination. The second numerical descriptor of errors in prediction can be introduced by considering what happens when you take Formula 12.4 and square it. You will find:

FORMULA 12.5
variance of errors in production

$$\sigma_{\hat{y}}^2 = \sigma_y^2(1 - r^2)$$

where $\sigma_{\hat{y}}^2$ is the variance of the errors in prediction, and σ_y^2 is the variance of the Y scores. Like the standard error of the estimate ($\sigma_{\hat{y}}$), the variance of the errors in prediction is a good measure of the accuracy of prediction. Formula 12.5 says that the variance of the errors in prediction is equal to the variance of the Y scores *reduced by a proportion equal to* r^2. Perhaps a better way to see this is to simplify Formula 12.5 further to $\sigma_{\hat{y}}^2 = \sigma_{\hat{y}}^2 - r^2\sigma_{\hat{y}}^2$. To get a better feeling for what Formula 12.5 indicates, consider the separate situations when $r = 0$ and when $r = \pm 1$.

When $r = 0$, you always predict the mean of Y, and the magnitudes of your errors are determined by how much the actual Y values vary above and below their mean. The variance of the errors in prediction will equal the variance of the Y scores. Once again, you see that if $r = 0$ there is no advantage in knowing the X value. You guess the mean of Y whether you know the X value or not. You can say that knowing the value of X does not *reduce the uncertainty* about the value of Y, since the variance of the actual Y values above and below the predicted $\hat{Y}$ values will equal the variance of the Y values above and below their mean. Your prediction will be equally inaccurate whether you know X or not.

When $r = \pm 1$, you can make a perfect prediction. The variance of the errors in prediction will equal 0 because there will be no errors (each $Y - \hat{Y}$ will be 0). Thus you have complete reduction of uncertainty about the value of Y. If you do not have a value of X from which to predict $\hat{Y}$, then you will guess the mean of Y and the variance of your errors in prediction will be σ_y^2. However, if you know the value of X, then you can perfectly predict $\hat{Y}$, and the variance of the errors in prediction will be reduced to zero. Intermediate values of r between 0 and ± 1 give intermediate degrees of reduction of uncertainty in prediction. In fact, the reduction in uncertainty of prediction is equal to r^2, because $\sigma_{\hat{y}}^2$, a measure of uncertainty in prediction, is equal to σ_y^2 multiplied by $(1 - r^2)$. Since σ_y^2 is a measure of the uncertainty in prediction when you do not have a X value and you must guess the mean of Y, the quantity r^2 tells you how much the knowledge of X reduces the uncertainty in prediction as compared to the situation in which you do not know X and you must guess the mean of Y. The quantity $\boldsymbol{r^2}$, often called the **coefficient of determination,** is very useful for determining the extent to which your knowledge of the values of X helps you in predicting the value of Y.

r^2

coefficient of determination

A useful way of conceptualizing the coefficient of determination is to consider it as the amount of variability (as measured by the variance) in the Y variable accounted for by knowing X. In this conceptual framework,

r^2 is converted to a percent by multiplying it by 100. Thus, if $r = .71$, then $r^2 = .50$, so multiplying .50 by 100 reveals that 50% of the variability in Y is accounted for, or explained, by differences in X. For example, if the correlation between IQ and academic achievement is .71, and you are using IQ to predict academic achievement, then you can say that differences in intelligence account for 50% of the variability in achievement. In other words, people differ in achievement, and difference in intelligence explains, or *determines,* 50% of the difference in achievement between individuals. The other 50% of the variability in achievement might be explained by differences in other factors, such as motivation, environmental stimulation, and so on. This interpretation explains why r^2 is referred to as the coefficient of determination.

You have seen that the usefulness of the correlation coefficient in making predictions can be measured by the reduction of uncertainty in prediction, which is proportional to r^2, not r. This is an important principle to remember—r^2 and not r is the relevant measure of usefulness in prediction. Therefore, intermediate values of r do not have directly proportional levels of usefulness in prediction. Many people think, for example, that an $r = .5$ implies a level of usefulness in prediction that is halfway between the levels of usefulness when $r = 0$ and when $r = \pm 1$. On the contrary, the level of usefulness is proportional to r^2, and not r. Thus, in terms of usefulness in prediction, an $r = 1$ is really four times as good as an $r = .5$, since $.5^2 = .25$, whereas $1^2 = 1$. This is why you cannot use the values of r to compare the proportionate strengths of two correlation coefficients, you must use r^2.

The above analysis of r^2 also reveals that low correlation coefficients are not very helpful in making predictions. For example, if $r = .2$, then $r^2 = .04$, which is very close to the case in which $r = 0$ and you are guessing the mean of Y. Thus trying to make predictions with a correlation coefficient of .2 is not much better than simply guessing the mean. To get an r^2 of .5, which is halfway between the usefulness afforded by correlations of 0 and 1, you need an $r = .71$, since $.71^2 = .5$.

It is important to realize that the usefulness in prediction is an entirely different concept than the *significance* of a sample r. For example, you may observe an $r = .2$ that is *statistically significant,* but you still have a correlation coefficient that is not very useful in making predictions since knowledge of X reduces uncertainty in prediction by only 4%. The only thing that a significant r indicates is that the correlation in the population is probably not zero. Significance gives no indication of the actual value of the population correlation coefficient, nor does it necessarily say that the r will be useful in making predictions; although to be useful in prediction an r should be significant since otherwise it could be due to sampling error and might vary widely from sample to sample. Given significance, usefulness in prediction is a function of r^2, the coefficient of determination. Thus r should be significant and r^2 should be of reasonable size.

KEY CONCEPTS

1 The regression line is the best-fitting straight line through the data points in a scatter plot. "Best fit" means that the sum of the squared errors in prediction is minimal. Symbolically, this is when $\Sigma(Y - \hat{Y})^2$ = minimum, where $\hat{Y}$ is the predicted score on the Y variable for a given predictor X score, and Y is the actual score seen in the scatter plot for that value of X.

2 The equation for the regression line for predicting $\hat{Y}$ from X is called the regression equation, and is:

$$\hat{Y} = r\frac{\sigma_y}{\sigma_x}(X - \bar{X}) + \bar{Y}$$

3 The regression equation can be rewritten in terms of z scores:

$$z_{\hat{y}} = rz_x$$

From this equation it can be demonstrated that whenever r is not equal to ± 1, the predicted $\hat{Y}$ score will be closer to the mean than the given predictor X score. This is called regression toward the mean. Whenever $r = 0$, the mean of the Y variable is the best prediction to make.

4 As the value of r approaches ± 1 the data points in the scatter plot lie closer to the regression line, and consequently the accuracy of prediction increases. An excellent measure of the accuracy of prediction is the standard error of the estimate, symbolized $\sigma_{\hat{y}}$. It is the standard deviation of the distribution of the errors in prediction. The definitional formula for $\sigma_{\hat{y}}$ is:

$$\sigma_{\hat{y}} = \sqrt{\frac{\Sigma(Y - \hat{Y})^2}{N}}$$

The computational formula for $\sigma_{\hat{y}}$ is:

$$\sigma_{\hat{y}} = \sigma_y\sqrt{1 - r^2}$$

5 The coefficient of determination, r^2, is a measure of the usefulness of knowing the X value when trying to predict the value for Y. r^2 also indicates the amount of variability in the Y variable explained or accounted for by differences in X variable.

EXERCISES

1 Construct a scatter plot for each of the data sets below. Pick two values of X at opposite ends of the abscissa and use the regression equation, Formula 12.1, to solve for $\hat{Y}$ and then draw the regression line in each scatter plot. The values

beneath each data set may be substituted into Formula 12.1.

(a)		(b)		(c)		(d)	
X	Y	X	Y	X	Y	X	Y
5	4	5	1	5	3	5	7
2	2	2	3	2	7	2	3
4	4	4	9	4	3	4	5
4	5	4	1	4	4	4	6
3	4	3	6	3	4	3	2
2	2	2	1	2	6	2	4
2	3	2	5	2	5	2	5
1	2	1	9	1	8	1	2
8	9	8	4	8	1	8	5
7	6	7	8	7	2	7	6
$\bar{X} = 3.8$		$\bar{X} = 3.8$		$\bar{X} = 3.8$		$\bar{X} = 3.8$	
$\sigma_x = 2.182$		$\sigma_x = 2.182$		$\sigma_x = 2.182$		$\sigma_x = 2.182$	
$\bar{Y} = 4.1$		$\bar{Y} = 4.7$		$\bar{Y} = 4.3$		$\bar{Y} = 4.5$	
$\sigma_y = 2.071$		$\sigma_y = 3.068$		$\sigma_y = 2.1$		$\sigma_y = 1.628$	
$r = .934$		$r = -.009$		$r = -.925$		$r = .619$	

2 Verbally define the standard error of the estimate.

3 Looking at the scatter plots constructed in Exercise 1, you can see that the data points for data sets (a) and (c) lie closest to the regression line, and that the data points for data set (b) generally lie farthest away from the regression line. The standard error of the estimate, $\sigma_{\hat{y}}$, is the standard deviation of these errors in prediction.

a. For each of the data sets in Exercise 1, calculate the standard error of the estimate using computational Formula 12.4.

b. Describe the effect that large errors in prediction have on the magnitude of the standard error of the estimate.

c. For each of the data sets in Exercise 1, calculate the value of $\hat{Y}$ that you would predict for someone who scored a 6 on the X variable. Use Formula 12.1.

4 Construct a scatter plot for each of the data sets below. Draw the regression line in each scatter plot.

(a)		(b)		(c)		(d)	
X	Y	X	Y	X	Y	X	Y
5	65	62	87	9.9	7.8	5	19.34
1	40	45	65	10.1	9.8	7	14.54
3	50	55	84	10.0	8.8	10	10.79
4	58	74	47	11.0	9.0	10	10.50
2	42	32	52	11.2	10.0	8	11.22

(a)		(b)		(c)		(d)	
X	Y	X	Y	X	Y	X	Y
		49	90	10.5	9.9	7	13.98
		88	31	10.3	8.7	9	11.30
		56	68			8	11.30
		73	76			6	15.98
		92	57			8	12.50
						9	12.89
						10	9.89
						8	12.89
$\bar{X} = 3$		$\bar{X} = 62.6$		$\bar{X} = 10.429$		$\bar{X} = 8.077$	
$\sigma_x = 1.414$		$\sigma_x = 18.122$		$\sigma_x = .465$		$\sigma_x = 1.492$	
$\bar{Y} = 51$		$\bar{Y} = 65.7$		$\bar{Y} = 9.143$		$\bar{Y} = 12.855$	
$\sigma_y = 9.466$		$\sigma_y = 18.188$		$\sigma_y = .744$		$\sigma_y = 2.514$	
$r = .986$		$r = -.397$		$r = .549$		$r = -.919$	

5 The coefficient of determination, r^2, is a measure of the amount of variability observed along the Y variable that can be explained by changes in X. Calculate the coefficient of determination for each of the data sets in Exercise 4. What percentage of the total variability in Y is accounted for by changes in X?

6 Assume that in a large sample of district sixth-graders, you found that "social maturity" and performance on a standardized achievement test were directly related, with $r = .774$. You also know the following values:

Social maturity (X)	*Achievement test scores (Y)*
$\bar{X} = 141.87$	$\bar{Y} = 526.06$
$\sigma_x = 37.287$	$\sigma_y = 94.879$

a. What score on the achievement test would you predict for someone scoring 150 in social maturity?

b. What score on the achievement test would you predict for someone scoring 115 in social maturity?

c. What percentage of the total variability in achievement test scores is accounted for by differences in social maturity?

d. If a person scores one standard deviation below the mean in social maturity, what is this person's predicted z score on the achievement test? Use Formula 12.2.

7 You are interested in the relationship between oxygen consumption and jogging performance. You asked a group of 20 joggers of normal body weight to jog as far as they could during a 50-minute period. Afterward you measured their average oxygen consumption per minute and the number of miles jogged. You obtained the following results:

Average oxygen consumption per minute (X)		*Miles jogged (Y)*
$\bar{X} = 264.7$		$\bar{Y} = 5.87$
$\sigma_x = 112.965$		$\sigma_y = 2.63$
	$r = .933$	

a. Construct a graph of the regression line for predicting the number of miles jogged from oxygen consumption.

b. How many miles would you predict someone could jog in 50 minutes if you knew the person's average oxygen consumption is 150 per minute while jogging?

c. What is the standard error of the estimate for predicting $\hat{Y}$ from X?

d. What percentage of the total variability in the number of miles jogged in 50 minutes is accounted for by knowing one's average oxygen consumption per minute?

e. Switch the X and Y labels for the two variables so that the number of miles jogged is labeled X and the average oxygen consumption per minute is Y. Construct a graph of the regression line for predicting average oxygen consumption per minute from the number of miles jogged.

f. What average oxygen consumption per minute would you predict for someone who jogs 4.5 miles in 50 minutes?

g. What average oxygen consumption per minute would you predict for someone who jogs 8 miles in 50 minutes?

h. What is the standard error of the estimate for predicting oxygen consumption from miles jogged?

13

ONE-WAY ANALYSIS OF VARIANCE (ANOVA)

CHAPTER GOALS

After studying this chapter you should be able to:

1. Describe the effect on the mean and on the variance when you add a constant to each score in a group of scores.
2. Define the between groups estimate of the variance of the sampling distribution of the mean ($s_{\bar{x}}^2$), and define the within-groups estimate of the variance of the sampling distribution of the mean.
3. Define F, and describe what the magnitude of F tells you about your sample means.
4. Perform a one-way ANOVA with the computational formulas.
5. Perform a one-way ANOVA with unequal sample sizes.
6. Construct a source table for the one-way ANOVA.
7. State the assumptions for the one-way ANOVA.

CHAPTER IN A NUTSHELL

Researchers frequently use multiple-sample or parametric designs in which samples of subjects differ either on some naturally occurring characteristics or on the level of the independent variable. The one-way analysis of variance (ANOVA) is a hypothesis-testing method enabling you to simultaneously compare the means of several different samples and decide whether an observed difference between the sample means is the result of a difference in population means, or the result of sampling error. The null hypothesis is that the means of the populations from which your samples were drawn are equal. H_0 therefore implies that the differences between your sample means are due to chance. The ANOVA involves comparing two estimates of the variance of the sampling distribution from which your sample means arose. The first estimate is computed from the actual values of your sample means, and the second estimate is calculated from the assumption that H_0 is true. The more the first estimate exceeds the second estimate, the more unlikely it is that H_0 is true. The comparison between the two estimates is made with the F ratio, which is simply the first estimate divided by the second estimate. The larger the F ratio, the more unlikely it is that H_0 is true. You can compare your F ratio with a distribution of F values that would occur by chance if H_0 is true. If the probability of observing an F value as large as yours is less than alpha, then you can reject H_0 and conclude that the population means differ.

Introduction to one-way analysis of variance

Suppose you are interested in examining the effect of motivation on performance in a learning task. You set up a study in which subjects learn a list of three-letter nonsense syllables such as GMB, TQH, MBS, HZT, and so on. You present the syllables one by one and repeat the list five times. The subject's task is to guess the next syllable in the list before it is actually presented. On the fifth presentation of the list, you measure the percentage of the syllables each subject correctly guesses. This "percentage correct" score is your measure of learning. You set up six groups and manipulate the level of motivation between the groups. One group is a control group whose subjects are simply instructed to learn the list as rapidly as possible. The subjects in each of the other five groups receive a different intensity of electric shock to the arm each time they make a wrong guess. The subjects in one of these groups receive very mild shock, those in another group receive more intense shock, and so on, up to the subjects in the last group, who receive very severe shock. You assume that motivation varies directly with shock intensity, so subjects who receive more severe shock are more motivated.

After the learning task you calculate the mean percentage correct for each group of subjects and plot each group mean as in Figure 13-1. You can see that the groups with higher levels of motivation performed better on the learning task, but only up to a point. After that point, it appears that further increases in motivation impair performance. These differences between the group means are very intriguing, but naturally you wonder whether your sample results are due to chance. Is there a true difference between the means of the populations from which your treatment groups were drawn, or are your results due to sampling error? In other words, if you manipulated motivation in the entire population of interest, would

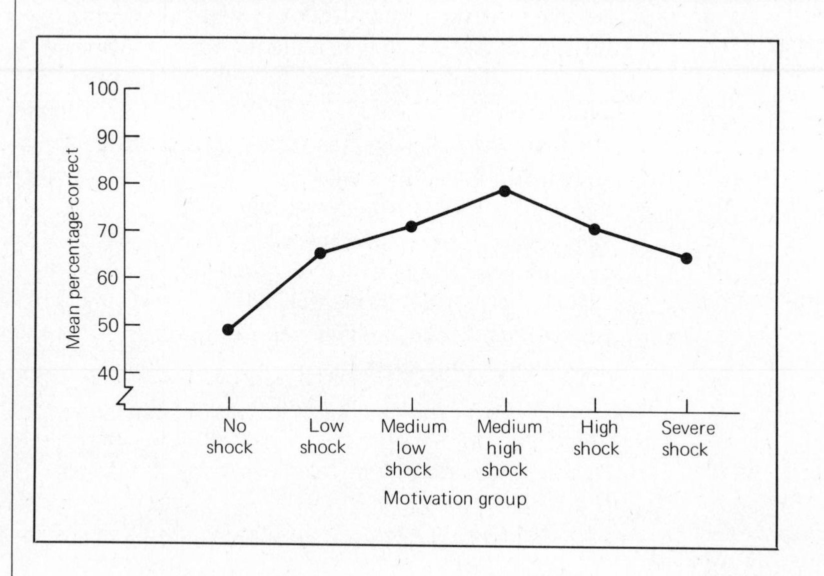

FIGURE 13-1
Mean percentage correct for each of the six motivational groups.

you actually change the population mean performance; or would your manipulation have no effect on the population mean?

Put simply, are your results due to chance or are they due to the effects of varying motivation? These two competing explanations for why your group means differ are illustrated in Figure 13-2. You can use the hypothesis-testing process to help you decide whether the differences between the group means are due to the effects of motivation or are due to sampling error.

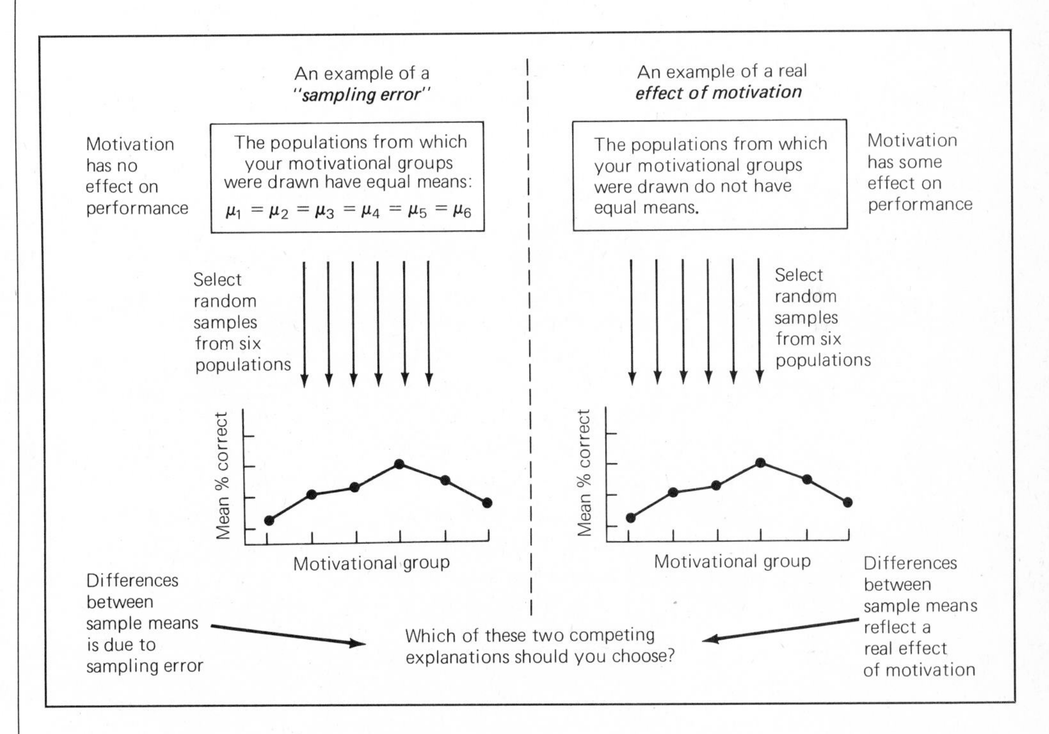

FIGURE 13-2
Two different explanations for the observed results in Figure 13-1.

So far, the only hypothesis tests that seem appropriate here are the two-sample independent t test and the Mann-Whitney U test. You could pick a pair of groups and test the difference between the two means to see whether the difference could have occurred by chance. Since you have six different groups, if you tested the difference between each possible pair you would have to perform 15 separate two-sample hypothesis tests. Needless to say, this would not be too much fun unless you had just purchased a new calculator and were looking for ways to put it to use! More important, you would not actually be answering the question you originally set out to resolve. That is, your experiment was designed to test whether changes in motivation over a range of levels have an effect on performance, but the t and U tests can only tell you whether changes in motivation between *two* levels have an effect on performance.

To see the difficulties you may run into if you initially use *t* or *U* tests when you have more than two groups, consider the following example. Suppose you test the difference in performance between Groups 3 and 5 and find that there is not a significant difference between the two means. If you look back at Figure 13-1, you can see that it would be very misleading to conclude that there is no effect of increasing motivation of Group 3 to the level of Group 5, because this ignores the possibility that increasing the motivation of Group 3 to the level of Group 4 may have actually increased performance. Another problem with performing many separate two-sample hypothesis tests on the same set of data is that you increase the likelihood of making at least one alpha error. For example, if $\alpha = .05$ in each of 20 separate two-sample hypothesis tests, then on the average you can expect to make one alpha error and erroneously report that a difference exists between population means. The more tests you perform with a given level of alpha in each test, the more likely it will be for you to make at least one alpha error.

It would be nice if you could use just one sweeping hypothesis test to determine whether changes in motivation over the entire range of motivation levels being tested have an effect on performance. This would avoid many problems associated with using many separate two-sample hypothesis tests. The analysis of variance **(ANOVA)** is a hypothesis test with the characteristics you desire. It compares the variability between all of the sample means with what you would expect to observe by chance if the population means were equal.

ANOVA

Suppose you assume the null hypothesis that the means (and variances) of the populations from which your samples were drawn are equal. That is, assume there is no treatment effect. If H_0 is true, then you would expect your sample means to differ only slightly between each other because any differences would be due only to sampling error. If you observe a very large difference between your sample means—one that is so large that you suspect it cannot be due to sampling error—then you should reject H_0 and conclude that the population means are not equal. The question is, how much difference between sample means do you need to observe before you can decide it is too much to be attributable to sampling error?

Since you have more than two sample means, an appropriate way to measure the "difference" between them is to calculate the variance of the sample means. The more the sample means differ from each other, the farther apart they are from each other, and hence the greater is their variance. The amount of variance between your sample means that you can expect to occur by chance depends on the population variance. If H_0 is true, and you are drawing your samples from populations with equal means and equal variances, then you can consider the populations to be indistinguishable on the attribute/trait being measured (the dependent variable), and it is the same as drawing the samples from one population having a given mean and variance. The greater the variance of the population, the more variance you can expect to observe between sample means. This is a consequence of the central limit theorem, which tells you the characteristics of the sampling distribution of the mean. Since the sampling distribution of the mean is the distribution of the means of samples drawn at random from a given population, it shows you what to expect from any particular set of sample means. You can consider your set of sample means to be a "sample" of sample means drawn from the sampling distribution which would occur if H_0 is true. The standard deviation of the sampling distribution of the mean is $\sigma_{\bar{x}} = \sigma/\sqrt{N}$, where σ is the population standard deviation and N is the sample size. Thus, the larger the population variance is, the larger $\sigma_{\bar{x}}$ will be. If you knew the population standard deviation, you could use the central limit theorem to infer the characteristics of the sampling distribution, and then you could determine whether or not the variance among your sample means is too large to be attributable to chance sampling error.

Unfortunately, you generally do not know the population variance. However, there is a method of estimating the population variance (and hence the standard deviation) that is unaffected by any differences in sample means or by changes in the independent variable. This method is based on the **principle of variance constancy.** The principle of variance constancy states that certain types of treatments will not affect estimates of the population variance that are made from samples having different levels of the treatment.

The principle of variance constancy

To illustrate the principle of variance constancy, suppose the scores 3, 4, 4, 6, 8 represent the observed number of errors for each of five children during a spelling test. Refer to these scores as the X_1 scores, and calculate their mean and variance:

X_1	X_1^2
3	9
4	16
4	16
6	36
8	64
$\Sigma X_1 = 25$	$\Sigma X_1^2 = 141$

$$\bar{X}_1 = \frac{\Sigma X_1}{N_1} = \frac{25}{5} = 5$$

$$(\Sigma X_1)^2 = 25^2 = 625$$

$$\sigma_1^{\,2} = \frac{\Sigma X_1^2 - \dfrac{(\Sigma X_1)^2}{N_1}}{N_1} = \frac{141 - \dfrac{625}{5}}{5}$$

$$= \frac{141 - 125}{5}$$

$$= \frac{16}{5} = 3.2$$

You find $\bar{X}_1 = 5$ and $\sigma_1^{\,2} = 3.2$.

Now suppose you add a constant to each of the X_1 scores. This could possibly occur in a research study if you purposely distracted the children while they were spelling the words. Suppose you distracted each child equally so you increased each X_1 score by 10. Refer to the resulting scores as the X_2 scores. What effect does distraction have on the mean and variance?

X_2 (or $X_1 + 10$)	$X_2^{\,2}$ [or $(X_1 + 10)^2$]
13	169
14	196
14	196
16	256
18	324
$\Sigma X_2 = 75$	$\Sigma X_2^{\,2} = 1141$

$$\bar{X}_2 = \frac{\Sigma X_2}{N_2} = \frac{75}{5} = 15$$

$$(\Sigma X_2)^2 = (75)^2$$

$$= 5625$$

$$\sigma_2^{\,2} = \frac{\Sigma X_2^{\,2} - \dfrac{(\Sigma X_2)^2}{N_2}}{N_2} = \frac{1141 - \dfrac{5625}{5}}{5} = \frac{1141 - 1125}{5}$$

$$= \frac{1141 - 1125}{5} = \frac{16}{5} = 3.2$$

Adding 10 to each score increases the mean by 10 but does not change the variance. This is an important principle that holds true whenever you increase each score in the group by a constant value.

principle of variance constancy

Adding a constant to each score in a distribution increases the mean by that constant but does not change the variance.

The principle of variance constancy becomes clearer when you consider graphically what happens to a distribution of scores when you add a constant to each score. Consider the distribution of scores in Figure 13-3. Imagine

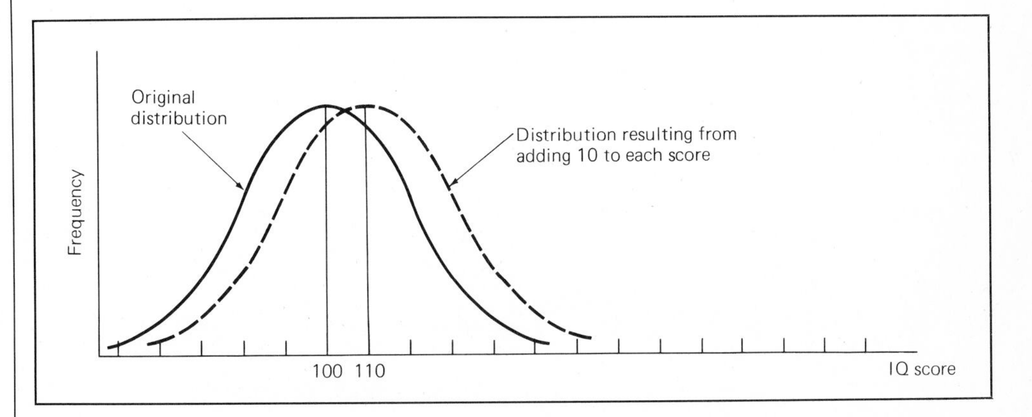

FIGURE 13-3
Adding a constant to each score in a distribution changes the mean but not the variance.

that this represents the frequency distribution of the IQs of a large group of people eating at a local hamburger stand. If you add a constant to each person's IQ, as might happen if you add a miracle IQ-enhancing secret sauce to the hamburger, you will move each person in the graph over to the right by an equal amount. For example, if the secret sauce has the effect of increasing everyone's IQ by 10 points, then everyone will shift to the right by 10 units. The mean is thus increased by 10 because the entire distribution will shift to the right by 10 units. Notice that the spread of the distribution remains the same. Since everyone is shifted over by the same amount, the distance in IQ points that separates one person from the next remains the same. For example, a person with an initial IQ of 100 becomes one with an IQ of 110, and a person with an initial IQ of 130 becomes one with an IQ of 140. These two people remain 30 points apart.

Although we illustrated the principle of variance constancy with the formula for calculating the population variance, σ^2, the principle remains the same if you have sample data and you estimate the population variance with the formula for s^2. Remember, the only difference between σ^2 and s^2 is that $N - 1$ appears in the denominator of the formula for s^2, whereas N alone appears in the denominator of the formula for σ^2. Neither σ^2 nor s^2 changes when a constant is added to each score because their formulas are based on the distances of the scores from the mean of the distribution, and these distances do not change when a constant is added to each score.

Applying the principle of variance constancy for measuring the strength of an effect

You can use the principle of variance constancy to estimate the population variance, which will then allow you to determine how much variability between sample means you should expect if H_0 is true. The principle of variance constancy tells you that treatments having a constant effect on each subject do not affect the variance within each population, nor do such treatments affect the variance within each sample drawn from the populations. Most important, treatments that have a constant effect on each subject do not affect the estimate of the population variance made from each sample. Thus you can use your samples to estimate the population variance, and this estimate will not be affected by any possible treatment effects. There is one small problem, however, which is easily overcome. For example, consider the motivation-performance study that was used to introduce this chapter. You have six different samples, and hence you can get six different estimates of the population variance (one from each sample). Since the sample sizes are all equal, each of these estimates is as good as another. Do you remember when you had a similar problem in estimating the population variance with the two-sample independent t test? There you had two independent samples from which to estimate σ. You can adopt the same solution here as you did there—you can average the estimates. The raw data for the motivation-performance study is given below, along with the steps to calculate σ^2. First use Formula 8.2 to calculate s^2, an estimate of the population variance from each sample. Then average these estimates to get an overall estimate of the population variance as follows:

Raw data

Group 1: No Shock X_1	*Group 2: Low Shock* X_2	*Group 3: Medium Low Shock* X_3	*Group 4: Medium High Shock* X_4	*Group 5: High Shock* X_5	*Group 6: Severe Shock* X_6
38	64	55	87	83	47
26	85	79	88	76	55
70	81	52	66	80	68
55	43	83	58	67	79
37	44	80	60	88	67
55	49	90	97	82	76
68	78	85	98	71	78
55	74	59	67	50	66
30	53	80	81	69	46
52	81	53	82	44	41
$\bar{X}_1 = 48.6$	$\bar{X}_2 = 65.2$	$\bar{X}_3 = 71.6$	$\bar{X}_4 = 78.4$	$\bar{X}_5 = 71.0$	$\bar{X}_6 = 62.3$

Now use Formula 8.2 to calculate s^2 for each group; that is,

$$s^2 = \frac{\Sigma X^2 - \dfrac{(\Sigma X)^2}{N}}{N - 1}$$

	Group 1: X_1	X_1^2	*Group 2:* X_2	X_2^2	*Group 3:* X_3	X_3^2	*Group 4:* X_4	X_4^2	*Group 5:* X_5	X_5^2	*Group 6:* X_6	X_6^2
	38	1444	64	4096	55	3025	87	7569	83	6889	47	2209
	26	676	85	7225	79	6241	88	7744	76	5776	55	3025
	70	4900	81	6561	52	2704	66	4356	80	6400	68	4624
	55	3025	43	1849	83	6889	58	3364	67	4489	79	6241
	37	1369	44	1936	80	6400	60	3600	88	7744	67	4489
	55	3025	49	2401	90	8100	97	9409	82	6724	76	5776
	68	4624	78	6084	85	7225	98	9604	71	5041	78	6084
	55	3025	74	5476	59	3481	67	4489	50	2500	66	4356
	30	900	53	2809	80	6400	81	6561	69	4761	46	2116
	52	2704	81	6561	53	2809	82	6724	44	1936	41	1681
$\Sigma X =$	486		652		716		784		710		623	
$(\Sigma X)^2 =$	236,196		425,104		512,656		614,656		504,100		388,129	
$\Sigma X^2 =$		25,692		44,998		53,274		63,420		52,260		40,601

For Group 1:

$$s_1^2 = \frac{X_1^2 - \dfrac{(\Sigma X_1)^2}{N_1}}{N_1 - 1} = \frac{25{,}692 - \dfrac{236{,}196}{10}}{10 - 1} = 230.27$$

Similarly, for Group 2:

$$s_2^2 = \frac{44{,}998 - \dfrac{425{,}104}{10}}{10 - 1} = 276.4$$

For Group 3:

$$s_3^2 = \frac{53{,}274 - \dfrac{512{,}656}{10}}{10 - 1} = 223.16$$

For Group 4:

$$s_4^2 = \frac{63{,}420 - \dfrac{614{,}656}{10}}{10 - 1} = 217.16$$

For Group 5:

$$s_5^2 = \frac{52{,}260 - \dfrac{504{,}100}{10}}{10 - 1} = 205.56$$

For Group 6:

$$s_6^2 = \frac{40{,}601 - \dfrac{388{,}129}{10}}{10 - 1} = 198.68$$

Now average the above six values of s^2 to get your overall estimate, symbolized s^2_{WG}:

$$s^2_{WG} = \frac{s_1^2 + s_2^2 + s_3^2 + s_4^2 + s_5^2 + s_6^2}{6}$$

$$= \frac{230.27 + 276.4 + 223.16 + 217.16 + 205.56 + 198.68}{6}$$

$$= \frac{1351.23}{6} = 225.2$$

within-groups estimate of σ^2

s^2_{WG}

The overall estimate, symbolized s^2_{WG}, is called the **within-groups estimate of σ^2** because it is calculated by finding the variance within each group and then averaging.

Once you have s^2_{WG} as an estimate of the population variance, you can estimate the variance of the sampling distribution of the mean. The central limit theorem tells you that the standard deviation of the sampling distribution, $\sigma_{\bar{x}}$, is equal to $\sigma/\sqrt{N}$. If you square both sides of the expression $\sigma_{\bar{x}} = \sigma/\sqrt{N}$, then you will find that the *variance* of the sampling distribution, $\sigma_{\bar{x}}^2$, is equal to σ^2/N. If you substitute s^2_{WG}, your estimate of σ^2, into this expression you get an estimate of $\sigma_{\bar{x}}^2$:

1. Given the expression $\sigma_{\bar{x}}^2 = \frac{\sigma^2}{N}$,

2. since the estimate of $\sigma^2 = s^2_{WG}$,

3. then the estimate of $\sigma_{\bar{x}}^2 = \frac{s^2_{WG}}{N}$.

within-groups estimate of $\sigma_{\bar{x}}^2$

We call this the **within-groups estimate of $\sigma_{\bar{x}}^2$** because it is calculated from the within-groups estimate of σ^2. For the motivation-performance example, since $s^2_{WG} = 225.2$ and N, the size of each group, is 10, then the within-groups estimate of $\sigma_{\bar{x}}^2$ is

$$\frac{s^2_{WG}}{N} = \frac{225.2}{10} = 22.52$$

Now that you have an estimate of the variance of the sampling distribution of the mean, what can you say about your observed sample means? Could they have arisen by chance from such a distribution? If there is considerable variability between your sample means, then it is reasonable to conclude that they did not come from the sampling distribution whose variance you just estimated by assuming H_0 that the population means are equal. This would lead to the rejection of H_0. On the other hand, if the sample means are reasonably close together, then they might have come from the sampling distribution specified under H_0, and you should retain H_0. How do you decide whether the sample means came from the specified sampling distribution?

One way to help you answer this question is to use your sample means to estimate the variance of the sampling distribution from which they came. If the value of this estimate is fairly close to your within-groups estimate of $\sigma_{\bar{x}}^2$ calculated by assuming H_0, then it is reasonable to conclude that your sample means came from the sampling distribution specified under H_0. Thus you can retain H_0 and conclude that there is no reason to say the population means differ. However, if your estimate of the variance of the sampling distribution is much larger than your within-groups estimate of $\sigma_{\bar{x}}^2$, then your sample means probably did not come from the specified sampling distribution and you should reject H_0 and conclude that the population means differ. How do you use your sample means to obtain such an estimate of the variance of the sampling distribution?

Consider that your six sample means for the motivation-performance example arose by chance from some sampling distribution of the mean. That is, they are a sample of sample means from the sampling distribution of the mean. This is suggested in Figure 13-4. You already know how

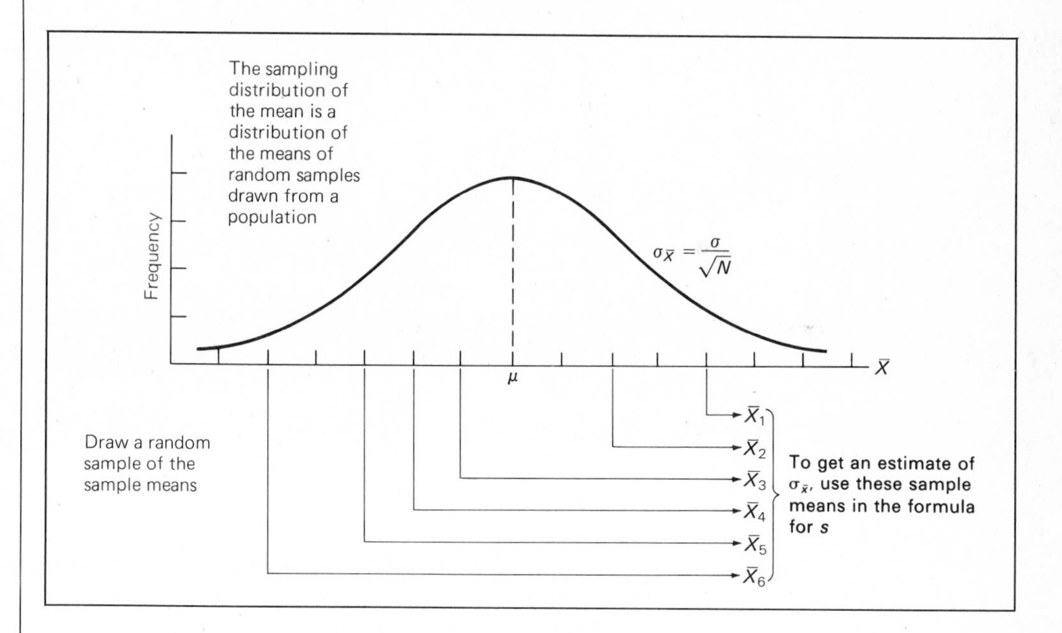

FIGURE 13-4
In deriving the between-groups variance estimate, group means are considered as a sample from the sampling distribution of the mean.

to use Formula 8.2 for s^2 to estimate the variance of a population when you have a sample of scores. You can similarly estimate the variance of the sampling distribution from your sample of sample means. All you have to do is adapt Formula 8.2. Instead of individual X scores, you generally will have K number of treatment groups and hence K sample means. Therefore substitute $\bar{X}$ for X and K for N in Formula 8.2 as shown below:

$s_{\bar{x}}^2$

$$s^2 = \frac{\Sigma X^2 - \frac{(\Sigma X)^2}{N}}{N - 1} \quad \text{becomes} \quad s_{\bar{x}}^2 = \frac{\Sigma \bar{X}^2 - \frac{(\Sigma \bar{X})^2}{K}}{K - 1}$$

between-groups estimate

$s_{\bar{x}}^2$ is called the **between-groups estimate** of the variance of the sampling distribution of the mean because it is calculated using the means of all the groups. You can now calculate $s_{\bar{x}}^2$ for the motivation-performance example. Simply take the six sample means and calculate the variance between them with the adapted form of Formula 8.2, just as you would do with individual raw scores. The computation is shown below:

	$\bar{X}$	$\bar{X}^2$
sample mean for Group 1 =	48.6	2361.96
sample mean for Group 2 =	65.2	4251.04
sample mean for Group 3 =	71.6	5126.56
sample mean for Group 4 =	78.4	6146.56
sample mean for Group 5 =	71.0	5041.00
sample mean for Group 6 =	62.3	3881.29
$\Sigma \bar{X}$ =	397.1	$\Sigma \bar{X}^2 = 26{,}808.41$

$$(\Sigma \bar{X})^2 = (397.1)^2 = 157{,}688.41$$

$$s_{\bar{x}}^2 = \frac{\Sigma \bar{X}^2 - \frac{(\Sigma \bar{X})^2}{K}}{K - 1} = \frac{26{,}808.41 - \frac{157{,}688.41}{6}}{6 - 1}$$

$$= \frac{26{,}808.41 - 26{,}281.40}{5}$$

$$= \frac{527.01}{5} = 105.4$$

Now you are in a position to compare your value of $s_{\bar{x}}^2$ with the within groups estimate of $\sigma_{\bar{x}}^2$ you calculated by assuming H_0 is true. If H_0 is true, the variance of the sampling distribution should be 22.52 according to your estimate. However, the estimate of the variance of the sampling distribution from which your sample means arose is 105.4. Does this immediately tell you that H_0 is false? No, because both of your figures are estimates, and so each is subject to error, namely, sampling error. Well, then, just how much error is possible, or likely? If the two different estimates were the same, or very close, you would probably dismiss any difference as due to sampling error and retain the H_0 that the populations from which your samples were drawn have the same means. But if the estimates are extremely different from one another, then you would be unwilling to attribute the difference to sampling error. Is the difference between 22.52 and 105.4 too extreme to be attributable to sampling error? How much discrepancy between the two estimates can there be before you should decide that the discrepancy is not due to sampling error?

There is an objective way of deciding how much discrepancy between estimates could reasonably be due to sampling error. This method involves comparing your two estimates of the variance of the sampling distribution in a ratio known as the ***F* ratio,** or simply ***F***. The F ratio is the between-groups estimate of $\sigma_{\bar{x}}^2$ divided by the within-groups estimate of $\sigma_{\bar{x}}^2$. Thus, the **definitional formula for *F*** is:

F* ratio, *F

FORMULA 13.1
definitional formula for *F*

$$F = \frac{\text{between-groups estimate of } \sigma_{\bar{x}}^2}{\text{within-groups estimate of } \sigma_{\bar{x}}^2} = \frac{s_{\bar{x}}^2}{\dfrac{s_{\text{WG}}^2}{N}}$$

Therefore the F ratio is your estimate of the variance of the sampling distribution from which your sample means came, divided by your estimate of the variance of the sampling distribution if H_0 is true. The estimate in the denominator is influenced only by sampling error and not by differences in the population means or by treatment effects.[1] On the other hand, the estimate in the numerator is influenced not only by sampling error but also by possible differences in the population means, or treatment effects. Thus F is comprised of an estimate in the numerator which is influenced by sampling error *and* treatment effects, and an estimate in the denominator which is influenced only by sampling error. When F is equal to 1, both estimates are the same, so you should retain H_0 and conclude that there is not sufficient evidence to say the population means differ. Your estimate of the variance of the sampling distribution from which your sample means came is the same as your estimate of the variance if H_0 is true. There is no reason to think there are any treatment effects. When F is less than 1, the within-groups estimate is greater than the between-groups estimate, and this also indicates that you should retain H_0, since there is no indication that there are any treatment effects. An

[1]A treatment effect is an effect on the dependent variable due to the introduction of different treatments, or levels of the independent variable, in the groups of a study. When there is a treatment effect, then the means of the populations from which your different treatment groups were drawn are different.

F value less than 1 indicates that the sample means are close together, and the logical inference is that the population means are not different. In contrast, only when F is greater than 1 is the between-groups estimate larger than the within-groups estimate. This may provide evidence that the population means differ. The more the between-groups estimate exceeds the within-groups estimate, the larger F will be and the larger is the possibility that there is some treatment effect. How much larger than 1 does F have to be before you can decide that it is too extreme to be due to differences in sampling error alone? Putting it another way, how much larger than 1 does F have to be before you can conclude that the population means differ? For the motivation-performance study, since $s_{\bar{x}}^2 = 105.4$ and $s_{WG}^2/N = 22.52$, then

$$F = \frac{s_{\bar{x}}^2}{\dfrac{s_{WG}^2}{N}} = \frac{105.4}{22.52} = 4.68.$$

What can you say about such an F value? As with all the other hypothesis-testing procedures you have examined so far, you can compare your F value with a distribution of F values that would occur by chance assuming H_0 is true.

F distribution

F^*

The distribution of F values which would occur by chance if H_0 is true is called the **F distribution.** The characteristics of the F distribution can be inferred mathematically. Critical values of F, symbolized **F^***, can be mathematically determined such that any F value that is greater than F^* has a probability of occurring by chance that is less than alpha. Three important factors determine the value of F^*: the level of alpha; the number of subjects in each group; and the number of groups. First, as the level of alpha is made more stringent by going from .05 to .01, F^* increases because you are requiring your sample results to be more extreme before you reject H_0. Second, as the number of subjects in each group increases, F^* decreases. This is reflected in the central limit theorem, which states that the larger the sample size is, the less variability there will be in the sampling distribution of the mean. Thus the larger the sample size, the less variability you should expect by chance between your sample means, and the smaller the values of F that occur by chance should be. Third, as the number of groups increases, the more sample means there are and hence the better an estimate of the sampling distribution of the mean you have. Thus you expect F^* to be smaller since you expect less chance variation.

The critical values of F are given in Table F^*. The critical values are listed according to *two* different calculations of *degrees of freedom,* which enables you to find the appropriate value of F^* for your particular number of subjects and number of groups and also allows you to use the table for different types of analyses of variance other than the particular

numerator $df = K - 1$

denominator $df = K(N - 1)$

one-way ANOVA discussed here. One calculation of degrees of freedom reflects the number of groups and relates to the numerator of the F ratio, your between-groups estimate $s_{\bar{x}}^2$. The other calculation of degrees of freedom reflects the number of subjects and relates to the denominator, your within-groups estimate of $\sigma_{\bar{x}}^2$. The **degrees of freedom for the numerator** are equal to $K - 1$, where K is the number of groups you have. The **degrees of freedom for the denominator** are $K(N - 1)$, where N is the number of subjects within each equal-sized group. You can see in Table F^* that the *df* values for the numerator are listed across the top of the table and that the *df* values for the denominator are listed down the extreme left-hand column. The relevant F^* value lies at the row-column intersection of the degrees of freedom for the numerator and denominator. Two critical values are listed at each row-column intersection. The one in lightface type is the critical value for $\alpha = .05$ and the one in boldface type is the critical value for $\alpha = .01$. To illustrate the use of Table F^*, consider the motivation-performance data. Since there are six groups, the *df* for the numerator is: $df = K - 1 = 6 - 1 = 5$. For the denominator, since there are ten subjects in each group: $df = K(N - 1) = 6(10 - 1) = 6(9) = 54$. Setting alpha at .05, look across the top of the table and find $df = 5$ for the numerator, and then go down the leftmost column and try to find $df = 54$ for the denominator. You will notice that $df = 54$ is not listed. Whenever this occurs, select the listed *df* value that is closest to your calculated *df* value, and which gives you the *larger* value for F^*. This is a standard conservative procedure that makes it harder to reject H_0 and thus reduces the chance of an alpha error. Therefore, you intersect your $df = 5$ with a $df = 50$, and the lightface value gives you $F^* = 2.40$. This means that when H_0 is true, if you were to replicate the study 100 times using new samples, you would observe a value of F greater than 2.40 only 5 times out of the 100 replications.

You can now compare your observed F of 4.68 in the motivation-performance study with the F^* value of 2.40. Since $F = 4.68$ is greater than $F^* = 2.40$, the probability of obtaining your F value by chance alone is less than .05. Therefore you can reject H_0 and conclude that changes in motivation do have an effect on performance.

When you reject H_0 as you have just done, you only know that the population means are probably not equal. It is possible that this is due to a single mean being different from all the others, which themselves are equal. Alternately, there may be two means that differ, or three, or they may all be different from each other. With the simple ANOVA you cannot tell which of these situations is most likely. There are procedures for answering the question of which mean or means are different from the others. These procedures involve comparing the means by pairs, as with several two-sample t tests. However, these procedures compensate for the increased tendency to make an alpha error when performing multiple paired comparisons. These procedures, such as Scheffe's procedure and

Tukey's procedure, are beyond the scope of this book. The interested reader is invited to study them in advanced texts on the ANOVA.

The hypothesis-testing procedure using the one-way ANOVA for equal size groups may be summarized:

1. Determine H_0 and H_A. (H_0 generally will be that your samples were drawn from populations having equal means. Symbolically, this is $\mu_1 = \mu_2 = \mu_3$, etc. H_A will then be that your samples were drawn from populations with different means.) Assume H_0 is true.
2. Set alpha and calculate $df = K - 1$ for the numerator and $df = K(N - 1)$ for the denominator, where K is the number of groups and N is the number of subjects in each group. For your alpha and df's, determine the critical value of F from Table F^*.
3. Calculate an F ratio for your data by using Formula 13.1:

$$F = \frac{\text{between-groups estimate of } \sigma_{\bar{x}}^2}{\text{within-groups estimate of } \sigma_{\bar{x}}^2} = \frac{s_{\bar{x}}^2}{\frac{s_{\text{WG}}^2}{N}}$$

In the numerator, $s_{\bar{x}}^2$ is an estimate of the variance of the sampling distribution from which your sample means arose. It is calculated using the formula below:

$$s_{\bar{x}}^2 = \frac{\Sigma\bar{X}^2 - \frac{(\Sigma\bar{X})^2}{K}}{K - 1}$$

This is the same formula as Formula 8.2 but with $\bar{X}$ substituted for X, and K substituted for N. (K is equal to the number of $\bar{X}$'s, just as N is equal to the number of X's.)

In the denominator of the F ratio, your within-groups estimate of $\sigma_{\bar{x}}^2$ is obtained by first calculating the population variance estimates within each group using the standard Formula 8.2:

$$s^2 = \frac{\Sigma X^2 - \frac{(\Sigma X)^2}{N}}{N - 1}$$

You average these separate estimates from each group to get a within-group estimate of σ^2, and then you divide this result by N to get the within-groups estimate of $\sigma_{\bar{x}}^2$:

$$\frac{\text{within-groups estimate of } \sigma^2}{N} = \frac{s_{\text{WG}}^2}{N}$$

4. If $F > F^*$, reject H_0 and conclude that the population means differ. If $F \leq F^*$, retain H_0 and conclude that there is not sufficient evidence to say your population means differ.

Assumptions of the one-way ANOVA

There are many different types of analysis of variance. The selection of the appropriate type depends primarily on the number of independent variables being manipulated and on how subjects are assigned to groups. For example, subjects in each group may be either independent or dependent (matched) subjects. The type of ANOVA presented in this chapter is called a one-way ANOVA because it is applied to research situations in which *one* independent variable is manipulated. Also, in this type of ANOVA the subjects in each group are independent. Consequently, this type of one-way ANOVA is appropriate in research situations in which the independent t test might be applied if you had only two independent groups. When more than two independent groups are being compared you should use the one-way ANOVA rather than several independent t tests.

The assumptions that should be met before the one-way ANOVA for independent samples is used are as follows:

1. The population distributions are normal.
2. The variances of the population distributions are homogeneous; that is,

$$\sigma_1^2 = \sigma_2^2 = \sigma_3^2 = \cdots$$

3. The subjects comprising each sample are selected randomly, and the samples are independent.
4. The means and variances of the populations must be independent; that is, any treatment effects must be approximately like adding a constant to each score in a group.

Assumptions 1, 2, and 3 are extensions of the assumptions for the independent t test. The first assumption requires that the populations be normal. This assumption allows you to compensate for the error introduced by your estimation of the variance of the sampling distribution of the mean. The error is compensated for in the tabled critical values of F. The second assumption of homogeneity of variance is necessary in order to use the separate variance estimates from each group and average them to provide a better estimate of the population variance. If the population variances are not equal, it would make little sense to average your sample variance estimates. The third assumption requires the subjects participating in your study to be randomly selected and randomly assigned to the different treatment groups. The subjects in each group must be different subjects and not matched on any trait. The fourth assumption requires the variance of your treatment populations to remain approximately unchanged, even though the treatment may cause an increase or decrease in the group means. This assumption assures you that any effect the treatment might have will be approximately like adding a constant.

robust

The assumptions for the one-way ANOVA are pliable and subject to a little bending. Like the t tests, the analysis of variance is **robust.** Robustness refers to Assumptions 1, 2, and 4 above, where it is common practice to apply the ANOVA even if your populations are not exactly normally distributed, or if the treatment groups do not have quite the same variance, or if your treatment is not quite like adding a constant. But you must not violate the third assumption. Your sampling must be random and the subjects in each sample must be different and not matched in any way.

Computational formulas for the one-way ANOVA

It is possible to derive computational formulas for the one-way ANOVA that make things a lot easier. Take Formula 13.1 for the F ratio:

$$F = \frac{s_{\bar{x}}^2}{\frac{s_{\mathrm{WG}}^2}{N}}$$

Now multiply the numerator and denominator of the F ratio by N to obtain the transformed F ratio:

$$F = \frac{Ns_{\bar{x}}^2}{s_{\mathrm{WG}}^2}$$

This transformed version of the F ratio gives a new interpretation to the ANOVA. In the numerator you actually have an estimate of the variance of the *population* from which your samples occurred. $Ns_{\bar{x}}^2$ is an estimate of the population variance, σ^2, because the central limit theorem states $\sigma_{\bar{x}} = \sigma/\sqrt{N}$. Squaring this yields $\sigma_{\bar{x}}^2 = \sigma^2/N$, and multiplying both sides by N gives $N\sigma_{\bar{x}}^2 = \sigma^2$. Since $s_{\bar{x}}^2$ is an estimate of $\sigma_{\bar{x}}^2$, then $Ns_{\bar{x}}^2$ is an estimate of σ^2. This estimate of σ^2 is based on the sample means. The greater the variance between the sample means, the greater this estimate will be. In the denominator of the transformed F ratio you have a within-groups estimate of σ^2. Remember, this is an estimate of σ^2 if H_0 is true. It is not based on the sample means and hence is not influenced by differences between sample means. Therefore, there is another interpretation of the ANOVA other than the one below that states you are estimating the variance of the *sampling distribution* from which your sample means were drawn:

$$F = \frac{\text{between-groups estimate of } \sigma_{\bar{x}}^2}{\text{within-groups estimate of } \sigma_{\bar{x}}^2} = \frac{s_{\bar{x}}^2}{\frac{s_{\mathrm{WG}}^2}{N}}$$

This new interpretation is that you are comparing an estimate of the variance of the *population* from which your samples occurred if H_0 is true with an estimate of the variance of the *population* which would give rise to your sample means:

$$F = \frac{\text{between-groups estimate of } \sigma^2}{\text{within-groups estimate of } \sigma^2}$$

These two interpretations are mathematically and logically equivalent.

In developing the computational formulas for the ANOVA we will use the transformed version of the F ratio and we will derive computational formulas for: the numerator, which we call the between-groups estimate of σ^2; and the denominator, which we will call the within-groups estimate of σ^2:

FORMULA 13.2
computational formula for F

$$F = \frac{\text{between-groups estimate of } \sigma^2}{\text{within-groups estimate of } \sigma^2} = \frac{Ns_{\bar{x}}^2}{s_{\text{WG}}^2}$$

When you previously calculated the within-groups estimate of σ^2 in the denominator of Formula 13.1 for the F ratio, the method you used was to sum the separate s^2 values computed within each group and average them. The general procedure for this can be written as a formula:

$$\text{within-groups estimate of } \sigma^2 = s_{\text{WG}}^2 = \frac{s_1^2 + s_2^2 + s_3^2 + \cdots}{K}$$

Here, K is the number of treatment groups, and s^2 for each group is given by Formula 8.2:

$$s^2 = \frac{\Sigma X^2 - \frac{(\Sigma X)^2}{N}}{N - 1}$$

If you take each s^2 term in the above formula for the within-groups estimate of σ^2, and replace s^2 with the right-hand term of Formula 8.2, you can eventually simplify things and produce the following computational formula for the denominator of the F ratio in Formula 13.2:

$$\text{within-groups estimate of } \sigma^2 = s_{\text{WG}}^2 = \frac{\Sigma\Sigma X^2 - \frac{\Sigma(\Sigma X)^2}{N}}{K(N - 1)}$$

The double summation sign, $\Sigma\Sigma$, indicates that you should add things up in two steps. The first summation sign is an instruction to "add up" the terms that follow it. Thus $\Sigma\Sigma X^2$ means to first "add up" the squared scores within each group to get as many ΣX^2 terms as you have groups and then to add up these sums. The $\Sigma(\Sigma X)^2$ term means to first "add up'' the scores within each group to get as many ΣX terms as you have groups, square these sums, and then add up the resulting squares. These verbal definitions of the double summation sign will become clearer with the numerical example below. With both the terms $\Sigma\Sigma X^2$ and $\Sigma(\Sigma X)^2$, you might say you are "summing" the sums of the "squared" terms. Consequently, the numerator of your within groups estimate of σ^2 is often called the **"sum of squares"** and is abbreviated SS.

sum of squares, SS

When you previously computed the numerator of the F ratio in Formula 13.1, you had to first compute $s_{\bar{x}}^2$ using the formula:

$$s_{\bar{x}}^2 = \frac{\Sigma \bar{X}^2 - \dfrac{(\Sigma \bar{X})^2}{K}}{K - 1}$$

When you take each $\bar{X}$ term in the above formula for $s_{\bar{x}}^2$, and replace $\bar{X}$ with the right-hand side of the formula for the mean, $\bar{X} = \Sigma X/N$, you can eventually simplify the formula for $s_{\bar{x}}^2$ to:

$$s_{\bar{x}}^2 = \frac{\Sigma(\Sigma X)^2}{N^2(K-1)} - \frac{(\Sigma\Sigma X)^2}{KN^2(K-1)}$$

To get the numerator of the transformed F ratio expressed in Formula 13.2, all you have to do is multiply $s_{\bar{x}}^2$ by N:

$$\text{between-groups estimate of } \sigma^2 = Ns_{\bar{x}}^2$$

$$= N\left[\frac{\Sigma(\Sigma X)^2}{N^2(K-1)} - \frac{(\Sigma\Sigma X)^2}{KN^2(K-1)}\right] = \frac{\Sigma(\Sigma X)^2}{N(K-1)} - \frac{(\Sigma\Sigma X)^2}{KN(K-1)}$$

After this derivation of the computational formulas for the between- and within-groups estimates of σ^2, you should be rightfully upset when you read the postscript, "however, in practice, these computational formulas are not used in this form." Actually, though, things get even better. If you look at the denominator of the computational formula for the within-groups estimate of σ^2, you will notice the term $K(N - 1)$, which is precisely the degrees of freedom associated with the denominator of the F ratio. Therefore, your within-groups estimate of σ^2 is equal to the "sum of squares" divided by the degrees of freedom. The computational formula which is used in practice is for only the "sum of squares," or the numerator of the within-groups estimate of σ^2. That is, the degrees of freedom are mathematically factored out. Since the within-groups computational formula you will use deals only with the numerator sum of squares, it is called the **sum of squares within groups,** and is abbreviated SS_{WG};

FORMULA 13.3
computational formula for sum of squares within groups, SS_{WG}

$$SS_{WG} = \Sigma\Sigma X^2 - \frac{\Sigma(\Sigma X)^2}{N}$$

If you use Formula 13.3 to compute SS_{WG}, then all you have to do to get the within-groups estimate of σ^2 is to divide SS_{WG} by *df:*

$$\text{within-groups estimate of } \sigma^2 = s^2_{WG}$$

$$= \frac{SS_{WG}}{df} = \frac{SS_{WG}}{K(N-1)} = \frac{\Sigma\Sigma X^2 - \dfrac{\Sigma(\Sigma X)^2}{N}}{K(N-1)}$$

You can use a similar procedure and factor out the degrees of freedom in the computational formula for the between-groups estimate of σ^2. The degrees of freedom, $K - 1$, associated with the numerator of the F ratio is a factor in the denominators of both terms in the between-groups estimate of σ^2. Factoring K − 1 out leaves the **sum of squares between groups, or SS_{BG}**, as the computational formula to use:

FORMULA 13.4
computational formula for sum of squares between groups, SS_{BG}

$$SS_{BG} = \frac{\Sigma(\Sigma X)^2}{N} - \frac{(\Sigma\Sigma X)^2}{KN}$$

After you have computed SS_{BG}, simply divide by *df* to derive the between-groups estimate of σ^2:

$$\text{between-groups estimate of } \sigma^2 = Ns_{\bar{x}}^2$$

$$= \frac{SS_{BG}}{df} = \frac{SS_{BG}}{K-1} = \frac{\Sigma(\Sigma X)^2}{N(K-1)} - \frac{(\Sigma\Sigma X)^2}{KN(K-1)}$$

The reason the degrees of freedom are mysteriously eliminated from the computational formulas for the within-groups and between-groups estimates of σ^2 is because the remaining terms, SS_{WG} and SS_{BG}, are encountered in the descriptions of many different types of analysis of variance, and these terms are employed in different ways for different purposes.

The use of computational Formulas 13.3 and 13.4 is illustrated below for the data concerning the effect of motivation on performance:

	Group 1:		Group 2:		Group 3:		Group 4:		Group 5:		Group 6:	
	X_1	X_1^2	X_2	X_2^2	X_3	X_3^2	X_4	X_4^2	X_5	X_5^2	X_6	X_6^2
	38	1444	64	4096	55	3025	87	7569	83	6889	47	2209
	26	676	85	7225	79	6241	88	7744	76	5776	55	3025
	70	4900	81	6561	52	2704	66	4356	80	6400	68	4624
	55	3025	43	1936	83	6889	58	3364	67	4489	79	6241
	37	1369	44	1936	80	6400	60	3600	88	7744	67	4489
	55	3025	49	2401	90	8100	97	9409	82	6724	76	5776
	68	4624	78	6084	85	7225	98	9604	71	5041	78	6084
	55	3025	74	5476	59	3481	67	4489	50	2500	66	4356
	30	900	53	2809	80	6400	81	6561	69	4761	46	2116
	52	2704	81	6561	53	2809	82	6724	44	1936	41	1681
$\Sigma X =$	486		652		716		784		710		623	
$(\Sigma X)^2 =$	236,196		425,104		512,656		614,656		504,100		388,129	
$\Sigma X^2 =$		25,692		44,998		53,274		63,420		52,260		40,601

$$\Sigma\Sigma X = 486 + 652 + 716 + 784 + 710 + 623 = 3971$$

$$(\Sigma\Sigma X)^2 = (3971)^2 = 15,768,841$$

$$\Sigma(\Sigma X)^2 = 236,196 + 425,104 + 512,656 + 614,656 + 504,100 + 388,129 = 2,680,841$$

$$\Sigma\Sigma X^2 = 25,692 + 44,998 + 53,274 + 63,420 + 52,260 + 40,601 = 280,245$$

The things you need to find to "plug in" values into the computational formulas for SS_{WG} and SS_{BG} are ΣX, $\Sigma\Sigma X$ and $(\Sigma\Sigma X)^2$, $(\Sigma X)^2$ and $\Sigma(\Sigma X)^2$, and ΣX^2 and $\Sigma\Sigma X^2$. First calculate ΣX by adding up all the scores within each individual group. You obtain six separate values of ΣX because you have six groups. $\Sigma\Sigma X$ then tells you to sum the six values of ΣX. You then get the term $(\Sigma\Sigma X)^2$ by squaring $\Sigma\Sigma X$. You can derive the term $\Sigma(\Sigma X)^2$ by first squaring each of the separate values of ΣX to get several $(\Sigma X)^2$ values, and then adding these up. Next, derive the several values of ΣX^2 by squaring each score and adding them up for each individual group. Then compute the term $\Sigma\Sigma X^2$ by summing the six values of ΣX^2. Now plug the values of these terms into the computational formulas to give:

$$SS_{BG} = \frac{\Sigma(\Sigma X)^2}{N} - \frac{(\Sigma\Sigma X)^2}{KN} = \frac{2,680,841}{10} - \frac{15,768,841}{6(10)}$$

$$= 268,084.1 - 262,814.02$$

$$= 5270.08$$

$$SS_{WG} = \Sigma\Sigma X^2 - \frac{\Sigma(\Sigma X)^2}{N} = 280,245 - \frac{2,680,841}{10}$$

$$= 280{,}245 - 268{,}084.1$$
$$= 12{,}160.9$$

source table

The usual way to use the computational formulas in the analysis of variance is to construct a **source table.** A source table is a convenient way to summarize the final stages of your data analysis, and is often printed in research reports. Here is an example of a one-way ANOVA source table for the study of the effect of motivation on performance:

Source	*SS*	*df*	$MS = \frac{SS}{df}$	$F = \frac{MS_{BG}}{MS_{WG}}$
BG	527.08	$K - 1 = 6 - 1 = 5$	$\frac{5270.08}{5} = 1054$	$\frac{1054}{225.2} = 4.68$
WG	12,160.9	$K(N - 1) = 6(10 - 1) = 54$	$\frac{12{,}160.9}{54} = 225.2$	

MS

mean squares

MS_{BG}

MS_{WG}

The first column lists the source of the variance estimates: the between-groups estimate (BG) and the within-groups estimate (WG). The between-groups estimate is listed first, at the top. The second column lists the sum of squares (SS) values for SS_{BG} and SS_{WG} obtained from the computational formulas. The third column lists the degrees of freedom for both sources. The fourth column, labeled **MS,** lists the between-groups and within-groups variance estimates of the population variance in the appropriate row. MS is the abbreviation for **mean squares,** which is merely a very common synonym for the "variance estimate." You obtain $\mathbf{MS_{BG}}$ and $\mathbf{MS_{WG}}$ by dividing the sums of squares by their respective degrees of freedom. Finally, the fifth column lists the value of F obtained by dividing MS_{BG} by MS_{WG} (which is the same as dividing your between-groups estimate of σ^2 by your within-groups estimate of σ^2:

$$F = \frac{\text{between-groups estimate of } \sigma^2}{\text{within-groups estimate of } \sigma^2} = \frac{\frac{SS_{BG}}{df}}{\frac{SS_{WG}}{df}} = \frac{MS_{BG}}{MS_{WG}}$$

In actual practice, source tables are constructed in a slightly more abbreviated fashion than the one above. The more abbreviated source table for the same study is:

Source	*SS*	*df*	*MS*	*F*
BG	527.08	5	1054	4.68
WG	12,160.9	54	225.2	

Dealing with unequal sample sizes

The computational Formulas 13.3 and 13.4 for the sum of squares are appropriate when the number of subjects in each group is the same. Occasionally, however, you must deal with situations in which the sample sizes are unequal. This will occur when you are unable to measure an equal number of subjects in each sample, but for a reason that leaves the samples *random*. An assumption for any ANOVA is that the samples are randomly selected. Thus you could not go through your samples and selectively eliminate certain scores on a systematic basis since this would make the samples no longer random. You could, however, *randomly* select samples of unequal size. With unequal sample sizes you could compute the within-groups estimate of the variance of the sampling distribution by taking a weighted average of the variance estimates computed within groups. A weighted average would reflect the fact that the larger samples give you a better estimate than the smaller samples. You could weight each variance estimate by a direct function of the sample size, much as you did with the two-sample independent t test. Likewise, you would have to compensate for the unequal sample size when you estimate the variance of the sampling distribution from your sample means. It is possible to develop computational formulas for the sums of squares when sample sizes are unequal. These are presented below:

FORMULA 13.5
computational formula for SS_{BG} for unequal sample sizes

$$SS_{BG} = \Sigma \frac{(\Sigma X)^2}{n_g} - \frac{(\Sigma\Sigma X)^2}{N_{total}}$$

FORMULA 13.6
computational formula for SS_{WG} for unequal sample sizes

$$SS_{WG} = \Sigma\Sigma X^2 - \Sigma \frac{(\Sigma X)^2}{n_g}$$

Here n_g stands for the number of subjects in a particular group, and N_{total} stands for the total number of subjects in the entire research study. In Formula 13.5, the first right-hand term tells you to sum up the scores in each group, ΣX, square these sums, $(\Sigma X)^2$, then divide each of these squares of sums by the number of scores in that group, $(\Sigma X)^2/n_g$, and finally add up all of the resultant quotients. The term $(\Sigma\Sigma X)^2/N_{total}$ tells you to add up the scores in each group, ΣX, add up these sums over all groups, $\Sigma\Sigma X$, then square this sum, $(\Sigma\Sigma X)^2$, and finally divide this by the total number of subjects in the study. In Formula 13.6, the term $\Sigma\Sigma X^2$ tells you to square all of the scores, X^2, add up the squares in each group, ΣX^2, and then add up these sums of squares, $\Sigma\Sigma X^2$. A numerical example using Formulas 13.5 and 13.6 should help clarify this verbal explanation.

Below are the data from the motivation-performance study, but with some scores randomly deleted from several groups.

	Group 1: X_1	X_1^2	Group 2: X_2	X_2^2	Group 3: X_3	X_3^2	Group 4: X_4	X_4^2	Group 5: X_5	X_5^2	Group 6: X_6	X_6^2
	38	1444	64	4096	55	3025	87	7569	83	6889	47	2209
	26	676	85	7225	79	6241	88	7744	76	5776	55	3025
	70	4900	85	6561	52	2704	66	4356	80	6400	68	4624
	55	3025	43	1849	83	6889	58	3364	67	4489	79	6241
	37	1369	44	1936	80	6400	60	3600	88	7744	67	4489
	55	3025	49	2401	90	8100	97	9409	82	6724		
	68	4624	78	6084	85	7225	98	9604	71	5041		
	55	3025	74	5476	59	3481			50	2500		
	30	900	53	2809					69	4761		
			81	6561								
$\Sigma X =$	434		652		583		554		666		316	
$(\Sigma X)^2 =$	188,356		425,104		339,889		306,916		443,556		99,856	
$\frac{(\Sigma X)^2}{n_g} =$	$\frac{188,356}{9}$		$\frac{425,104}{10}$		$\frac{339,889}{8}$		$\frac{306,916}{7}$		$\frac{443,556}{9}$		$\frac{99,856}{5}$	
$=$	20,928.44		= 42,510.4		= 42,486.12		= 43,845.14		= 49,284		= 19,971.2	
$\Sigma X^2 =$		22,988		44,998		44.065		45,646		50,324		20,588

$$\Sigma \frac{(\Sigma X)^2}{n_g} = 20{,}928.44 + 42{,}510.4 + 42{,}486.12 + 43{,}845.14 + 49{,}284 + 19{,}971.2$$

$$= 219{,}025.3$$

$$\Sigma\Sigma X = 434 + 652 + 583 + 554 + 666 + 316 = 3205$$

$$(\Sigma\Sigma X)^2 = (3205)^2 = 10{,}272{,}025$$

$$N_{\text{total}} = 9 + 10 + 8 + 7 + 9 + 5 = 48$$

$$\frac{(\Sigma\Sigma X)^2}{N_{\text{total}}} = \frac{10{,}272{,}025}{48} = 214{,}000.52$$

$$\Sigma\Sigma X^2 = 22{,}988 + 44{,}998 + 44{,}065 + 45{,}646 + 50{,}324 + 20{,}588$$

$$= 228{,}609$$

$$SS_{BG} = \Sigma \frac{(\Sigma X)^2}{n_g} - \frac{(\Sigma\Sigma X)^2}{N_{\text{total}}} = 219{,}025.3 - 214{,}000.52 = 5024.78$$

$$SS_{WG} = \Sigma\Sigma X^2 - \Sigma \frac{(\Sigma X)^2}{n_g} = 228{,}609 - 219{,}025.3 = 9583.7$$

Now that you have calculated SS_{BG} and SS_{WG}, you can set up a source table just as you did when you had equal-size groups, except that you must adjust the degrees of freedom in the denominator of the F ratio. In the numerator of the F ratio, for SS_{BG}, $df = K - 1$, which is the same as when you have equal-size groups. However, for SS_{WG}, in the denominator of the F ratio, $df = N_{total} - K$. Actually this formula for the degrees of freedom associated with the denominator of the F ratio will also work when you have equal-size groups, since N_{total} will then be the number of scores in each group times the number of groups. Thus for equal-size groups $N_{total} = KN$, so $df = N_{total} - K = KN - K = K(N - 1)$, which is the same formula for *df* given in the previous section for equal-sized groups.

For unequal-size groups, the source table in general will look like this:

Source	*SS*	*df*	*MS*	*F*
BG	SS_{BG}	$K - 1$	$\frac{SS_{BG}}{K - 1}$	$\frac{MS_{BG}}{MS_{WG}}$
WG	SS_{WG}	$N_{total} - K$	$\frac{SS_{WG}}{N_{total} - K}$	

For our specific example, the source table is:

Source	*SS*	*df*	*MS*	*F*
BG	5024.78	$6 - 1 = 5$	$\frac{5024.78}{5} = 1004.96$	$\frac{1004.96}{228.18} = 4.40$
WG	9583.7	$48 - 6 = 42$	$\frac{9583.7}{42} = 228.18$	

If $\alpha = .05$, you can look in Table F^* to find that for $df = 5$ and 42, $F^* = 2.44$. Since your calculated F of 4.40 is greater than F^*, you should reject H_0 and conclude that changes in motivation affect performance.

Journal form

The journal form for the ANOVA is very similar to that used to report the results of *t* tests. There is one major difference, however. This difference lies in the fact that there are two degrees of freedom to calculate for each F ratio you calculate (*df* for the numerator of the F ratio, MS_{BG}, and *df* for the denominator, MS_{WG}). For example, for the study concerning the effects of motivation on performance, you might report the results of your ANOVA as follows: "The level of motivation has a significant effect on subjects' performance, $F(5, 54) = 4.68$, $p < .05$." Notice that both values of *df* are listed in parentheses. Conventionally, the *df* for the numerator is listed first, followed by the *df* for the denominator.

KEY CONCEPTS

1 Analysis of variance (ANOVA) is a parametric hypothesis-testing procedure useful in many research situations that employ more than two groups of subjects who are tested under different treatment conditions. There are many types of ANOVA. The one-way ANOVA is appropriate for problems involving three or more equal-size groups of independent subjects tested under different levels of one independent variable, or measured under different naturally occurring conditions.

2 The assumptions that must be met before applying the one-way ANOVA are: (1) the population distributions are normal; (2) the populations have the same variance; (3) the subjects in each sample are selected at random and the samples are independent; and (4) the means and variances of the populations are independent of each other.

3 The principle of variance constancy refers to the fact that adding a constant to every score in a distribution affects the mean but not the variance. Therefore, any treatment that has a constant effect on each subject's performance will change the mean but leave the variance the same.

4 The ANOVA proceeds by assuming that the populations from which your treatment groups were drawn have equal means. Under this assumption you can estimate the variance of the sampling distribution of means. You can then estimate the variance of the sampling distribution from which your observed sample means came, and then compare your two estimates with an F ratio to see how similar they are. If your estimate for the variance of the sampling distribution from which your sample means came is much larger than your estimate calculated under H_0, the F ratio will be much larger than 1, and you can reject H_0.

5 The method of estimating the variance of the sampling distribution which would occur if H_0 is true involves first estimating the variance of the population from which your samples were drawn. You find the estimate of the population variance separately for each sample and then average these variance estimates. The result is called the within-groups estimate of σ^2, and is abbreviated s^2_{WG}:

$$s^2_{\text{WG}} = \frac{s_1^2 + s_2^2 + s_3^2 + \cdots}{K}$$

Because of the principle of variance constancy, s^2_{WG} is unaffected by any treatment effects (or differences in the means of the populations from which the samples were drawn). You can then estimate the variance of the sampling distribution by dividing s^2_{WG} by the size of each sample, N:

$$\text{within-groups estimate of } \sigma_{\bar{x}}^2 = \frac{s_{WG}^2}{N}$$

6 You can estimate the variance of the sampling distribution from which your sample means arose by treating your sample means as a sample of means from the sampling distribution of the mean, and using the formula to estimate the population variance from a sample. The formula adapted to the present case is:

$$s_{\bar{x}}^2 = \frac{\Sigma \bar{X}^2 - \dfrac{(\Sigma \bar{X})^2}{K}}{K - 1}$$

7 The definitional formula for the F ratio is the ratio of your two estimates of the variance of the sampling distribution:

$$F = \frac{\text{between-groups estimate of } \sigma_{\bar{x}}^2}{\text{within-groups estimate of } \sigma_{\bar{x}}^2} = \frac{s_{\bar{x}}^2}{\dfrac{s_{WG}^2}{N}}$$

Values of F that are much larger than 1 indicate that your sample means vary more than you expect on the basis of chance if H_0 is true.

8 With the one-way ANOVA you test the H_0 that the sample means came from populations with equal means, or $\mu_1 = \mu_2 = \mu_3 = \cdots$ by comparing your calculated F value with a distribution of F values that would occur by chance assuming H_0 is true. The hypothesis-testing procedure is:

1. State H_0 and H_A. Assume H_0 is true.
2. Set α and calculate $df = K - 1$ for the numerator of the F ratio, and $df = K(N - 1)$ for the denominator of the F ratio if the sample sizes are equal (or $df = N_{total} - K$ if the sample sizes are unequal). Here, K is the number of groups, N is the number of subjects in each group, and N_{total} is the total number of subjects in the study. Determine F^* from Table F^* for your level of alpha and df's.
3. Calculate an F ratio with either the definitional or the computational formulas. The definitional formula for F is:

$$F = \frac{\text{between-groups estimate of } \sigma_{\bar{x}}^2}{\text{Within-groups estimate of } \sigma_{\bar{x}}^2} = \frac{s_{\bar{x}}^2}{\dfrac{s_{WG}^2}{N}}$$

where

$$s_{\bar{x}}^2 = \frac{\Sigma \bar{X}^2 - \dfrac{(\Sigma \bar{X})^2}{K}}{K - 1} \quad \text{and} \quad s_{WG}^2 = \frac{s_1^2 + s_2^2 + s_3^2 + \cdots}{K}$$

The computational formula for F is:

$$F = \frac{\text{between-groups estimate of } \sigma^2}{\text{within-groups estimate of } \sigma^2} = \frac{N s_{\bar{x}}^2}{s_{\text{WG}}^2} = \frac{\text{MS}_{\text{BG}}}{\text{MS}_{\text{WG}}}$$

where

$$\text{MS}_{\text{WG}} = \frac{\text{SS}_{\text{WG}}}{df} = \frac{\text{SS}_{\text{WG}}}{K(N-1)} \quad \text{for equal-size samples, or}$$

$$\text{MS}_{\text{WG}} = \frac{\text{SS}_{\text{WG}}}{N_{\text{total}} - K} \quad \text{for unequal-size samples; and where}$$

$$\text{SS}_{\text{WG}} = \Sigma\Sigma X^2 - \frac{\Sigma(\Sigma X)^2}{N} \quad \text{for equal-size samples, or}$$

$$\text{SS}_{\text{WG}} = \Sigma\Sigma X^2 - \Sigma\frac{(\Sigma X)^2}{n_g} \quad \text{for unequal-size samples.}$$

In this computational formula for F,

$$\text{MS}_{\text{BG}} = \frac{\text{SS}_{\text{BG}}}{df} = \frac{\text{SS}_{\text{BG}}}{K-1}$$

and

$$\text{SS}_{\text{BG}} = \frac{\Sigma(\Sigma X)^2}{N} - \frac{(\Sigma\Sigma X)^2}{KN} \quad \text{for equal-size samples, or}$$

$$\text{SS}_{\text{BG}} = \Sigma\frac{(\Sigma X)^2}{n_g} - \frac{(\Sigma\Sigma X)^2}{N_{\text{total}}} \quad \text{for unequal-size samples.}$$

4. If $F > F^*$, reject H_0. If $F \leq F^*$, retain H_0.

EXERCISES

1 A source table was constructed for each of six experiments employing a one-way ANOVA. The source tables are:

A.

Source	*SS*	*df*	*MS*	*F*
BG	137.057	2	68.528	1.225
WG	1846.501	33	55.954	

B.

Source	*SS*	*df*	*MS*	*F*
BG	407.532	3	135.844	4.853
WG	1735.446	62	27.991	

C.

Source	*SS*	*df*	*MS*	*F*
BG	3.852	2	1.926	.203
WG	227.778	24	9.491	

D.

Source	*SS*	*df*	*MS*	*F*
BG	50.8	2	25.4	2.436
WG	156.373	15	10.425	

E.

Source	*SS*	*df*	*MS*	*F*
BG	843.621	4	210.905	130.997
WG	28.977	18	1.61	

F.

Source	*SS*	*df*	*MS*	*F*
BG	14,732.895	3	4910.965	1076.706
WG	164.2	36	4.561	

a. Which experiment used the largest number of groups? How many groups were used in this experiment?
b. Which experiment used the largest total number of subjects?
c. Setting $\alpha = .05$, determine the critical value of F for each experiment. Setting $\alpha = .01$, determine the critical value of F for each experiment.
d. In which experiments would the null hypothesis that the samples were drawn from populations having the same mean be rejected at the .05 level of significance?

2 You are interested in determining whether the type of reward given to third-grade children who are introduced to multiplication will affect their ultimate scores on a speed test of their mastery of the basic multiplication tables. You randomly divide 40 third-graders into four groups of ten subjects each. The control group receives no specific reward while an adult tutor practices individually with the children one hour a day for 10 days. The children in the "social rewards" group receive verbal praise, smiles of approval, and so on, while an adult tutor practices individually with the children one hour per day for 10 days. The children in the "candy" group receive candy rewards for daily achievement from an adult tutor after each one-hour daily session for 10 days. The children in the "token" group receive plastic chips from an adult tutor as rewards for daily achievement after each one-hour daily session for 10 days. Children in this group can cash in their tokens for toys after the 10-day period. On the eleventh day all children in each group were given a five-minute timed multiplication test. The number of problems each child got correct on the speed test are given below, along with the group means and various group sums.

No Rewards Subject	X_1	Social Rewards Subject	X_2	Candy Rewards Subject	X_3	Token Rewards Subject	X_4
S_1	119	S_{11}	111	S_{21}	130	S_{31}	140
S_2	112	S_{12}	123	S_{22}	117	S_{32}	139
S_3	114	S_{13}	110	S_{23}	136	S_{33}	138
S_4	100	S_{14}	126	S_{24}	131	S_{34}	124
S_5	115	S_{15}	124	S_{25}	125	S_{35}	140
S_6	97	S_{16}	124	S_{26}	115	S_{36}	116
S_7	124	S_{17}	132	S_{27}	123	S_{37}	131
S_8	127	S_{18}	140	S_{28}	103	S_{38}	139
S_9	105	S_{19}	131	S_{29}	134	S_{39}	147
S_{10}	109	S_{20}	135	S_{30}	118	S_{40}	126
$\bar{X} =$	112.2		125.6		123.2		134.0
$\Sigma X =$	1122		1256		1232		1340
$(\Sigma X)^2 =$	1,258,884		1,577,536		1,517,824		1,795,600
$\Sigma X^2 =$	126,746		158,588		152,714		180,364

a. Within each group, use computational Formula 8.2 and solve for s^2. This gives you four separate estimates of the population variance.
b. Calculate s^2_{WG}, the within-groups estimate of the population population variance, by computing the mean of the variance estimates from within each group in (a).
c. Calculate the within-groups estimate of the variance of the sampling distribution of the mean by dividing the within-groups estimate of the population variance computed in (b) by N.
d. Calculate $s_{\bar{x}}^2$, the between-groups estimate of the variance of the sampling distribution of the mean, by adapting Formula 8.2 using the group means in place of individual raw scores.
e. Setting $\alpha = .05$, use the definitional Formula 13.1 for the one-way ANOVA to test the null hypothesis that your samples were drawn from populations having the same mean; that is, the type of reward given children by tutors does not affect mastery of basic multiplication. Use the hypothesis-testing procedure.
f. Put your results in journal form.

3 Using the data in Exercise 2, do the following:
a. Calculate the F ratio using Formula 13.2, the computational version of the F ratio.
b. Compute SS_{WG}, the sum of the squares within groups using the computational Formula 13.3. Compute SS_{BG}, the sum of squares between groups using Formula 13.4.
c. Compute the within-groups estimate of σ^2 by dividing SS_{WG} by its associated degrees of freedom, $K(N - 1)$. Compute the between-groups estimate of σ^2 by dividing SS_{BG} by its associated degrees of freedom, $K - 1$.
d. Construct a source table.

4 You are interested in examining recognition memory for English comparative adjectives (such as "bigger," "thinner," and "smarter") used in written

passages. You assume that the amount of time between reading the passages and the test of adjective recognition will influence memory for the specific adjectives used in the passages. That is, the retention interval affects recognition memory for adjectives. To test this assumption, 30 subjects were randomly assigned to one of three groups such that ten subjects comprised each group. All subjects were allowed to read passages of fiction and then tested later by giving them a brief summary of the characters and numerous contexts, and then asking them to choose from five adjectives the one used in a particular context. In one group, the subjects were tested a half hour after reading; in the second group, the subjects were tested one week later; and in the third group, the subjects were tested three weeks later. You obtain the following scores of the number of adjectives correctly recognized from 20 contexts:

1/2 hour Subject	X_1	*1 week* Subject	X_2	*3 weeks* Subject	X_3
S_1	14	S_{11}	9	S_{21}	11
S_2	14	S_{12}	14	S_{22}	8
S_3	15	S_{13}	15	S_{23}	9
S_4	16	S_{14}	13	S_{24}	11
S_5	13	S_{15}	12	S_{25}	10
S_6	17	S_{16}	10	S_{26}	12
S_7	15	S_{17}	12	S_{27}	14
S_8	12	S_{18}	13	S_{28}	11
S_9	14	S_{19}	13	S_{29}	9
S_{10}	16	S_{20}	10	S_{30}	12

a. Setting $\alpha = .05$, use computational Formulas 13.3 and 13.4 and complete the steps of the hypothesis-testing procedure for the one-way ANOVA. What can you conclude about the effect of the retention interval on memory for comparative adjectives?

b. Construct a source table.

c. Put your results in journal form.

d. Using the values for K, N, and N_{total} derived from this study, show by example that when you have equal-size groups the degrees of freedom associated with the denominator can be calculated with either $df = K(N - 1)$ or $df = N_{\text{total}} - K$.

5 You have developed a do-it-yourself desensitization program that has been demonstrated to reduce math anxiety in students enrolling in an introductory statistics class. Now you wish to examine whether or not five modes of presenting the desensitization program will differentially affect the reduction of math anxiety. To examine the mode of presentation, all students in a large introductory statistics class were asked to answer the following question on the first day of class, "Which one of the following phrases best describes your feelings about math: I enjoy math very much; I enjoy math; I am neutral in my feelings about math; I fear math; or I fear math very much." The 32 students who answered that they feared math very much participated in the study. These students were randomly assigned to receive one of the following five presentation modes: written transcript; audiotape cassette; written transcript plus audiotape cassette; audiovideotape cassette; and written transcript plus audiovideotape cassette. One week after being presented with the desensitization program the students were given a test designed to measure math anxiety. Low scores signify low math anxiety. You obtained the following results:

Written Subject	X_1	*Audiotape* Subject	X_2	*Written and Audiotape* Subject	X_3	*Audiovideotape* Subject	X_4	*Written and Audiovideotape* Subject	X_5
S_1	12	S_8	8	S_{15}	7	S_{21}	10	S_{27}	6
S_2	7	S_9	12	S_{16}	4	S_{22}	8	S_{28}	6
S_3	12	S_{10}	7	S_{17}	5	S_{23}	6	S_{29}	2
S_4	11	S_{11}	8	S_{18}	3	S_{24}	4	S_{30}	10
S_5	11	S_{12}	9	S_{19}	2	S_{25}	3	S_{31}	7
S_6	6	S_{13}	7	S_{20}	8	S_{26}	4	S_{32}	9
S_7	8	S_{14}	5						

a. Setting $\alpha = .05$, use computational Formulas 13.5 and 13.6 for unequal sample sizes and complete the steps of the hypothesis-testing procedure. What can you conclude about the effect that mode of presentation has on the reduction of math anxiety?

b. Construct a source table.

c. Put your results in journal form.

6 Although relaxation training is widely done in treatment centers for hypertension, there are very few data to suggest which methods of relaxation training are most appropriate with these patients. You design a study to compare the effectiveness of four different relaxation strategies. Forty married men who were medically diagnosed as having essential hypertension volunteered to participate. The subjects were randomly assigned to one of the following relaxation conditions: progressive relaxation; autogenic relaxation; transcendental meditation; and metronome-induced relaxation. After one month, you gather follow-up data by asking each subject's wife to evaluate her husband's recuperation by filling out a "behavior checklist" designed to measure hypertension. The behavior checklist scores signify the total number of hypertension-related behaviors demonstrated by subjects during the month. You obtain the following data:

Progressive Relaxation Subject	X_1	*Autogenic Relaxation* Subject	X_2	*Transcendental Meditation* Subject	X_3	*Metronome Relaxation* Subject	X_4
S_1	36	S_{11}	36	S_{21}	36	S_{31}	38
S_2	35	S_{12}	34	S_{22}	30	S_{32}	31
S_3	37	S_{13}	35	S_{23}	37	S_{33}	35
S_4	35	S_{14}	35	S_{24}	35	S_{34}	37
S_5	38	S_{15}	39	S_{25}	31	S_{35}	33
S_6	34	S_{16}	36	S_{26}	35	S_{36}	34
S_7	35	S_{17}	29	S_{27}	34	S_{37}	35
S_8	28	S_{18}	35	S_{28}	37	S_{38}	32
S_9	36	S_{19}	37	S_{29}	40	S_{39}	35
S_{10}	36	S_{20}	39	S_{30}	33	S_{40}	34

a. Setting $\alpha = .01$, complete the hypothesis-testing procedure using the appropriate computational formulas for SS_{BG} and SS_{WG}. What can you conclude about the effect of method of relaxation on hypertension scores?
b. Construct a source table.
c. Put your results in journal form.

14

The purpose of this short chapter is to organize and summarize the tools that have been presented in this text. Figure 14-1 is a schematic summary of the various techniques. The following outline briefly describes the content of this schematic.

I. *Descriptive statistics* are tools used to summarize and describe groups of data.

A. *Frequency distributions and graphs* are methods of summarizing all the data at once, so you can get an overall picture of how the data stack up. Frequency distributions can be either of the "regular" type or of the cumulative type. A regular *frequency distribution* gives you immediate information about how often a given score occurs. A *cumulative frequency distribution* tells you how many scores occur at or below a given score. Frequency distributions and graphs may also be either ungrouped or grouped. Whether you use an ungrouped or a grouped type of presentation depends on which one most clearly represents the important information in the data. Roughly speaking, *histograms* and *bar graphs* are constructed with bars, and *frequency polygons* and *cumulative frequency polygons* are constructed with points connected by straight lines. A *scatter plot* reveals the relationship (correlation) between two variables. Each point in a scatter plot corresponds to an individual's combination of values on the two variables represented on the abscissa and the ordinate.

B. *Measures of central tendency* are single numbers that represent the general location (magnitude) of a distribution of scores. The *mode* is the most frequently occurring score, the *median* is the middle score in the distribution, and the *mean* is the arithmetic average. Unlike the mode

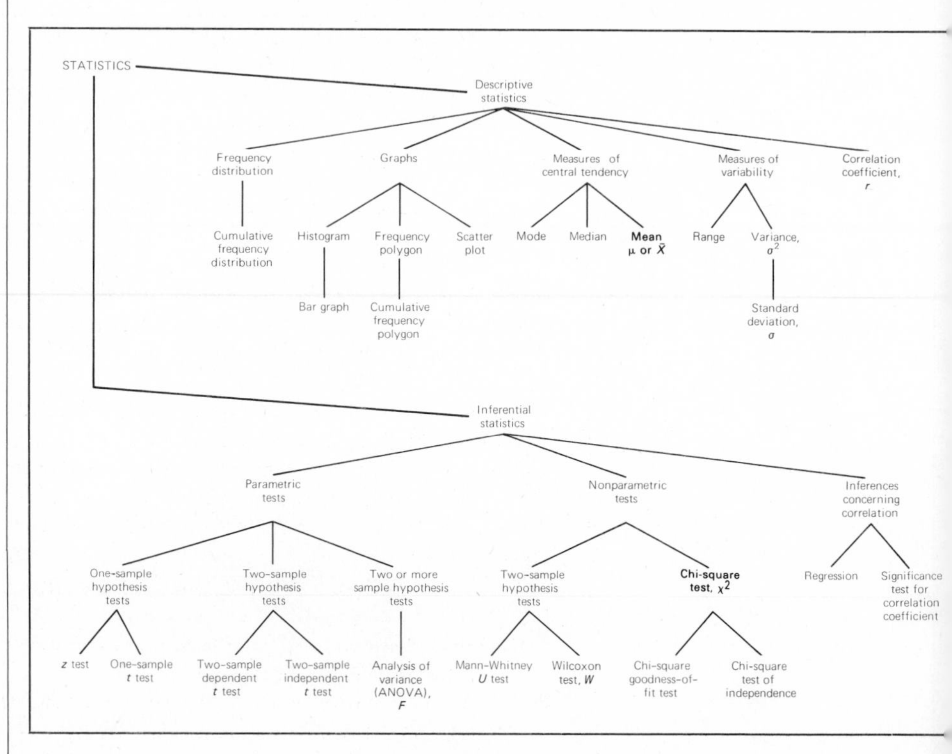

FIGURE 14-1

Schematic summary of tools in Basic Statistics for the Behavioral Sciences.

and the median, whose computation relies on the value of only one or two scores in the distribution, the calculation of the mean takes into account the value of every score.

C. *Measures of variability* are single numbers that represent how spread out the scores in a distribution are. The *range* is one plus the difference between the highest and the lowest scores. Unlike the range, whose computation relies on only the two extreme scores in the distribution, computation of the variance and the standard deviation takes into account the value of every score. The *variance* is the mean squared deviation from the mean, and the *standard deviation* is the square root of the variance.

D. The *correlation coefficient* (r) is a single number between -1 and $+1$ which describes the linear (straight-line) relationship between two variables. The sign of r tells whether the variables vary directly (positive sign) or inversely (negative sign). The magnitude of r tells how closely the points lie to a straight line in a scatter plot.

II. *Inferential Statistics* are tools for making inferences about populations from samples.

A. *Parametric tests* are hypothesis-testing procedures that enable you to determine whether a difference between sample means occurred by chance, provided that fairly stringent assumptions about the population parameters are met (such as that the population distributions should be normal).

1. The *z test* helps you decide whether a sample with a given mean could have occurred by chance from a population with a known mean and standard deviation.
2. The *one-sample t test* is the same as the z test except that with the one-sample t test the population standard deviation is not known and must be estimated from the sample scores.
3. The *two-sample independent t test* helps you decide if independent samples with given means could have occurred by chance from populations with equal means.
4. The *two-sample dependent t test* is the same as the two-sample independent t test except that with the dependent t test the samples are matched pairwise, or are composed of the same subjects tested under two different conditions.
5. The *one-way analysis of variance* (ANOVA) is an extension of the t test that enables you to compare several sample means at once. It helps you decide whether samples with known means could have arisen from populations with equal means.

B. *Nonparametric tests* are hypothesis-testing procedures that make much less stringent assumptions about the population parameters than do the parametric tests. As a consequence, however, nonparametric tests are less powerful than their corresponding parametric tests.

1. The *Mann-Whitney U test* is the nonparametric counterpart of the two-sample independent t test.
2. The *Wilcoxon test* is the nonparametric counterpart of the two-sample dependent t test.

3. The chi-square (χ^2) tests allow hypothesis testing when your data consist of frequencies of occurrence or scores on a nominal scale of measurement.
    a. The *chi-square goodness-of-fit test* helps you decide whether a sample with a given frequency distribution could have occurred by chance from a population with a known frequency distribution.
    b. The *chi-square test of independence* helps you decide from a sample contingency table whether two variables are dependent.

C. *Inferences concerning correlation* involve either using the correlation coefficient to make predictions from known values of one variable to unknown values of a related variable, or testing hypotheses about the value of the population correlation coefficient.
1. *Regression* is a technique that allows you to make a prediction from a known value on one variable to an unknown value on another variable, provided the two variables are linearly related.
2. The *significance test for the correlation coefficient* helps you decide whether a sample with a given correlation coefficient could have occurred by chance from a population in which there is no correlation between the two variables in question.

APPENDIXES

I

A BRIEF REVIEW OF MATHEMATICS AND THE USE OF STATISTICAL FORMULAS

Mathematical symbols and formulas are exceedingly useful in statistics. Mathematics is a language with which researchers can express various elements of data and manipulations of these data elements in a shorthand fashion.

A statistical formula indicates three important things to the researcher:

1. The elements of data to be manipulated.
2. The types of manipulations to perform (e.g., add, subtract, multiply, divide, square, etc.).
3. The order of manipulation (e.g., multiply two elements together and then add a third element).

To use a statistical formula you need to know how elements of data are symbolized and how manipulations are symbolized. You also need to know the rules that prescribe the order in which manipulations should be performed.

Symbols for data elements

Since the data collected usually vary from subject to subject, as well as over time, the elements of data are called *variables.* Variables are usually symbolized by arabic letters such as X, Y, or Z. Thus you might use the letter "X" to represent any value of a particular attribute.

Symbols for manipulations

Addition is symbolized by "+", so $Y + Z$ says to add Y and Z together. Thus, if $Y = 2$ and $Z = 10$, then $Y + Z$ means to add $2 + 10$ together to get 12.

Summation is repeated addition, and is symbolized by the capital Greek letter sigma, Σ. Thus ΣX says to add up all the values of X, one by one. If X refers to each raw score in a collection of data, such as the raw scores 5, 9, 7, 6, 9, 2, then ΣX means to add $5 + 9 + 7 + 6 + 9 + 2$ to get 38.

Subtraction is symbolized by "−", so $Y - Z$ says to subtract Z from

Y. If Y = 15 and Z = 7, then Y − Z means to subtract 7 from 15 to get 8.

Multiplication is symbolized by "×", "·", or by juxtaposition, either with or without parentheses. Because the symbol × resembles the X used to symbolize a variable or a raw score, you will most often see either the dot or juxtaposition used to symbolize multiplication. Therefore $Y \cdot Z$, YZ, (Y)(Z), and (Y)Z, and Y(Z) all mean to multiply Y by Z. Thus if Y = 5 and Z = 3, all of the preceding say to multiply 5 by 3 to get 15.

Division is symbolized by "÷", "—", or "/", so Y ÷ Z, $\frac{Y}{Z}$, and Y/Z, all mean to divide Y by Z. The number to be divided, in this case Y, is called the *numerator*. The number doing the dividing, in this case Z, is called the *denominator*. Thus, if Y = 12 and Z = 3, all of the above mean to divide 12 by 3 to get 4.

Squaring means to multiply a number by itself, so "X squared" means $X \cdot X$. A special symbol that tells you to square X is X^2, read as "X squared." Thus, if X = 8, then $X^2 = 8^2$ which means to multiply 8 by 8 to get 64.

Square root is a manipulation that tells you to find a number which, when squared, will yield the given number. Thus the square root of 16 is 4, because 4 squared is 16. The square root of X is symbolized by $\sqrt{X}$. Thus, if X = 36, then $\sqrt{X} = \sqrt{36}$ which is 6.

Some special numbers and symbols

There are two numbers that have special properties when they are subjected to the manipulations summarized above. These special numbers are 0 and 1.

When 0 is added to or subtracted from any other number, you get the original number. Thus X + 0 is X, and X − 0 is X. When 0 is multiplied by any other number, you get 0. Thus $X \cdot 0$ is 0. When 0 is divided by any other number you get zero. Thus 0/X is 0. For example, if X = 8, then $X + 0 = 8 + 0 = 8$, $X - 0 = 8 - 0 = 8$, $X \cdot 0 = 8 \cdot 0 = 0$, and $0/X = 0/8 = 0$. The rules of mathematics forbid you from ever *dividing* by 0.

If you multiply or divide any other number by one, you will get the original number. Thus $X \cdot 1$ is X, and X/1 is also X. For example, if X = 6, then $X \cdot 1 = 6 \cdot 1 = 6$ and $X/1 = 6/1 = 6$.

There are certain numbers used in statistical formulas that do not change as the collected data change, so they are called *constants*. Constants are usually symbolized by arabic letters such as *a, b, c,* or *k*.

Order of manipulations

The general rule for the order of performing various manipulations is:

1. First perform any special operations, such as squaring or finding the square root.
2. Second, perform multiplication or division.
3. Last, perform addition or subtraction.

For example, $X^2Y + Z$ says to first square X, then multiply the answer by Y, and finally add Z. You would be violating the rules if you first added Y and Z before multiplying Y by X^2.

Parentheses are used whenever it is desired to circumvent the general rules above, or when it is necessary to eliminate ambiguity. The rule for the use of parentheses is:

1. Perform the manipulations inside the parentheses first.
2. Then perform the manipulations indicated outside the parentheses.

For example, $(X + Y)/Z$ says first add X and Y, and then divide the answer by Z. Another example is

$$\frac{\left(X + \frac{Y^2}{Z}\right)^2}{(a + b)c}$$

First perform the operations inside the parentheses, using the general rule for the priority of manipulations. Since you have one set of parentheses in the numerator and another in the denominator, first do the work inside both sets of parentheses before going on. It does not matter which set of parentheses you start with. In the numerator, first square Y and then divide the result by Z. Add X to the result. In the denominator, add together a and b, and then multiply by c. Now square the result in the parentheses of the numerator, and then divide by the result in the denominator.

Solving equations

An equation is created wherever you use the equal sign, =, to state that two expressions are equal to one another.

For example, $X = Y$ says that the values of X and Y are equal. Some equations are much more complex than this simple example, however. You might have something like

$$\frac{3X^2 + Y}{2} = 10Z$$

To solve such an equation for one of the variables, say X, you must transform the equation into a form in which X appears by itself on one side of the equation. Thus you want to achieve a form that looks like "$X =$ something." Since an equation is a statement that two things are equal, you may perform any manipulation you please to *both* sides of an equation, and the two sides will remain equal. You may do anything you please to one side of the equation and the two sides will remain equal if you perform the same manipulation to the other side of the equation. For example, if you add a number to one side of the equation, you can add the same number to the other side of the equation and the two sides will remain equal. You can use this rule to transform any equation to the form you desire. Thus in the example given above,

$$\frac{3X^2 + Y}{2} = 10Z$$

suppose you want to solve for X. That is, you want to get X alone on the left side of the equation. Since the entire left side of the equation that contains X is divided by 2, you can eliminate this 2 by multiplying the left side by 2. When you multiply by 2 you "undo" the division, since any quantity multiplied and divided by the same number remains the original quantity. To keep both sides of the equation equal you must also multiply the right side by 2, as follows:

$$2 \cdot \frac{3X^2 + Y}{2} = 2 \cdot 10Z$$

Thus you obtain

$$3X^2 + Y = 20Z$$

Now you can get rid of the Y on the left side by subtracting Y from both sides:

$$3X^2 + Y - Y = 20Z - Y$$

Since $Y - Y = 0$, and adding 0 to any number does not change it, you have

$$3X^2 = 20Z - Y$$

Since the left side is multiplied by 3, you can undo this by dividing both sides by 3:

$$\frac{3X^2}{3} = \frac{20Z - Y}{3}$$

or

$$X^2 = \frac{20Z - Y}{3}$$

Now you can undo the squaring on the left side by taking the square root of both sides (the square root of a number squared is the number itself)

$$\sqrt{X^2} = \sqrt{\frac{20Z - Y}{3}}$$

and therefore

$$X = \sqrt{\frac{20Z - Y}{3}}$$

Inequalities

The symbol "$<$" means "less than," so "$X < Y$" says that X is less than Y. Similarly, "$>$" means "greater than," so "$X > Y$" says that X is greater than Y. The equality and inequality symbols can also be combined, for example "$\leq$" means "less than or equal to" and "$\geq$" means "greater than or equal to."

II

GLOSSARY OF SYMBOLS AND ABBREVIATIONS

*	The asterisk is used as a superscript to denote critical values, as in z^*.
α	Alpha, the criterion for significance. Alpha specifies a value for judging what is too unlikely to occur by chance.
C	The column total in a contingency table of observed values in the χ^2 test of independence.
cf	Cumulative frequency
χ^2	Read "chi-square," this is a measure of how much the observed frequency scores differ from the expected frequency scores in the chi-square test.
χ^{2*}	Read "chi-square critical," this is a criterion for judging the significance of an observed value of χ^2 in the χ^2 test.
D	A difference score, the difference between pretest and posttest scores in the two-sample dependent t test.
$\bar{D}$	Read "D bar," this signifies the mean of the D scores in the two-sample dependent t test.
df	Degrees of freedom.
E	Expected value in the χ^2 test.
F	A ratio of two different variance estimates. Used in the analysis of variance as a measure of the magnitude of a treatment effect.
F^*	Read "F critical," this is a criterion for judging the significance of an observed F value.
H_A	The alternate hypothesis.
H_0	The null hypothesis.
$>$	Greater than. For example, $X > Y$ states that X is greater than Y.
$\geq$	Greater than or equal to. For example, "$X \geq Y$" states that X is greater than or equal to Y.
$<$	Less than. For example, $X < Y$ states that X is less than Y.
$\leq$	Less than or equal to. For example, $X \leq Y$ states that X is less than or equal to Y.
K	The number of groups in the analysis of variance.

K_c	The number of categories of observations for the χ^2 goodness-of-fit test, and the number of columns in the contingency table for the χ^2 test of independence.
K_r	The number of rows in the contingency table for the χ^2 test of independence.
μ	Pronounced "mew," this signifies the mean of a population of scores.
μ_x	The mean of the variable labeled X.
μ_y	The mean of the variable labeled Y.
MS_{BG}	The mean square between groups, which is the between-groups estimate of the population variance in the analysis of variance.
MS_{WG}	The mean square within groups, which is the within-groups estimate of the population variance in the analysis of variance.
N	The number of scores in a group, or the numbers of pairs of scores in a correlational study or a two-sample dependent design.
n_g	The number of scores in each group in the analysis of variance.
N_{total}	The total number of scores in all groups in the analysis of variance.
N.S.	Not significant.
O	Observed value in χ^2 test.
p	The probability of an event occurring.
R	The row total in a contingency table of observed values in the χ^2 test of independence.
R_1	Sum of ranks in Group 1 in the Mann-Whitney U test.
R_2	Sum of ranks in Group 2 in the Mann-Whitney U test.
r	The Pearson product-moment correlation coefficient, which describes the linear relationship between two variables.
r^*	Read "r critical," this is a criterion for judging the significance of an observed r value.
r^2	The coefficient of determination, which specifies the amount of the variance of a variable that is accounted for by knowledge of another variable.
ρ	Rho, Pearson product-moment correlation coefficient in the population.
s	The sample standard deviation, which is an estimate of the standard deviation of the population from which the sample came when doing t tests.
s_D	The standard deviation of the difference scores in a sample. Used to estimate σ_D in the two-sample dependent t test.
$s^{2'}$	The sample variance, which is an estimate of the variance of the population from which the sample came.
s^2_{pooled}	The pooled variance estimate, a weighted average of the estimates of the variance of the population from which two independent samples came.
s^2_{WG}	The within-groups variance estimate, an estimate of the population variance calculated by averaging the variance estimates within each treatment group in the analysis of variance.
$s_{\bar{x}}^2$	An estimate of the variance of the sampling distribution of the mean, calculated from the sample means in the analysis of variance.
σ	Read "sigma," this signifies the standard deviation of a population.

σ^2	Read "sigma squared," this signifies the variance of a population.
$\sigma_{\bar{x}}^2$	The variance of the sampling distribution of the mean.
σ_x	The standard deviation of the variable labeled X.
$\sigma_{\bar{x}}$	The standard error of the mean (the standard deviation of the sampling distribution of the mean).
$\sigma_{\bar{x}_1-\bar{x}_2}$	The standard error of the difference (the standard deviation of the sampling distribution of the difference between independent sample means).
σ_y	The standard deviation of the variable labeled Y.
$\sigma_{\hat{y}}$	The standard error of the estimate (the standard deviation of the errors in prediction when using the regression line).
Σ	Sigma, the summation sign. This directs you to add up the values of the symbol that follows the sign. For example "ΣX" is read "the sum of X" and signifies the sum of all the values of X.
ΣX	The sum of X. To calculate this, add up all of the individual values of X.
$(\Sigma X)^2$	Read "the sum of X, quantity squared." Calculated by first summing the values of X and then squaring the sum.
$\Sigma(\Sigma X)^2$	The sum of the sum of X, quantity squared. Calculated by first summing the X values separately in each group, then squaring these sums, and finally summing these squared sums.
ΣX^2	The sum of the squared values of X. Calculated by first squaring each X value and then summing.
$\Sigma\Sigma X^2$	The double sum of the squared values of X. Calculated by first squaring each X value and summing in each treatment group in the analysis of variance and then summing these sums of squared values.
$\Sigma\Sigma X$	The double sum of X, calculated by summing the scores in each treatment group in the analysis of variance, and then summing these sums.
$(\Sigma\Sigma X)^2$	The double sum of X, quantity squared. Calculated by first summing the X values in each group, then summing these sums, and finally squaring the result.
SS_{BG}	The sum of squares between groups, used in calculating the between-groups estimate of the population variance in the computational formula for the analysis of variance.
SS_{WG}	The sum of squares within groups, used in calculating the within-groups estimate of the population variance in the computational formula for the analysis of variance.
t	The estimated number of standard deviations that a sample result lies away from the mean of the sampling distribution.
$\|t\|$	The absolute value of the calculated t in the t test.
t^*	Read "t critical," this is a value of t that delimits the region(s) of rejection in a t test.
U	A measure of the difference between two independent samples in the Mann-Whitney U test.
U^*	Read "U critical," this is a criterion for judging the significance of a calculated U value in the Mann-Whitney U test.

W	A measure of the difference between two dependent samples in the Wilcoxon matched-pairs signed-ranks test.
W^*	Read "W critical," this is a criterion for judging the significance of a calculated W value in the Wilcoxon test.
X	This usually signifies a raw score or the values of some variable.
$\bar{X}$	Read "X bar," this signifies the mean of a sample of scores.
$X - \mu$	The difference, or deviation of a score from the mean of a population.
$\hat{Y}$	Read "Y hat," this signifies a value of Y that is predicted from knowledge of a correlated variable X.
z	The number of standard deviations that a score lies away from the mean of a distribution of scores.
$\lvert z \rvert$	The absolute value of the calculated z in the z test.
z_x	Denotes z scores for the variable labeled X.
z_y	Denotes z scores for the variable labeled Y.
z^*	Read "z critical," this is a value of z that delimits the region(s) of rejection in a z test.

III

GLOSSARY OF FORMULAS

FORMULA 2.1 (page 16)

$$\text{midpoint of class interval} = \frac{\text{lowest score} + \text{highest score}}{2}$$

FORMULA 3.1 (page 29)

mode = value of score occurring most frequently

FORMULA 3.2 (page 32)

median (N is odd) = the middle score

FORMULA 3.3 (page 32)

median (N is even) = value halfway between the middle scores

FORMULA 3.4 (page 34)

$$\text{mean} = \frac{\Sigma X}{N}$$

Formula for calculating the population mean, μ, or the sample mean, $\bar{X}$

FORMULA 3.5 (page 38)

range = highest score − lowest score + 1

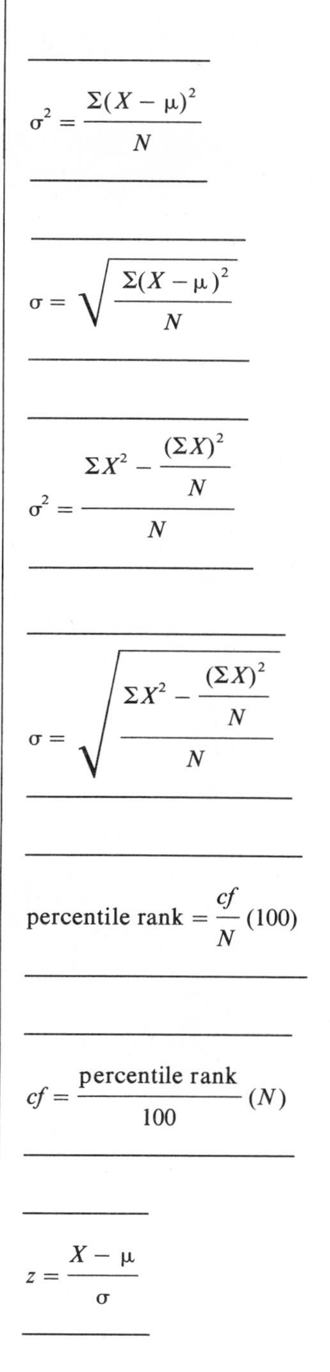

FORMULA 3.6 (page 40)

$$\sigma^2 = \frac{\Sigma(X - \mu)^2}{N}$$

Definitional formula for calculating the population variance

FORMULA 3.7 (page 40)

$$\sigma = \sqrt{\frac{\Sigma(X - \mu)^2}{N}}$$

Definitional formula for calculating the population standard deviation

FORMULA 3.8 (page 43)

$$\sigma^2 = \frac{\Sigma X^2 - \frac{(\Sigma X)^2}{N}}{N}$$

Computational formula for calculating the population variance

FORMULA 3.9 (page 44)

$$\sigma = \sqrt{\frac{\Sigma X^2 - \frac{(\Sigma X)^2}{N}}{N}}$$

Computational formula for calculating the population standard deviation

FORMULA 4.1 (page 52)

$$\text{percentile rank} = \frac{cf}{N}(100)$$

For transforming raw scores into percentile ranks

FORMULA 4.2 (page 52)

$$cf = \frac{\text{percentile rank}}{100}(N)$$

Used in transforming percentile ranks into raw scores

FORMULA 4.3 (page 53)

$$z = \frac{X - \mu}{\sigma}$$

For transforming raw scores into z scores

FORMULA 4.4 (page 54)

$$X = \mu + z\sigma$$

For transforming z scores into raw scores

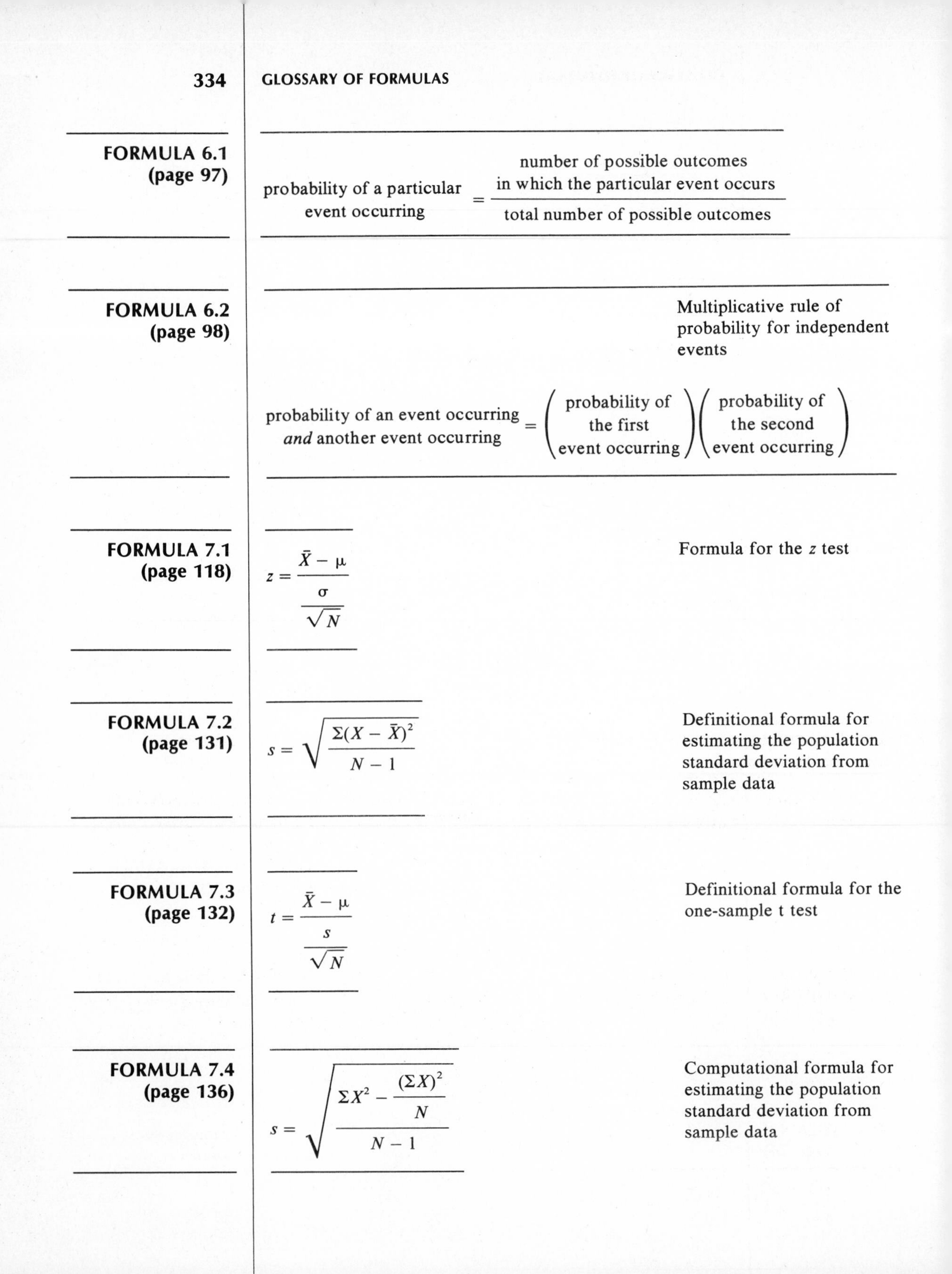

FORMULA 6.1
(page 97)

$$\text{probability of a particular event occurring} = \frac{\text{number of possible outcomes in which the particular event occurs}}{\text{total number of possible outcomes}}$$

FORMULA 6.2
(page 98)

Multiplicative rule of probability for independent events

$$\text{probability of an event occurring } \textit{and} \text{ another event occurring} = \left(\text{probability of the first event occurring}\right)\left(\text{probability of the second event occurring}\right)$$

FORMULA 7.1
(page 118)

Formula for the z test

$$z = \frac{\bar{X} - \mu}{\dfrac{\sigma}{\sqrt{N}}}$$

FORMULA 7.2
(page 131)

Definitional formula for estimating the population standard deviation from sample data

$$s = \sqrt{\frac{\Sigma(X - \bar{X})^2}{N - 1}}$$

FORMULA 7.3
(page 132)

Definitional formula for the one-sample t test

$$t = \frac{\bar{X} - \mu}{\dfrac{s}{\sqrt{N}}}$$

FORMULA 7.4
(page 136)

Computational formula for estimating the population standard deviation from sample data

$$s = \sqrt{\frac{\Sigma X^2 - \dfrac{(\Sigma X)^2}{N}}{N - 1}}$$

FORMULA 7.5 (page 137)

$$t = \frac{\bar{X} - \mu}{\sqrt{\dfrac{\Sigma X^2 - \dfrac{(\Sigma X)^2}{N}}{N(N-1)}}}$$

Computational formula for the one-sample t test

FORMULA 8.1 (page 155)

$$s^2 = \frac{\Sigma(X - \bar{X})^2}{N - 1}$$

Definitional formula for estimating the population variance from sample data

FORMULA 8.2 (page 155)

$$s^2 = \frac{\Sigma X^2 - \dfrac{(\Sigma X)^2}{N}}{N - 1}$$

Computational formula for estimating the population variance from sample data

FORMULA 8.3 (page 156)

$$s^2_{\text{pooled}} = \frac{(N_1 - 1)s_1^2 + (N_2 - 1)s_2^2}{N_1 + N_2 - 2}$$

Definitional formula for estimating the population variance by combining the estimates from two independent samples

FORMULA 8.4 (page 156)

$$\text{estimate of } \sigma_{\bar{x}_1 - \bar{x}_2} = \sqrt{s^2_{\text{pooled}}\left(\frac{1}{N_1} + \frac{1}{N_2}\right)}$$

Formula for estimating the standard deviation (standard error) of the sampling distribution of the difference for independent samples

FORMULA 8.5 (page 157)

$$\text{independent } t = \frac{(\bar{X}_1 - \bar{X}_2) - 0}{\sqrt{s^2_{\text{pooled}}\left(\dfrac{1}{N_1} + \dfrac{1}{N_2}\right)}}$$

$$= \frac{(\bar{X}_1 - \bar{X}_2) - 0}{\sqrt{\dfrac{(N_1 - 1)s_1^2 + (N_2 - 1)s_2^2}{N_1 + N_2 - 2}\left(\dfrac{1}{N_1} + \dfrac{1}{N_2}\right)}}$$

Definitional formula for the two-sample independent t test

FORMULA 8.6 (page 169)

$$s_D = \sqrt{\frac{\Sigma(D - \bar{D})^2}{N - 1}}$$

Definitional formula for estimating the standard deviation of the population of difference scores

FORMULA 8.7 (page 169)

$$s_D = \sqrt{\frac{\Sigma D^2 - \frac{(\Sigma D)^2}{N}}{N - 1}}$$

Computational formula for estimating the standard deviation of the population of difference scores

FORMULA 8.8 (page 169)

$$\text{dependent } t = \frac{\bar{D} - 0}{\frac{s_D}{\sqrt{N}}}$$

Definitional formula for the two-sample dependent t test

FORMULA 8.9 (page 175)

Computational formula for two-sample independent t test

$$\text{independent } t = \frac{(\bar{X}_1 - \bar{X}_2)}{\sqrt{\frac{\Sigma X_1^2 - \frac{(\Sigma X_1)^2}{N_1} + \Sigma X_2^2 - \frac{(\Sigma X_2)^2}{N_2}}{N_1 + N_2 - 2}\left(\frac{1}{N_1} + \frac{1}{N_2}\right)}}$$

FORMULA 8.10 (page 177)

$$\text{dependent } t = \frac{\bar{D}}{\sqrt{\frac{\Sigma D^2 - \frac{(\Sigma D)^2}{N}}{N(N - 1)}}}$$

Computational formula for the two-sample dependent t test

FORMULA 9.1 (page 195)

$$U = R_1 - \frac{(N_1)(N_1 + 1)}{2}$$

Definitional formula for the Mann-Whitney U test

FORMULA 9.2 (page 200)

$$z = \frac{U - \frac{(N_1)(N_2)}{2}}{\sqrt{\frac{(N_1)(N_2)(N_1 + N_2 + 1)}{12}}}$$

Formula for converting U to z in large samples

FORMULA 10.1 (page 220)

$$\chi^2 = \sum \frac{(0 - E)^2}{E}$$

Definitional formula for the χ^2 test

FORMULA 10.2 (page 226)

$$\text{expected frequency of a particular cell} = E = \frac{(R)(C)}{N}$$

Used for calculating expected frequencies in the χ^2 test of independence

FORMULA 11.1 (page 244)

$$r = 1 - \left(\frac{1}{2}\right)\left(\frac{\Sigma(z_x - z_y)^2}{N}\right)$$

Definitional formula for the correlation coefficient

FORMULA 11.2 (page 246)

$$r = \frac{\Sigma z_x z_y}{N}$$

z score computational formula for the correlation coefficient

FORMULA 11.3 (page 247)

$$r = \frac{N\Sigma XY - \Sigma X \Sigma Y}{\sqrt{[N\Sigma X^2 - (\Sigma X)^2]\,[N\Sigma Y^2 - (\Sigma Y)^2]}}$$

Raw-score computational formula for the correlation coefficient

FORMULA 12.1 (page 266)

$$Y = r\frac{\sigma_y}{\sigma_x}(X - \bar{X}) + \bar{Y}$$

Raw-score formula for the regression line

FORMULA 12.2 (page 269)

$$z_{\hat{y}} = r z_x$$

z score formula for the regression line

FORMULA 12.3
(page 274)

$$\sigma_{\hat{y}} = \sqrt{\frac{\Sigma(Y - \hat{Y})^2}{N}}$$

Definitional formula for the standard error of the estimate

FORMULA 12.4
(page 275)

$$\sigma_{\hat{y}} = \sigma_y \sqrt{1 - r^2}$$

Computational formula for the standard error of the estimate

FORMULA 12.5
(page 277)

$$\sigma_{\hat{y}}^{\ 2} = \sigma_y^{\ 2}(1 - r^2)$$

Formula for the variance of the errors in prediction

FORMULA 13.1
(page 295)

$$F = \frac{\text{between-groups estimate of } \sigma_{\bar{x}}^{\ 2}}{\text{within-groups estimate of } \sigma_{\bar{x}}^{\ 2}} = \frac{s_{\bar{x}}^{\ 2}}{\frac{s_{\text{WG}}^2}{N}}$$

Definitional formula for the F ratio

FORMULA 13.2
(page 301)

$$F = \frac{\text{between-groups estimate of } \sigma^2}{\text{within-groups estimate of } \sigma^2} = \frac{Ns_{\bar{x}}^{\ 2}}{s_{\text{WG}}^2}$$

Computational formula for the F ratio

FORMULA 13.3
(page 303)

$$\text{SS}_{\text{WG}} = \Sigma\Sigma X^2 - \frac{\Sigma(\Sigma X)^2}{N}$$

Formula for the sum of squares within groups

FORMULA 13.4
(page 303)

$$\text{SS}_{\text{BG}} = \frac{\Sigma(\Sigma X)^2}{N} - \frac{(\Sigma\Sigma X)^2}{KN}$$

Formula for the sum of squares between groups

FORMULA 13.5 (page 306)

$$SS_{BG} = \Sigma \frac{(\Sigma X)^2}{n_g} - \frac{(\Sigma\Sigma X)^2}{N_{total}}$$

Formula for the sum of squares between groups for unequal sample sizes

FORMULA 13.6 (page 306)

$$SS_{WG} = \Sigma\Sigma X^2 - \Sigma \frac{(\Sigma X)^2}{n_g}$$

Formula for the sum of squares within groups for unequal sample sizes

IV

GLOSSARY OF TERMS

Abscissa The horizontal axis of a graph.

Absolute value The magnitude of a number expressed as a positive value.

Alpha Symbolized α, this is a criterion for deciding what is "too unlikely to occur by chance." Any event with a probability less than alpha is judged "too unlikely to have occurred by chance."

Alpha error A Type I error; rejecting a true H_0.

Alpha level The particular value of alpha selected in a hypothesis-testing procedure.

Alternate hypothesis Symbolized H_A, this generally asserts that your research results are not due to chance sampling error.

Analysis of variance See "one-way analysis of variance."

ANOVA The abbreviation for "analysis of variance."

Argument by contradiction An argument to demonstrate the falsity of a statement. You assume the statement is true, apply logic, and try to find a contradiction.

Axes The horizontal and vertical lines in a graph, upon which values of a measurement or the corresponding frequencies are plotted.

Bar graph A graphical technique of descriptive statistics that uses the heights of separated bars to show how often each score occurs.

Between-groups variance estimate Refers to the numerator of the F ratio, where you have $s_{\bar{x}}^2$ as an estimate of the variance of the sampling distribution of the mean calculated from each sample mean in the one-way analysis of variance.

Beta error A Type II error; retaining a false H_0.

Bimodal Refers to a distribution with two modes.

Causality A relationship of cause and effect; the effect will invariably occur when the cause is present. In behavioral science, causality is usually statistical. That is, changes in the causal variable will alter values of the affected variable on the average.

Cell A compartment in a matrix or table.

Central limit theorem States that with a sufficiently large sample size ($N \geq 25$) the sampling distribution of the mean will be approximately normal with a mean equal to the population mean, μ, and a standard deviation equal to the population standard deviation, σ, divided by the square root of the sample size.

Chi-square Symbolized χ^2, this is a nonparametric technique of inferential statistics that is useful with nominal data. Chi-square measures the discrepancies between expected and observed frequencies. It is defined by Formula 10.1:

$$\chi^2 = \sum \frac{(O - E)^2}{E}$$

Chi-square distribution A distribution of chi-square values that would occur by chance if H_0 is true.

Chi-square goodness-of-fit test A technique of inferential statistics used to decide whether a sample with a given frequency distribution could have occurred by chance from a population with a known frequency distribution (or known percentage composition).

Chi-square test of independence A technique of inferential statistics used with a sample to decide whether two variables represented in a contingency table are dependent.

Class interval The interval between the highest and lowest value in each category of a grouped frequency distribution or histogram.

Coefficient of determination Symbolized r^2, the square of the correlation coefficient, signifies how much you can reduce the uncertainty of prediction if you know the value of a variable that is correlated with another variable.

Condition A treatment condition. Often used synonymously with "group" and "sample" when discussing research designs.

Confounding This occurs when the effects of an extraneous variable cannot be separated from the effects of the variable of interest. The effects of the extraneous variable thus confound the interpretation of the research results.

Contingency table A table showing the joint frequency distribution of two nominal variables. The contingency table shows how often each combination of levels on each variable occurs.

Continuous measurement Measurement in which a potentially infinite number of different values can occur in any given interval.

Control group In an experimental design, the control group is a group in which the independent variable is left unchanged. The control group serves as a reference to compare the effect of manipulating the independent variable in the experimental group(s).

Correlation coefficient Symbolized r, this is a measure of the degree of linear relationship between two variables.

Critical value A value of a test statistic (z, t, χ^2, etc.) that demarcates the region of rejection, and which is thus used as a criterion for significance in hypothesis testing.

Cumulative frequency Abbreviated *cf*, the number of scores falling at or below a given score.

Cumulative frequency distribution A frequency distribution which gives the number of scores occurring at or below each value of the dependent variable.

Cumulative frequency polygon A frequency polygon that shows how often scores occur at or below each value of the dependent variable.

Data The numbers, or scores, generated by a research study. "Data" is plural, and "datum" is singular.

Degrees of freedom Abbreviated *df*, this is a number related to the sample size in a way that depends on the particular statistical technique employed. In many hypothesis tests *df* is used to look up critical values.

Dependent events Events that influence the probability of occurrence of each other.

Dependent *t* test A hypothesis-testing procedure used to decide whether two given dependent samples could have occurred by chance from populations with equal means.

Dependent variable The variable that is measured in a research study.

Descriptive statistics The branch of statistics concerned with describing and summarizing data.

Deviation from the mean The distance of a score from the mean, symbolized $X - \mu$ or $X - \bar{X}$.

Direct relationship A relationship between variables such that both variables either increase together or decrease together; also called a "positive relationship."

Discrete measurement Measurement that can generate only certain values that are separated by discrete intervals.

Distribution The pattern of frequency of occurrence of scores.

Empirical sampling distribution A sampling distribution generated by actually taking random samples and measuring each sample's characteristics.

Experimental group In an experimental design, a group in which the independent variable is manipulated.

Extraneous variable A variable that is of no interest in itself to the experimenter, but might affect the dependent variable and thus confound the interpretation of a research study.

***F* ratio** The between-group estimates of the variance of the sampling distribution of the mean divided by the within-groups estimate; or, equivalently, the between-groups estimate of the population variance divided by the within-groups estimate. The F ratio is a measure of the strength of a treatment effect.

Frequency distribution A technique of descriptive statistics that shows how often each score occurs.

Frequency polygon A graphical technique of descriptive statistics that uses the height of connected dots to show how often each score occurs.

Group A group of subjects. The term "group" is often used synonymously with "sample" and "condition" when discussing research designs.

Grouped cumulative frequency distribution An extension of a grouped frequency distribution that shows how often scores occur in or below each interval.

Grouped frequency distribution A frequency distribution that shows how often scores occur in given intervals.

Grouped frequency polygon A frequency polygon that shows how often scores occur in given intervals.

Grouped histogram A histogram that shows how often scores occur in given intervals.

Histogram A graphical technique of descriptive statistics that uses the heights of adjoining bars to show how often each score occurs.

Homogeneity of variance An assumption (of the two-sample independent t test and of the one-way ANOVA) that the populations from which your samples were drawn have equal variances.

Hypothesis A prediction of the results of a subsequent observation, or a proposed explanation for those results.

Hypothesis testing A technique of inferential statistics which helps you decide whether research results are due to chance sampling error.

Independent events Events which do not influence the probability of occurrence of one another.

Independent samples Samples drawn in such a way that the particular subjects chosen for one sample have no influence on which subjects are chosen for the other sample.

Independent t test A hypothesis-testing procedure used to decide whether two given independent samples could have occurred by chance from populations with equal means.

Independent variable The variable that is manipulated in an experiment.

Inferential statistics The branch of statistics concerned with inferring the characteristics of populations from the characteristics of samples.

Interval scale A scale of measurement in which objects, events, or processes are assigned to ordered categories that are separated by equal intervals.

Inverse relationship A relationship between two variables such that one variable increases as the other variable decreases, or vice versa; also called "negative relationship."

Inversely Two variables vary inversely if one increases as the other decreases, or vice versa.

Line of best fit See "regression line."

Mann-Whitney U test A nonparametric hypothesis-testing procedure used to decide whether two given independent samples would have arisen by chance from identically distributed populations.

Matching Pairing up subjects in different groups on the basis of scores on the dependent variable, or on some related variable.

Mean Symbolized $\bar{X}$ or μ; the arithmetic average of a group of scores. It is defined in Formula 3.4:

$$\text{Mean} = \frac{\Sigma X}{N}$$

Mean squared deviation from the mean The variance, symbolized σ^2. Defined by Formula 3.6:

$$\sigma^2 = \frac{\Sigma(X - \mu)^2}{N}$$

Mean squared difference between z scores Symbolized $\Sigma(z_x - z_y)^2/N$, this is a measure of the degree of relationship between two variables. The calculation of the correlation coefficient, r, derives from this.

Mean squares between groups Symbolized MS_{BG}, this is the between-groups estimate of the population variance in the analysis of variance.

Mean squares within groups Symbolized MS_{WG}, this is the within-groups estimate of the population variance in the analysis of variance.

Measure of central tendency A single number that describes the location, or relative magnitude, of a group of scores; synonymous with "average." The mode, median, and mean are measures of central tendency.

Measure of variability A single number that describes how spread out a group of scores are. The range, variance, and standard deviation are measures of variability.

Median The middle score in a group of scores.

Midpoint of class interval Obtained by averaging the lowest and highest scores of a class interval.

Mode The most frequently occurring score in a group of scores.

Multiple-group design An experimental design with one control group and several experimental groups.

Multiplicative rule A rule of probability that allows you to determine the probability of two independent events occurring together. Formula 6.2 demonstrates the multiplicative rule:

$$\begin{array}{c}\text{probability of an event}\\ \text{occurring } \textit{and} \text{ another}\\ \text{event occurring}\end{array} = \left(\begin{array}{c}\text{probability of}\\ \text{the first}\\ \text{event occurring}\end{array}\right)\left(\begin{array}{c}\text{probability of}\\ \text{the second}\\ \text{event occurring}\end{array}\right)$$

Negatively skewed distribution A distribution whose frequency polygon has an elongated tail extending toward the lower scores.

Negative relationship See "inverse relationship."

Nominal scale A scale of measurement in which objects, events, or processes are assigned to categories having no inherent order.

Nonparametric tests Hypothesis-testing procedures that do not make stringent assumptions about population parameters. The Mann-Whitney U test, Wilcoxon test, and χ^2 tests are examples of nonparametric tests.

Normal distribution A symmetrical, bell-shaped distribution which often arises when a trait is composed of a large number of random, independent factors.

Null hypothesis Symbolized H_0, this generally asserts that your research results are due to chance sampling error.

One-sample design An experimental design in which a random sample is selected from a population and the independent variable is manipulated in the sample.

One-sample *t* test A hypothesis-testing procedure used to decide whether a given sample could have occurred by chance from a population with a given mean, μ, but with an unknown variance.

One-tailed test A hypothesis test in which only high values or only low values of a sample result can be significant.

One-way analysis of variance A hypothesis test used to decide whether two or more samples could have occurred by chance from populations with equal means.

Operational definition A definition written in terms of the operations that must be performed to measure that which is being defined.

Ordinal scale A scale of measurement in which objects, events, or processes are assigned to ordered categories.

Ordinate The vertical axis of a graph.

Origin The point of a graph at which the abscissa and ordinate intersect.

Outcome A possible result of an experiment or observation.

Parameters The characteristics of a population.

Parametric design See "multiple-group design."

Parametric tests Hypothesis-testing procedures that make relatively stringent assumptions about population parameters. The *z* test, *t* tests, and analysis of variance are examples of parametric tests.

Pearson product-moment correlation coefficient See "correlation coefficient."

Percent Synonymous with "in 100," or the number of cases out of one hundred.

Percentile rank A transformed score that tells you the percentage of scores falling at or below a given score.

Perfect relationship A relationship, either direct or inverse, in which there is perfect predictability between two variables; when all points in the scatter plot lie exactly on the regression line.

Placebo An inactive substance given to subjects in drug studies to control for the effects of expectation.

Pooled variance estimate Symbolized s^2_{pooled}, this is an estimate of the population variance obtained by forming a weighted average of the population variance estimates from each sample in the two-sample independent *t* test.

Population All the scores of interest.

Population distribution A distribution of the scores in a population.

Positive relationship See "direct relationship."

Positively skewed distribution A distribution whose frequency polygon has an elongated tail extending toward the higher scores.

Power The probability of rejecting a false H_0.

Predicted variable The Y variable in the regression equation. Given values of X are used to predict $\hat{Y}$ values.

Predictor variable The X variable in the regression equation. Given values of this variable are used to predict values on the Y variable.

Principle of variance constancy This refers to the fact that adding a constant to each score in a group of scores does not affect the variance.

Probability Symbolized p, a measure of likelihood; can be defined as the number of outcomes in which an event can occur divided by the total number of possible outcomes.

Proportion A fraction of one.

Psychotropic Having some psychological effect. Used in reference to drugs.

Random Chance; with no systematic method other than that of chance.

Random sample A sample drawn by a method that ensures that each subject in the population has an equal chance of being included in the sample.

Randomization A technique of random assignment of subjects to groups. Randomization has the effect of making the groups equal, on the average, on all variables.

Range One plus the difference between the highest and lowest scores in a group.

Ratio scale A scale of measurement in which objects, events, or processes are assigned to ordered categories which are separated with equal intervals, and where the zero point is not arbitrary.

Raw score A score directly obtained by measuring some characteristic of an object, event, or process in a research study.

Regions of rejection Areas in the tail(s) of the sampling distribution such that any sample result falling into these areas has a probability of occurring by chance that is less than alpha.

Regression Moving toward the mean.

Regression equation Formula 12.1:

$$\hat{Y} = r\frac{\sigma_y}{\sigma_x}(X - \bar{X}) + \bar{Y}$$

or Formula 12.2:

$$z_{\hat{y}} = rz_x$$

These are the equations for the regression line. They are used to calculate predicted values on the Y variable from given values of the X variable.

Regression line The line of best fit in a scatter plot. It is used to predict values of the Y variable from values of the X variable, and it minimizes the sum of the squared errors in prediction.

Rho Symbolized ρ, this is the population correlation coefficient.

Robustness This refers to the property that certain hypothesis-testing procedures yield the same decision regardless of whether all assumptions for the test are strictly satisfied.

Sample A subset of a population often used synonymously with "group" and "condition" when discussing research designs.

Sample distribution A distribution of the scores in a sample.

Sampling distribution A distribution of the characteristics of a large number of random samples.

Sampling distribution of correlation A distribution of r values that would occur in random samples, when the population correlation coefficient, ρ, is zero.

Sampling distribution of the difference for independent samples A distribution of $\bar{X}_1 - \bar{X}_2$ values (differences between sample means) for pairs of independent samples drawn randomly from populations with equal means.

Sampling distribution of the mean A distribution of the means of random samples.

Sampling distribution of U A distribution of U values for samples drawn randomly from identically distributed populations.

Sampling error Refers to cases when sample statistics differ by chance from population parameters.

Scales of measurement Different ways of categorizing or assigning numbers to objects, events, or processes.

Scatter plot A graphical representation of the relationship between two variables.

Significant Judged too unlikely to have occurred by chance sampling error.

Source table Used in the computational version of the analysis of variance, it is a table that lists the SS, *df,* and MS for both the within- and the between-groups sources of variance, and also gives the F ratio.

Spurious Occurring by chance.

Standard deviation Symbolized σ, the square root of the variance. Defined by Formula 3.7:

$$\sigma = \sqrt{\frac{\Sigma(X - \mu)^2}{N}}$$

Standard error of the difference This refers to the standard deviation of the sampling distribution of the difference for independent samples.

Standard error of the estimate Symbolized $\sigma_{\hat{y}}$, this is the standard deviation of the errors in prediction. $\sigma_{\hat{y}}$ thus serves as a measure of the accuracy in prediction. Defined by Formula 12.3: 12.3:

$$\sigma_{\hat{y}} = \sqrt{\frac{\Sigma(Y - \hat{Y})^2}{N}}$$

Standard error of the mean Symbolized $\sigma_{\bar{x}}$, this is the standard deviation of the sampling distribution of the mean.

Statistics In comparison to the term "parameters," statistics refers to the characteristics of a sample.

Subject A participant in a research study.

Subset A part of a set.

Sum of squares between groups Symbolized SS_{BG}; this is the between-groups estimate of the population variance, but without the degrees of freedom in the denominator. For equal-sized samples it is calculated using Formula 13.4:

$$SS_{BG} = \frac{\Sigma(\Sigma X)^2}{N} - \frac{(\Sigma\Sigma X)^2}{KN}$$

Sum of squares within groups Symbolized SS_{WG}, this is the within-groups estimate of the population variance, but without the degrees of freedom in the denominator. For equal-sized samples it is calculated using Formula 13.3:

$$SS_{WG} = \Sigma\Sigma X^2 - \frac{\Sigma(\Sigma X)^2}{N}$$

Symmetrical distribution A distribution with a frequency polygon whose left and right sides will coincide if the graph is folded in the middle along a vertical line.

***t* score** The estimated number of standard deviations that a sample result lies away from the mean of the sampling distribution.

t test Used to refer to one of the three hypothesis tests in which t scores are calculated: one-sample t test; dependent t test; independent t test.

Theoretical sampling distribution A sampling distribution generated by making assumptions and applying mathematics. The theoretical sampling distribution can be used as an approximation to empirical sampling distributions.

Theory A hypothetical system proposed to explain observable phenomena.

Transformed score A score that allows you to tell at a glance where it falls in a distribution of scores.

Treatment effect An effect on the dependent variable that is caused by changes in the independent variable.

Trend A relationship between two variables.

Two-group dependent design An experimental design with two matched groups.

Two-group independent design An experimental design with two independent randomized groups.

Two-group matched design See "two-group dependent design."

Two-tailed test A hypothesis test in which either high or low values of a sample result could be significant.

Type I error An alpha error; rejecting a true H_0.

Type II error A beta error; retaining a false H_0.

U distribution See "sampling distribution of U."

U test See "Mann-Whitney U test."

U statistic A measure of how much the distributions of two independent samples differ from one another. Computed using Formula 9.1:

$$U = R_1 - \frac{(N_1)(N_1 + 1)}{2}$$

Unimodal Refers to a distribution with only one mode.

Unstable Refers to a statistic that greatly changes value from sample to sample.

Variable Any attribute whose value, or level, can change.

Variance Symbolized σ^2; the mean squared deviation from the mean. Defined by Formula 3.6:

$$\sigma^2 = \frac{\Sigma(X - \mu)^2}{N}$$

Variance constancy See "principle of variance constancy."

Wilcoxon matched-pairs signed-ranks test A nonparametric hypothesis-testing procedure used to decide whether two given dependent samples could have occurred by chance from identically distributed populations.

Within-groups variance estimate Refers to the denominator of the F ratio, where you have s^2_{WG} as an estimate of the population variance calculated by averaging the variance estimates within each treatment condition in the one-way analysis of variance.

***X* variable** The variable plotted on the abscissa of a scatter plot, and the "predictor" variable (used to predict the Y variable) in regression.

***Y* variable** The variable plotted on the ordinate in a scatter plot, and the "predicted" variable (predicted from the X variable) in regression.

***z* score** Symbolized z; a transformed score that tells you how many standard deviations a score lies away from the mean of a distribution.

***z* test** A hypothesis-testing procedure used to decide whether a given sample could have occurred by chance from a population with a given mean, μ, and known standard deviation, σ.

V

TABLES

TABLE A

Table of squares, square roots, and reciprocals of numbers from 1 to 1000

N	N^2	$\sqrt{N}$	$1/N$	N	N^2	$\sqrt{N}$	$1/N$
1	1	1.0000	1.000000	26	676	5.0990	.038462
2	4	1.4142	.500000	27	729	5.1962	.037037
3	9	1.7321	.333333	28	784	5.2915	.035714
4	16	2.0000	.250000	29	841	5.3852	.034483
5	25	2.2361	.200000	30	900	5.4772	.033333
6	36	2.4495	.166667	31	961	5.5678	.032258
7	49	2.6458	.142857	32	1024	5.6569	.031250
8	64	2.8284	.125000	33	1089	5.7446	.030303
9	81	3.0000	.111111	34	1156	5.8310	.029412
10	100	3.1623	.100000	35	1225	5.9161	.028571
11	121	3.3166	.090909	36	1296	6.0000	.027778
12	144	3.4641	.083333	37	1369	6.0828	.027027
13	169	3.6056	.076923	38	1444	6.1644	.026316
14	196	3.7417	.071429	39	1521	6.2450	.025641
15	225	3.8730	.066667	40	1600	6.3246	.025000
16	256	4.0000	.062500	41	1681	6.4031	.024390
17	289	4.1231	.058824	42	1764	6.4807	.023810
18	324	4.2426	.055556	43	1849	6.5574	.023256
19	361	4.3589	.052632	44	1936	6.6332	.022727
20	400	4.4721	.050000	45	2025	6.7082	.022222
21	441	4.5826	.047619	46	2116	6.7823	.021739
22	484	4.6904	.045455	47	2209	6.8557	.021277
23	529	4.7958	.043478	48	2304	6.9282	.020833
24	576	4.8990	.041667	49	2401	7.0000	.020408
25	625	5.0000	.040000	50	2500	7.0711	.020000

TABLE A

Table of squares, square roots, and reciprocals of numbers from 1 to 1000—Continued

N	N^2	$\sqrt{N}$	$1/N$	N	N^2	$\sqrt{N}$	$1/N$
51	2601	7.1414	.019608	91	8281	9.5394	.010989
52	2704	7.2111	.019231	92	8464	9.5917	.010870
53	2809	7.2801	.018868	93	8649	9.6437	.010753
54	2916	7.3485	.018519	94	8836	9.6954	.010638
55	3025	7.4162	.018182	95	9025	9.7468	.010526
56	3136	7.4833	.017857	96	9216	9.7980	.010417
57	3249	7.5498	.017544	97	9409	9.8489	.010309
58	3364	7.6158	.017241	98	9604	9.8995	.010204
59	3481	7.6811	.016949	99	9801	9.9499	.010101
60	3600	7.7460	.016667	100	10000	10.0000	.010000
61	3721	7.8102	.016393	101	10201	10.0499	.00990099
62	3844	7.8740	.016129	102	10404	10.0995	.00980392
63	3969	7.9373	.015873	103	10609	10.1489	.00970874
64	4096	8.0000	.015625	104	10816	10.1980	.00961538
65	4225	8.0623	.015385	105	11025	10.2470	.00952381
66	4356	8.1240	.015152	106	11236	10.2956	.00943396
67	4489	8.1854	.014925	107	11449	10.3441	.00934579
68	4624	8.2462	.014706	108	11664	10.3923	.00925926
69	4761	8.3066	.014493	109	11881	10.4403	.00917431
70	4900	8.3666	.014286	110	12100	10.4881	.00909091
71	5041	8.4261	.014085	111	12321	10.5357	.00900901
72	5184	8.4853	.013889	112	12544	10.5830	.00892857
73	5329	8.5440	.013699	113	12769	10.6301	.00884956
74	5476	8.6023	.013514	114	12996	10.6771	.00877193
75	5625	8.6603	.013333	115	13225	10.7238	.00869565
76	5776	8.7178	.013158	116	13456	10.7703	.00862069
77	5929	8.7750	.012987	117	13689	10.8167	.00854701
78	6084	8.8318	.012821	118	13924	10.8628	.00847458
79	6241	8.8882	.012658	119	14161	10.9087	.00840336
80	6400	8.9443	.012500	120	14400	10.9545	.00833333
81	6561	9.0000	.012346	121	14641	11.0000	.00826446
82	6724	9.0554	.012195	122	14884	11.0454	.00819672
83	6889	9.1104	.012048	123	15129	11.0905	.00813008
84	7056	9.1652	.011905	124	15376	11.1355	.00806452
85	7225	9.2195	.011765	125	15625	11.1803	.00800000
86	7396	9.2736	.011628	126	15876	11.2250	.00793651
87	7569	9.3274	.011494	127	16129	11.2694	.00787402
88	7744	9.3808	.011364	128	16384	11.3137	.00781250
89	7921	9.4340	.011236	129	16641	11.3578	.00775194
90	8100	9.4868	.011111	130	16900	11.4018	.00769231

TABLE A

Table of squares, square roots, and reciprocals of numbers from 1 to 1000—Continued

N	N^2	$\sqrt{N}$	$1/N$	N	N^2	$\sqrt{N}$	$1/N$
131	17161	11.4455	.00763359	171	29241	13.0767	.00584795
132	17424	11.4891	.00757576	172	29584	13.1149	.00581395
133	17689	11.5326	.00751880	173	29929	13.1529	.00578035
134	17956	11.5758	.00746269	174	30276	13.1909	.00574713
135	18225	11.6190	.00740741	175	30625	13.2288	.00571429
136	18496	11.6619	.00735294	176	30976	13.2665	.00568182
137	18769	11.7047	.00729927	177	31329	13.3041	.00564972
138	19044	11.7473	.00724638	178	31684	13.3417	.00561798
139	19321	11.7898	.00719424	179	32041	13.3791	.00558659
140	19600	11.8322	.00714286	180	32400	13.4164	.00555556
141	19881	11.8743	.00709220	181	32761	13.4536	.00552486
142	20164	11.9164	.00704225	182	33124	13.4907	.00549451
143	20449	11.9583	.00699301	183	33489	13.5277	.00546448
144	20736	12.0000	.00694444	184	33856	13.5647	.00543478
145	21025	12.0416	.00689655	185	34225	13.6015	.00540541
146	21316	12.0830	.00684932	186	34596	13.6382	.00537634
147	21609	12.1244	.00680272	187	34969	13.6748	.00534759
148	21904	12.1655	.00675676	188	35344	13.7113	.00531915
149	22201	12.2066	.00671141	189	35721	13.7477	.00529101
150	22500	12.2474	.00666667	190	36100	13.7840	.00526316
151	22801	12.2882	.00662252	191	36481	13.8203	.00523560
152	23104	12.3288	.00657895	192	36864	13.8564	.00520833
153	23409	12.3693	.00653595	193	37249	13.8924	.00518135
154	23716	12.4097	.00649351	194	37636	13.9284	.00515464
155	24025	12.4499	.00645161	195	38025	13.9642	.00512821
156	24336	12.4900	.00641026	196	38416	14.0000	.00510204
157	24649	12.5300	.00636943	197	38809	14.0357	.00507614
158	24964	12.5698	.00632911	198	39204	14.0712	.00505051
159	25281	12.6095	.00628931	199	39601	14.1067	.00502513
160	25600	12.6491	.00625000	200	40000	14.1421	.00500000
161	25921	12.6886	.00621118	201	40401	14.1774	.00497512
162	26244	12.7279	.00617284	202	40804	14.2127	.00495050
163	26569	12.7671	.00613497	203	41209	14.2478	.00492611
164	26896	12.8062	.00609756	204	41616	14.2829	.00490196
165	27225	12.8452	.00606061	205	42025	14.3178	.00487805
166	27556	12.8841	.00602410	206	42436	14.3527	.00485437
167	27889	12.9228	.00598802	207	42849	14.3875	.00483092
168	28224	12.9615	.00595238	208	43264	14.4222	.00480769
169	28561	13.0000	.00591716	209	43681	14.4568	.00478469
170	28900	13.0384	.00588235	210	44100	14.4914	.00476190

TABLE A

Table of squares, square roots, and reciprocals of numbers from 1 to 1000—Continued

N	N^2	$\sqrt{N}$	$1/N$	N	N^2	$\sqrt{N}$	$1/N$
211	44521	14.5258	.00473934	251	63001	15.8430	.00398406
212	44944	14.5602	.00471698	252	63504	15.8745	.00396825
213	45369	14.5945	.00469484	253	64009	15.9060	.00395257
214	45796	14.6287	.00467290	254	64516	15.9374	.00393701
215	46225	14.6629	.00465116	255	65025	15.9687	.00392157
216	46656	14.6969	.00462963	256	65536	16.0000	.00390625
217	47089	14.7309	.00460829	257	66049	16.0312	.00389105
218	47524	14.7648	.00458716	258	66564	16.0624	.00387597
219	47961	14.7986	.00456621	259	67081	16.0935	.00386100
220	48400	14.8324	.00454545	260	67600	16.1245	.00384615
221	48841	14.8661	.00452489	261	68121	16.1555	.00383142
222	49284	14.8997	.00450450	262	68644	16.1864	.00381679
223	49729	14.9332	.00448430	263	69169	16.2173	.00380228
224	50176	14.9666	.00446429	264	69696	16.2481	.00378788
225	50625	15.0000	.00444444	265	70225	16.2788	.00377358
226	51076	15.0333	.00442478	266	70756	16.3095	.00375940
227	51529	15.0665	.00440529	267	71289	16.3401	.00374532
228	51984	15.0997	.00438596	268	71824	16.3707	.00373134
229	52441	15.1327	.00436681	269	72361	16.4012	.00371747
230	52900	15.1658	.00434783	270	72900	16.4317	.00370370
231	53361	15.1987	.00432900	271	73441	16.4621	.00369004
232	53824	15.2315	.00431034	272	73984	16.4924	.00367647
233	54289	15.2643	.00429185	273	74529	16.5227	.00366300
234	54756	15.2971	.00427350	274	75076	16.5529	.00364964
235	55225	15.3297	.00425532	275	75625	16.5831	.00363636
236	55696	15.3623	.00423729	276	76176	16.6132	.00362319
237	56169	15.3948	.00421941	277	76729	16.6433	.00361011
238	56644	15.4272	.00420168	278	77284	16.6733	.00359712
239	57121	15.4596	.00418410	279	77841	16.7033	.00358423
240	57600	15.4919	.00416667	280	78400	16.7332	.00357143
241	58081	15.5242	.00414938	281	78961	16.7631	.00355872
242	58564	15.5563	.00413223	282	79524	16.7929	.00354610
243	59049	15.5885	.00411523	283	80089	16.8226	.00353357
244	59536	15.6205	.00409836	284	80656	16.8523	.00352113
245	60025	15.6525	.00408163	285	81225	16.8819	.00350877
246	60516	15.6844	.00406504	286	81796	16.9115	.00349650
247	61009	15.7162	.00404858	287	82369	16.9411	.00348432
248	61504	15.7480	.00403226	288	82944	16.9706	.00347222
249	62001	15.7797	.00401606	289	83521	17.0000	.00346021
250	62500	15.8114	.00400000	290	84100	17.0294	.00344828

TABLE A

Table of squares, square roots, and reciprocals of numbers from 1 to 1000—Continued

N	N^2	$\sqrt{N}$	$1/N$	N	N^2	$\sqrt{N}$	$1/N$
291	84681	17.0587	.00343643	331	109561	18.1934	.00302115
292	85264	17.0880	.00342466	332	110224	18.2209	.00301205
293	85849	17.1172	.00341297	333	110889	18.2483	.00300300
294	86436	17.1464	.00340136	334	111556	18.2757	.00299401
295	87025	17.1756	.00338983	335	112225	18.3030	.00298507
296	87616	17.2047	.00337838	336	112896	18.3303	.00297619
297	88209	17.2337	.00336700	337	113569	18.3576	.00296736
298	88804	17.2627	.00335570	338	114244	18.3848	.00295858
299	89401	17.2916	.00334448	339	114921	18.4120	.00294985
300	90000	17.3205	.00333333	340	115600	18.4391	.00294118
301	90601	17.3494	.00332226	341	116281	18.4662	.00293255
302	91204	17.3781	.00331126	342	116964	18.4932	.00292398
303	91809	17.4069	.00330033	343	117649	18.5203	.00291545
304	92416	17.4356	.00328947	344	118336	18.5472	.00290698
305	93025	17.4642	.00327869	345	119025	18.5742	.00289855
306	93636	17.4929	.00326797	346	119716	18.6011	.00289017
307	94249	17.5214	.00325733	347	120409	18.6279	.00288184
308	94864	17.5499	.00324675	348	121104	18.6548	.00287356
309	95481	17.5784	.00323625	349	121801	18.6815	.00286533
310	96100	17.6068	.00322581	350	122500	18.7083	.00285714
311	96721	17.6352	.00321543	351	123201	18.7350	.00284900
312	97344	17.6635	.00320513	352	123904	18.7617	.00284091
313	97969	17.6918	.00319489	353	124609	18.7883	.00283286
314	98596	17.7200	.00318471	354	125316	18.8149	.00282486
315	99225	17.7482	.00317460	355	126025	18.8414	.00281690
316	99856	17.7764	.00316456	356	126736	18.8680	.00280899
317	100489	17.8045	.00315457	357	127449	18.8944	.00280112
318	101124	17.8326	.00314465	358	128164	18.9209	.00279330
319	101761	17.8606	.00313480	359	128881	18.9473	.00278552
320	102400	17.8885	.00312500	360	129600	18.9737	.00277778
321	103041	17.9165	.00311526	361	130321	19.0000	.00277008
322	103684	17.9444	.00310559	362	131044	19.0263	.00276243
323	104329	17.9722	.00309598	363	131769	19.0526	.00275482
324	104976	18.0000	.00308642	364	132496	19.0788	.00274725
325	105625	18.0278	.00307692	365	133225	19.1050	.00273973
326	106276	18.0555	.00306748	366	133956	19.1311	.00273224
327	106929	18.0831	.00305810	367	134689	19.1572	.00272480
328	107584	18.1108	.00304878	368	135424	19.1833	.00271739
329	108241	18.1384	.00303951	369	136161	19.2094	.00271003
330	108900	18.1659	.00303030	370	136900	19.2354	.00270270

TABLE A

Table of squares, square roots, and reciprocals of numbers from 1 to 1000—Continued

N	N^2	$\sqrt{N}$	$1/N$	N	N^2	$\sqrt{N}$	$1/N$
371	137641	19.2614	.00269542	411	168921	20.2731	.00243309
372	138384	19.2873	.00268817	412	169744	20.2978	.00242718
373	139129	19.3132	.00268097	413	170569	20.3224	.00242131
374	139876	19.3391	.00267380	414	171396	20.3470	.00241546
375	140625	19.3649	.00266667	415	172225	20.3715	.00240964
376	141376	19.3907	.00265957	416	173056	20.3961	.00240385
377	142129	19.4165	.00265252	417	173889	20.4206	.00239808
378	142884	19.4422	.00264550	418	174724	20.4450	.00239234
379	143641	19.4679	.00263852	419	175561	20.4695	.00238663
380	144400	19.4936	.00263158	420	176400	20.4939	.00238095
381	145161	19.5192	.00262467	421	177241	20.5183	.00237530
382	145924	19.5448	.00261780	422	178084	20.5426	.00236967
383	146689	19.5704	.00261097	423	178929	20.5670	.00236407
384	147456	19.5959	.00260417	424	179776	20.5913	.00235849
385	148225	19.6214	.00259740	425	180625	20.6155	.00235294
386	148996	19.6469	.00259067	426	181476	20.6398	.00234742
387	149769	19.6723	.00258398	427	182329	20.6640	.00234192
388	150544	19.6977	.00257732	428	183184	20.6882	.00233645
389	151321	19.7231	.00257069	429	184041	20.7123	.00233100
390	152100	19.7484	.00256410	430	184900	20.7364	.00232558
391	152881	19.7737	.00255754	431	185761	20.7605	.00232019
392	153664	19.7990	.00255102	432	186624	20.7846	.00231481
393	154449	19.8242	.00254453	433	187489	20.8087	.00230947
394	155236	19.8494	.00253807	434	188356	20.8327	.00230415
395	156025	19.8746	.00253165	435	189225	20.8567	.00229885
396	156816	19.8997	.00252525	436	190096	20.8806	.00229358
397	157609	19.9249	.00251889	437	190969	20.9045	.00228833
398	158404	19.9499	.00251256	438	191844	20.9284	.00228311
399	159201	19.9750	.00250627	439	192721	20.9523	.00227790
400	160000	20.0000	.00250000	440	193600	20.9762	.00227273
401	160801	20.0250	.00249377	441	194481	21.0000	.00226757
402	161604	20.0499	.00248756	442	195364	21.0238	.00226244
403	162409	20.0749	.00248139	443	196249	21.0476	.00225734
404	163216	20.0998	.00247525	444	197136	21.0713	.00225225
405	164025	20.1246	.00246914	445	198025	21.0950	.00224719
406	164836	20.1494	.00246305	446	198916	21.1187	.00224215
407	165649	20.1742	.00245700	447	199809	21.1424	.00223714
408	166464	20.1990	.00245098	448	200704	21.1660	.00223214
409	167281	20.2237	.00244499	449	201601	21.1896	.00222717
410	168100	20.2485	.00243902	450	202500	21.2132	.00222222

TABLE A

Table of squares, square roots, and reciprocals of numbers from 1 to 1000—Continued

N	N^2	$\sqrt{N}$	$1/N$	N	N^2	$\sqrt{N}$	$1/N$
451	203401	21.2368	.00221729	491	241081	22.1585	.00203666
452	204304	21.2603	.00221239	492	242064	22.1811	.00203252
453	205209	21.2838	.00220751	493	243049	22.2036	.00202840
454	206116	21.3073	.00220264	494	244036	22.2261	.00202429
455	207025	21.3307	.00219780	495	245025	22.2486	.00202020
456	207936	21.3542	.00219298	496	246016	22.2711	.00201613
457	208849	21.3776	.00218818	497	247009	22.2935	.00201207
458	209764	21.4009	.00218341	498	248004	22.3159	.00200803
459	210681	21.4243	.00217865	499	249001	22.3383	.00200401
460	211600	21.4476	.00217391	500	250000	22.3607	.00200000
461	212521	21.4709	.00216920	501	251001	22.3830	.00199601
462	213444	21.4942	.00216450	502	252004	22.4054	.00199203
463	214369	21.5174	.00215983	503	253009	22.4277	.00198807
464	215296	21.5407	.00215517	504	254016	22.4499	.00198413
465	216225	21.5639	.00215054	505	255025	22.4722	.00198020
466	217156	21.5870	.00214592	506	256036	22.4944	.00197628
467	218089	21.6102	.00214133	507	257049	22.5167	.00197239
468	219024	21.6333	.00213675	508	258064	22.5389	.00196850
469	219961	21.6564	.00213220	509	259081	22.5610	.00196464
470	220900	21.6795	.00212766	510	260100	22.5832	.00196078
471	221841	21.7025	.00212314	511	261121	22.6053	.00195695
472	222784	21.7256	.00211864	512	262144	22.6274	.00195312
473	223729	21.7486	.00211416	513	263169	22.6495	.00194932
474	224676	21.7715	.00210970	514	264196	22.6716	.00194553
475	225625	21.7945	.00210526	515	265225	22.6936	.00194175
476	226576	21.8174	.00210084	516	266256	22.7156	.00193798
477	227529	21.8403	.00209644	517	267289	22.7376	.00193424
478	228484	21.8632	.00209205	518	268324	22.7596	.00193050
479	229441	21.8861	.00208768	519	269361	22.7816	.00192678
480	230400	21.9089	.00208333	520	270400	22.8035	.00192308
481	231361	21.9317	.00207900	521	271441	22.8254	.00191939
482	232324	21.9545	.00207469	522	272484	22.8473	.00191571
483	233289	21.9773	.00207039	523	273529	22.8692	.00191205
484	234256	22.0000	.00206612	524	274576	22.8910	.00190840
485	235225	22.0227	.00206186	525	275625	22.9129	.00190476
486	236196	22.0454	.00205761	526	276676	22.9347	.00190114
487	237169	22.0681	.00205339	527	277729	22.9565	.00189753
488	238144	22.0907	.00204918	528	278784	22.9783	.00189394
489	239121	22.1133	.00204499	529	279841	23.0000	.00189036
490	240100	22.1359	.00204082	530	280900	23.0217	.00188679

TABLE A

Table of squares, square roots, and reciprocals of numbers from 1 to 1000—Continued

N	N^2	$\sqrt{N}$	$1/N$	N	N^2	$\sqrt{N}$	$1/N$
531	281961	23.0434	.00188324	571	326041	23.8956	.00175131
532	283024	23.0651	.00187970	572	327184	23.9165	.00174825
533	284089	23.0868	.00187617	573	328329	23.9374	.00174520
534	285156	23.1084	.00187266	574	329476	23.9583	.00174216
535	286225	23.1301	.00186916	575	330625	23.9792	.00173913
536	287296	23.1517	.00186567	576	331776	24.0000	.00173611
537	288369	23.1733	.00186220	577	332929	24.0208	.00173310
538	289444	23.1948	.00185874	578	334084	24.0416	.00173010
539	290521	23.2164	.00185529	579	335241	24.0624	.00172712
540	291600	23.2379	.00185185	580	336400	24.0832	.00172414
541	292681	23.2594	.00184843	581	337561	24.1039	.00172117
542	293764	23.2809	.00184502	582	338724	24.1247	.00171821
543	294849	23.3024	.00184162	583	339889	24.1454	.00171527
544	295936	23.3238	.00183824	584	341056	24.1661	.00171233
545	297025	23.3452	.00183486	585	342225	24.1868	.00170940
546	298116	23.3666	.00183150	586	343396	24.2074	.00170648
547	299209	23.3880	.00182815	587	344569	24.2281	.00170358
548	300304	23.4094	.00182482	588	345744	24.2487	.00170068
549	301401	23.4307	.00182149	589	346921	24.2693	.00169779
550	302500	23.4521	.00181818	590	348100	24.2899	.00169492
551	303601	23.4734	.00181488	591	349281	24.3105	.00169205
552	304704	23.4947	.00181159	592	350464	24.3311	.00168919
553	305809	23.5160	.00180832	593	351649	24.3516	.00168634
554	306916	23.5372	.00180505	594	352836	24.3721	.00168350
555	308025	23.5584	.00180180	595	354025	24.3926	.00168067
556	309136	23.5797	.00179856	596	355216	24.4131	.00167785
557	310249	23.6008	.00179533	597	356409	24.4336	.00167504
558	311364	23.6220	.00179211	598	357604	24.4540	.00167224
559	312481	23.6432	.00178891	599	358801	24.4745	.00166945
560	313600	23.6643	.00178571	600	360000	24.4949	.00166667
561	314721	23.6854	.00178253	601	361201	24.5153	.00166389
562	315844	23.7065	.00177936	602	362404	24.5357	.00166113
563	316969	23.7276	.00177620	603	363609	24.5561	.00165837
564	318096	23.7487	.00177305	604	364816	24.5764	.00165563
565	319225	23.7697	.00176991	605	366025	24.5967	.00165289
566	320356	23.7908	.00176678	606	367236	24.6171	.00165017
567	321489	23.8118	.00176367	607	368449	24.6374	.00164745
568	322624	23.8328	.00176056	608	369664	24.6577	.00164474
569	323761	23.8537	.00175747	609	370881	24.6779	.00164204
570	324900	23.8747	.00175439	610	372100	24.6982	.00163934

TABLE A

Table of squares, square roots, and reciprocals of numbers from 1 to 1000—Continued

N	N^2	$\sqrt{N}$	$1/N$	N	N^2	$\sqrt{N}$	$1/N$
611	373321	24.7184	.00163666	651	423801	25.5147	.00153610
612	374544	24.7386	.00163399	652	425104	25.5343	.00153374
613	375769	24.7588	.00163132	653	426409	25.5539	.00153139
614	376996	24.7790	.00162866	654	427716	25.5734	.00152905
615	378225	24.7992	.00162602	655	429025	25.5930	.00152672
616	379456	24.8193	.00162338	656	430336	25.6125	.00152439
617	380689	24.8395	.00162075	657	431649	25.6320	.00152207
618	381924	24.8596	.00161812	658	432964	25.6515	.00151976
619	383161	24.8797	.00161551	659	434281	25.6710	.00151745
620	384400	24.8998	.00161290	660	435600	25.6905	.00151515
621	385641	24.9199	.00161031	661	436921	25.7099	.00151286
622	386884	24.9399	.00160772	662	438244	25.7294	.00151057
623	388129	24.9600	.00160514	663	439569	25.7488	.00150830
624	389376	24.9800	.00160256	664	440896	25.7682	.00150602
625	390625	25.0000	.00160000	665	442225	25.7876	.00150376
626	391876	25.0200	.00159744	666	443556	25.8070	.00150150
627	393129	25.0400	.00159490	667	444889	25.8263	.00149925
628	394384	25.0599	.00159236	668	446224	25.8457	.00149701
629	395641	25.0799	.00158983	669	447561	25.8650	.00149477
630	396900	25.0998	.00158730	670	448900	25.8844	.00149254
631	398161	25.1197	.00158479	671	450241	25.9037	.00149031
632	399424	25.1396	.00158228	672	451584	25.9230	.00148810
633	400689	25.1595	.00157978	673	452929	25.9422	.00148588
634	401956	25.1794	.00157729	674	454276	25.9615	.00148368
635	403225	25.1992	.00157480	675	455625	25.9808	.00148148
636	404496	25.2190	.00157233	676	456976	26.0000	.00147929
637	405769	25.2389	.00156986	677	458329	26.0192	.00147710
638	407044	25.2587	.00156740	678	459684	26.0384	.00147493
639	408321	25.2784	.00156495	679	461041	26.0576	.00147275
640	409600	25.2982	.00156250	680	462400	26.0768	.00147059
641	410881	25.3180	.00156006	681	463761	26.0960	.00146843
642	412164	25.3377	.00155763	682	465124	26.1151	.00146628
643	413449	25.3574	.00155521	683	466489	26.1343	.00146413
644	414736	25.3772	.00155280	684	467856	26.1534	.00146199
645	416025	25.3969	.00155039	685	469225	26.1725	.00145985
646	417316	25.4165	.00154799	686	470596	26.1916	.00145773
647	418609	25.4362	.00154560	687	471969	26.2107	.00145560
648	419904	25.4558	.00154321	688	473344	26.2298	.00145349
649	421201	25.4755	.00154083	689	474721	26.2488	.00145138
650	422500	25.4951	.00153846	690	476100	26.2679	.00144928

TABLE A

Table of squares, square roots, and reciprocals of numbers from 1 to 1000—Continued

N	N^2	$\sqrt{N}$	$1/N$	N	N^2	$\sqrt{N}$	$1/N$
691	477481	26.2869	.00144718	731	534361	27.0370	.00136799
692	478864	26.3059	.00144509	732	535824	27.0555	.00136612
693	480249	26.3249	.00144300	733	537289	27.0740	.00136426
694	481636	26.3439	.00144092	734	538756	27.0924	.00136240
695	483025	26.3629	.00143885	735	540225	27.1109	.00136054
696	484416	26.3818	.00143678	736	541696	27.1293	.00135870
697	485809	26.4008	.00143472	737	543169	27.1477	.00135685
698	487204	26.4197	.00143266	738	544644	27.1662	.00135501
699	488601	26.4386	.00143062	739	546121	27.1846	.00135318
700	490000	26.4575	.00142857	740	547600	27.2029	.00135135
701	491401	26.4764	.00142653	741	549081	27.2213	.00134953
702	492804	26.4953	.00142450	742	550564	27.2397	.00134771
703	494209	26.5141	.00142248	743	552049	27.2580	.00134590
704	495616	26.5330	.00142045	744	553536	27.2764	.00134409
705	497025	26.5518	.00141844	745	555025	27.2947	.00134228
706	498436	26.5707	.00141643	746	556516	28.3130	.00134048
707	499849	26.5895	.00141443	747	558009	28.3313	.00133869
708	501264	26.6083	.00141243	748	559504	28.3496	.00133690
709	502681	26.6271	.00141044	749	561001	28.3679	.00133511
710	504100	26.6458	.00140845	750	562500	27.3861	.00133333
711	505521	26.6646	.00140647	751	564001	27.4044	.00133156
712	506944	26.6833	.00140449	752	565504	27.4226	.00132979
713	508369	26.7021	.00140252	753	567009	27.4408	.00132802
714	509796	26.7208	.00140056	754	568516	27.4591	.00132626
715	511225	26.7395	.00139860	755	570025	27.4773	.00132450
716	512656	26.7582	.00139665	756	571536	27.4955	.00132275
717	514089	26.7769	.00139470	757	573049	27.5136	.00132100
718	515524	26.7955	.00139276	758	574564	27.5318	.00131926
719	516961	26.8142	.00139082	759	576081	27.5500	.00131752
720	518400	26.8328	.00138889	760	577600	27.5681	.00131579
721	519841	26.8514	.00138696	761	579121	27.5862	.00131406
722	521284	26.8701	.00138504	762	580644	27.6043	.00131234
723	522729	26.8887	.00138313	763	582169	27.6225	.00131062
724	524176	26.9072	.00138122	764	583696	27.6405	.00130890
725	525625	26.9258	.00137931	765	585225	27.6586	.00130719
726	527076	26.9444	.00137741	766	586756	27.6767	.00130548
727	528529	26.9629	.00137552	767	588289	27.6948	.00130378
728	529984	26.9815	.00137363	768	589824	27.7128	.00130208
729	531441	27.0000	.00137174	769	591361	27.7308	.00130039
730	532900	27.0185	.00136986	770	592900	27.7489	.00129870

TABLE A

Table of squares, square roots, and reciprocals of numbers from 1 to 1000—Continued

N	N^2	$\sqrt{N}$	$1/N$	N	N^2	$\sqrt{N}$	$1/N$
771	594441	27.7669	.00129702	811	657721	28.4781	.00123305
772	595984	27.7849	.00129534	812	659344	28.4956	.00123153
773	597529	27.8029	.00129366	813	660969	28.5132	.00123001
774	599076	27.8209	.00129199	814	662596	28.5307	.00122850
775	600625	27.8388	.00129032	815	664225	28.5482	.00122699
776	602176	27.8568	.00128866	816	665856	28.5657	.00122549
777	603729	27.8747	.00128700	817	667489	28.5832	.00122399
778	605284	27.8927	.00128535	818	669124	28.6007	.00122249
779	606841	27.9106	.00128370	819	670761	28.6182	.00122100
780	608400	27.9285	.00128205	820	672400	28.6356	.00121951
781	609961	27.9464	.00128041	821	674041	28.6531	.00121803
782	611524	27.9643	.00127877	822	675684	28.6705	.00121655
783	613089	27.9821	.00127714	823	677329	28.6880	.00121507
784	614656	28.0000	.00127551	824	678976	28.7054	.00121359
785	616225	28.0179	.00127389	825	680625	28.7228	.00121212
786	617796	28.0357	.00127226	826	682276	28.7402	.00121065
787	619369	28.0535	.00127065	827	683929	28.7576	.00120919
788	620944	28.0713	.00126904	828	685584	28.7750	.00120773
789	622521	28.0891	.00126743	829	687241	28.7924	.00120627
790	624100	28.1069	.00126582	830	688900	28.8097	.00120482
791	625681	28.1247	.00126422	831	690561	28.8271	.00120337
792	627264	28.1425	.00126263	832	692224	28.8444	.00120192
793	628849	28.1603	.00126103	833	693889	28.8617	.00120048
794	630436	28.1780	.00125945	834	695556	28.8791	.00119904
795	632025	28.1957	.00125786	835	697225	28.8964	.00119760
796	633616	28.2135	.00125628	836	698896	28.9137	.00119617
797	635209	28.2312	.00125471	837	700569	28.9310	.00119474
798	636804	28.2489	.00125313	838	702244	28.9482	.00119332
799	638401	28.2666	.00125156	839	703921	28.9655	.00119190
800	640000	28.2843	.00125000	840	705600	28.9828	.00119048
801	641601	28.3019	.00124844	841	707281	29.0000	.00118906
802	643204	28.3196	.00124688	842	708964	29.0172	.00118765
803	644809	28.3373	.00124533	843	710649	29.0345	.00118624
804	646416	28.3549	.00124378	844	712336	29.0517	.00118483
805	648025	28.3725	.00124224	845	714025	29.0689	.00118343
806	649636	28.3901	.00124069	846	715716	29.0861	.00118203
807	651249	28.4077	.00123916	847	717409	29.1033	.00118064
808	652864	28.4253	.00123762	848	719104	29.1204	.00117925
809	654481	28.4429	.00123609	849	720801	29.1376	.00117786
810	656100	28.4605	.00123457	850	722500	29.1548	.00117647

TABLE A

Table of squares, square roots, and reciprocals of numbers from 1 to 1000—Continued

N	N^2	$\sqrt{N}$	$1/N$	N	N^2	$\sqrt{N}$	$1/N$
851	724201	29.1719	.00117509	891	793881	29.8496	.00112233
852	725904	29.1890	.00117371	892	795664	29.8664	.00112108
853	727609	29.2062	.00117233	893	797449	29.8831	.00111982
854	729316	29.2233	.00117096	894	799236	29.8998	.00111857
855	731025	29.2404	.00116959	895	801025	29.9166	.00111732
856	732736	29.2575	.00116822	896	802816	29.9333	.00111607
857	734449	29.2746	.00116686	897	804609	29.9500	.00111483
858	736164	29.2916	.00116550	898	806404	29.9666	.00111359
859	737881	29.3087	.00116414	899	808201	29.9833	.00111235
860	739600	29.3258	.00116279	900	810000	30.0000	.00111111
861	741321	29.3428	.00116144	901	811801	30.0167	.00110988
862	743044	29.3598	.00116009	902	813604	30.0333	.00110865
863	744769	29.3769	.00115875	903	815409	30.0500	.00110742
864	746496	29.3939	.00115741	904	817216	30.0666	.00110619
865	748225	29.4109	.00115607	905	819025	30.0832	.00110497
866	749956	29.4279	.00115473	906	820836	30.0998	.00110375
867	751689	29.4449	.00115340	907	822649	30.1164	.00110254
868	753424	29.4618	.00115207	908	824464	30.1330	.00110132
869	755161	29.4788	.00115075	909	826281	30.1496	.00110011
870	756900	29.4958	.00114943	910	828100	30.1662	.00109890
871	758641	29.5127	.00114811	911	829921	30.1828	.00109769
872	760384	29.5296	.00114679	912	831744	30.1993	.00109649
873	762129	29.5466	.00114548	913	833569	30.2159	.00109529
874	763876	29.5635	.00114416	914	835396	30.2324	.00109409
875	765625	29.5804	.00114286	915	837225	30.2490	.00109290
876	767376	29.5973	.00114155	916	839056	30.2655	.00109170
877	769129	29.6142	.00114025	917	840889	30.2820	.00109051
878	770884	29.6311	.00113895	918	842724	30.2985	.00108932
879	772641	29.6479	.00113766	919	844561	30.3150	.00108814
880	774400	29.6648	.00113636	920	846400	30.3315	.00108696
881	776161	29.6816	.00113507	921	848241	30.3480	.00108578
882	777924	29.6985	.00113379	922	850084	30.3645	.00108460
883	779689	29.7153	.00113250	923	851929	30.3809	.00108342
884	781456	29.7321	.00113122	924	853776	30.3974	.00108225
885	783225	29.7489	.00112994	925	855625	30.4138	.00108108
886	784996	29.7658	.00112867	926	857476	30.4302	.00107991
887	786769	29.7825	.00112740	927	859329	30.4467	.00107875
888	788544	29.7993	.00112613	928	861184	30.4631	.00107759
889	790321	29.8161	.00112486	929	863041	30.4795	.00107643
890	792100	29.8329	.00112360	930	864900	30.4959	.00107527

TABLE A

Table of squares, square roots, and reciprocals of numbers from 1 to 1000—Continued

N	N^2	$\sqrt{N}$	$1/N$	N	N^2	$\sqrt{N}$	$1/N$
931	866761	30.5123	.00107411	966	933156	31.0805	.00103520
932	868624	30.5287	.00107296	967	935089	31.0966	.00103413
933	870489	30.5450	.00107181	968	937024	31.1127	.00103306
934	872356	30.5614	.00107066	969	938961	31.1288	.00103199
935	874225	30.5778	.00106952	970	940900	31.1448	.00103093
936	876096	30.5941	.00106838	971	942841	31.1609	.00102987
937	877969	30.6105	.00106724	972	944784	31.1769	.00102881
938	879844	30.6268	.00106610	973	946729	31.1929	.00102775
939	881721	30.6431	.00106496	974	948676	31.2090	.00102669
940	883600	30.6594	.00106383	975	950625	31.2250	.00102564
941	885481	30.6757	.00106270	976	952576	31.2410	.00102459
942	887364	30.6920	.00106157	977	954529	31.2570	.00102354
943	889249	30.7083	.00106045	978	956484	31.2730	.00102249
944	891136	30.7246	.00105932	979	958441	31.2890	.00102145
945	893025	30.7409	.00105820	980	960400	31.3050	.00102041
946	894916	30.7571	.00105708	981	962361	31.3209	.00101937
947	896809	30.7734	.00105597	982	964324	31.3369	.00101833
948	898704	30.7896	.00105485	983	966289	31.3528	.00101729
949	900601	30.8058	.00105374	984	968256	31.3688	.00101626
950	902500	30.8221	.00105263	985	970225	31.3847	.00101523
951	904401	30.8383	.00105152	986	972196	31.4006	.00101420
952	906304	30.8545	.00105042	987	974169	31.4166	.00101317
953	908209	30.8707	.00104932	988	976144	31.4325	.00101215
954	910116	30.8869	.00104822	989	978121	31.4484	.00101112
955	912025	30.9031	.00104712	990	980100	31.4643	.00101010
956	913936	30.9192	.00104603	991	982081	31.4802	.00100908
957	915849	30.9354	.00104493	992	984064	31.4960	.00100806
958	917764	30.9516	.00104384	993	986049	31.5119	.00100705
959	919681	30.9677	.00104275	994	988036	31.5278	.00100604
960	921600	30.9839	.00104167	995	990025	31.5436	.00100503
961	923521	31.0000	.00104058	996	992016	31.5595	.00100402
962	925444	31.0161	.00103950	997	994009	31.5753	.00100301
963	927369	31.0322	.00103842	998	996004	31.5911	.00100200
964	929296	31.0483	.00103734	999	998001	31.6070	.00100100
965	931225	31.0644	.00103627	1000	1000000	31.6228	.00100000

TABLE χ^2

Critical values of χ^2 for the chi-square tests

df	α = .05	α = .01
1	3.841	6.635
2	5.991	9.210
3	7.815	11.345
4	9.488	13.277
5	11.070	15.086
6	12.592	16.812
7	14.067	18.475
8	15.507	20.090
9	16.919	21.666
10	18.307	23.209
11	19.675	24.725
12	21.026	26.217
13	22.362	27.688
14	23.685	29.141
15	24.996	30.578
16	26.296	32.000
17	27.587	33.409
18	28.869	34.805
19	30.144	36.191
20	31.410	37.566
21	32.671	38.932
22	33.924	40.289
23	35.172	41.638
24	36.415	42.980
25	37.652	44.314
26	38.885	45.642
27	40.113	46.963
28	41.337	48.278
29	42.557	49.588
30	43.773	50.892
32	46.194	53.486
34	48.602	56.061
36	50.999	58.619
38	53.384	61.162
40	55.759	63.691
42	58.124	66.206
44	60.481	68.710
46	62.830	71.201
48	65.171	73.683
50	67.505	76.154
52	69.832	78.616
54	72.153	81.069
56	74.468	83.513
58	76.778	85.950
60	79.082	88.379
62	81.381	90.802
64	83.675	93.217
66	85.965	95.626
68	88.250	98.028
70	90.531	100.425

If your calculated χ^2 is greater than χ^{2*}, reject H_0. If your value of *df* is not listed, use χ^{2*} for the next larger value of *df*.

From Table IV of Fisher and Yates: *Statistical Tables of Biological, Agricultural and Medical Research,* 6th edition, 1974, published by Longman Group Ltd., London. (previously published by Oliver and Boyd, Edinburgh), and by permission of the authors and publishers.

TABLE F*

Critical values of F for the analysis of variance

Degrees of freedom for denominator	Degrees of freedom for numerator																						
	1	2	3	4	5	6	7	8	9	10	11	12	14	16	20	24	30	40	50	75	100	200	500	∞
1	161	200	216	225	230	234	237	239	241	242	243	244	245	246	248	249	250	251	252	253	253	254	254	254
	4052	**4999**	**5403**	**5625**	**5764**	**5859**	**5928**	**5981**	**6022**	**6056**	**6082**	**6106**	**6142**	**6169**	**6208**	**6234**	**6258**	**6286**	**6302**	**6323**	**6334**	**6352**	**6361**	**6366**
2	18.51	19.00	19.16	19.25	19.30	19.33	19.36	19.37	19.38	19.39	19.40	19.41	19.42	19.43	19.44	19.45	19.46	19.47	19.47	19.48	19.49	19.49	19.50	19.50
	98.49	**99.01**	**99.17**	**99.25**	**99.30**	**99.33**	**99.34**	**99.36**	**99.38**	**99.40**	**99.41**	**99.42**	**99.43**	**19.44**	**99.45**	**99.46**	**99.47**	**99.48**	**99.48**	**99.49**	**99.49**	**99.49**	**99.50**	**99.50**
3	10.13	9.55	9.28	9.12	9.01	8.94	8.88	8.84	8.81	8.78	8.76	8.74	8.71	8.69	8.66	8.64	8.62	8.60	8.58	8.57	8.56	8.54	8.54	8.53
	34.12	**30.81**	**29.46**	**28.71**	**28.24**	**27.91**	**27.67**	**27.49**	**27.34**	**27.23**	**27.13**	**27.05**	**26.92**	**26.83**	**26.69**	**26.60**	**26.50**	**26.41**	**26.30**	**26.27**	**26.23**	**26.18**	**26.14**	**26.12**
4	7.71	6.94	6.59	6.39	6.26	6.16	6.09	6.04	6.00	5.96	5.93	5.91	5.87	5.84	5.80	5.77	5.74	5.71	5.70	5.68	5.66	5.65	5.64	5.53
	21.20	**18.00**	**16.69**	**15.98**	**15.52**	**15.21**	**14.98**	**14.80**	**14.66**	**14.54**	**14.45**	**14.37**	**14.24**	**14.15**	**14.02**	**13.93**	**13.83**	**13.74**	**13.69**	**13.61**	**13.57**	**13.52**	**13.48**	**13.46**
5	6.61	5.79	5.41	5.19	5.05	4.95	4.88	4.82	4.78	4.74	4.70	4.68	4.64	4.60	4.56	4.53	4.50	4.46	4.44	4.42	4.40	4.38	4.37	4.36
	16.26	**13.27**	**12.06**	**11.39**	**10.97**	**10.67**	**10.45**	**10.27**	**10.15**	**10.05**	**9.96**	**9.89**	**9.77**	**9.68**	**9.55**	**9.47**	**9.38**	**9.29**	**9.24**	**9.17**	**9.13**	**9.07**	**9.04**	**9.02**
6	5.99	5.14	4.76	4.53	4.39	4.28	4.21	4.15	4.10	4.06	4.03	4.00	3.96	3.92	3.87	3.84	3.81	3.77	3.75	3.72	3.71	3.69	3.68	3.67
	13.74	**10.92**	**9.78**	**9.15**	**8.75**	**8.47**	**8.26**	**8.10**	**7.98**	**7.87**	**7.79**	**7.72**	**7.60**	**7.52**	**7.39**	**7.31**	**7.23**	**7.14**	**7.09**	**7.02**	**6.99**	**6.94**	**6.90**	**6.88**
7	5.59	4.74	4.35	4.12	3.97	3.87	3.79	3.73	3.68	3.63	3.60	3.57	3.52	3.49	3.44	3.41	3.38	3.34	3.32	3.29	3.28	3.25	3.24	3.23
	12.25	**9.55**	**8.45**	**7.85**	**7.46**	**7.19**	**7.00**	**6.84**	**6.71**	**6.62**	**6.54**	**6.47**	**6.35**	**6.27**	**6.15**	**6.07**	**5.98**	**5.90**	**5.85**	**5.78**	**5.75**	**5.70**	**5.67**	**5.65**
8	5.32	4.46	4.07	3.84	3.69	3.58	3.50	3.44	3.39	3.34	3.31	3.28	3.21	3.20	3.15	3.12	3.08	3.05	3.03	3.00	2.98	2.96	2.94	2.93
	11.26	**8.65**	**7.59**	**7.01**	**6.63**	**6.37**	**6.19**	**6.03**	**5.91**	**5.82**	**5.74**	**5.67**	**5.56**	**5.48**	**5.36**	**5.28**	**5.20**	**5.11**	**5.06**	**5.00**	**4.96**	**4.91**	**4.88**	**4.86**
9	5.12	4.26	3.86	3.63	3.48	3.37	3.29	3.23	3.18	3.13	3.10	3.07	3.02	2.98	2.93	2.90	2.86	2.82	2.80	2.77	2.76	2.73	2.72	2.71
	10.56	**8.02**	**6.99**	**6.42**	**6.06**	**5.80**	**5.62**	**5.47**	**5.35**	**5.26**	**5.18**	**5.11**	**5.00**	**4.92**	**4.80**	**4.73**	**4.64**	**4.56**	**4.51**	**4.45**	**4.41**	**4.36**	**4.33**	**4.31**
10	4.96	4.10	3.71	3.48	3.33	3.22	3.14	3.07	3.02	2.97	2.94	2.91	2.86	2.82	2.77	2.74	2.70	2.67	2.64	2.61	2.59	2.56	2.55	2.54
	10.04	**7.56**	**6.55**	**5.99**	**5.64**	**5.39**	**5.21**	**5.06**	**4.95**	**4.85**	**4.78**	**4.71**	**4.60**	**4.52**	**4.41**	**4.33**	**4.25**	**4.17**	**4.12**	**4.05**	**4.01**	**3.96**	**3.93**	**3.91**
11	4.84	3.98	3.59	3.36	3.20	3.09	3.01	2.95	2.90	2.86	2.82	2.79	2.74	2.70	2.65	2.61	2.57	2.53	2.50	2.47	2.45	2.42	2.41	2.40
	9.65	**7.20**	**6.22**	**5.67**	**5.32**	**5.07**	**4.88**	**4.74**	**4.63**	**4.54**	**4.46**	**4.40**	**4.29**	**4.21**	**4.10**	**4.02**	**3.94**	**3.86**	**3.80**	**3.74**	**3.70**	**3.66**	**3.62**	**3.60**
12	4.75	3.88	3.49	3.26	3.11	3.00	2.92	2.85	2.80	2.76	2.72	2.69	2.64	2.60	2.54	2.50	2.46	2.42	2.40	2.36	2.35	2.32	2.31	2.30
	9.33	**6.93**	**5.95**	**5.41**	**5.06**	**4.82**	**4.65**	**4.50**	**4.39**	**4.30**	**4.22**	**4.16**	**4.05**	**3.98**	**3.86**	**3.78**	**3.70**	**3.61**	**3.56**	**3.49**	**3.46**	**3.41**	**3.38**	**3.36**

Degrees of freedom for denominator																								
13	4.67	3.80	3.41	3.18	3.02	2.92	2.84	2.77	2.72	2.67	2.63	2.60	2.55	2.51	2.46	2.42	2.38	2.34	2.32	2.28	2.26	2.24	2.22	2.21
	9.07	**6.70**	**5.74**	**5.20**	**4.86**	**4.62**	**4.44**	**4.30**	**4.19**	**4.10**	**4.02**	**3.96**	**3.85**	**3.78**	**3.67**	**3.59**	**3.51**	**3.42**	**3.37**	**3.30**	**3.27**	**3.21**	**3.18**	**3.16**
14	4.60	3.74	3.34	3.11	2.96	2.85	2.77	2.70	2.65	2.60	2.56	2.53	2.48	2.44	2.39	2.35	2.31	2.27	2.24	2.21	2.19	2.16	2.14	2.13
	8.86	**6.51**	**5.56**	**5.03**	**4.69**	**4.46**	**4.28**	**4.14**	**4.03**	**3.94**	**3.86**	**3.80**	**3.70**	**3.62**	**3.51**	**3.43**	**3.34**	**3.26**	**3.21**	**3.14**	**3.11**	**3.06**	**3.02**	**3.00**
15	4.54	3.68	3.29	3.06	2.90	2.79	2.70	2.64	2.59	2.55	2.51	2.48	2.43	2.39	2.33	2.29	2.25	2.21	2.18	2.15	2.12	2.10	2.08	2.07
	8.68	**6.36**	**5.42**	**4.89**	**4.56**	**4.32**	**4.14**	**4.00**	**3.89**	**3.80**	**3.73**	**3.67**	**3.56**	**3.48**	**3.36**	**3.29**	**3.20**	**3.12**	**3.07**	**3.00**	**2.97**	**2.92**	**2.89**	**2.87**
16	4.49	3.63	3.24	3.01	2.85	2.74	2.66	2.59	2.54	2.49	2.45	2.42	2.37	2.33	2.28	2.24	2.20	2.16	2.13	2.09	2.07	2.04	2.02	2.01
	8.53	**6.23**	**5.29**	**4.77**	**4.44**	**4.20**	**4.03**	**3.89**	**3.78**	**3.69**	**3.61**	**3.55**	**3.45**	**3.37**	**3.25**	**3.18**	**3.10**	**3.01**	**2.96**	**2.89**	**2.86**	**2.80**	**2.77**	**2.75**
17	4.45	3.59	3.20	2.96	2.81	2.70	2.62	2.55	2.50	2.45	2.41	2.38	2.33	2.29	2.23	2.19	2.15	2.11	2.08	2.04	2.02	1.99	1.97	1.96
	8.40	**6.11**	**5.18**	**4.67**	**4.34**	**4.10**	**3.93**	**3.79**	**3.68**	**3.59**	**3.52**	**3.45**	**3.35**	**3.27**	**3.16**	**3.08**	**3.00**	**2.92**	**2.86**	**2.79**	**2.76**	**2.70**	**2.67**	**2.65**
18	4.41	3.55	3.16	2.93	2.77	2.66	2.58	2.51	2.46	2.41	2.37	2.34	2.29	2.25	2.19	2.15	2.11	2.07	2.04	2.00	1.98	1.95	1.93	1.92
	8.28	**6.01**	**5.09**	**4.58**	**4.25**	**4.01**	**3.85**	**3.71**	**3.60**	**3.51**	**3.44**	**3.37**	**3.27**	**3.19**	**3.07**	**3.00**	**2.91**	**2.83**	**2.78**	**2.71**	**2.68**	**2.62**	**2.59**	**2.57**
19	4.38	3.52	3.13	2.90	2.74	2.63	2.55	2.48	2.43	2.38	2.34	2.31	2.26	2.21	2.15	2.11	2.07	2.02	2.00	1.96	1.94	1.91	1.90	1.88
	8.18	**5.93**	**5.01**	**4.50**	**4.17**	**3.94**	**3.77**	**3.63**	**3.52**	**3.43**	**3.36**	**3.30**	**3.19**	**3.12**	**3.00**	**2.92**	**2.84**	**2.76**	**2.70**	**2.63**	**2.60**	**2.54**	**2.51**	**2.49**
20	4.35	3.49	3.10	2.87	2.71	2.60	2.52	2.45	2.40	2.35	2.31	2.28	2.23	2.18	2.12	2.08	2.04	1.99	1.96	1.92	1.90	1.87	1.85	1.84
	8.10	**5.85**	**4.94**	**4.43**	**4.10**	**3.87**	**3.71**	**3.56**	**3.45**	**3.37**	**3.30**	**3.23**	**3.13**	**3.05**	**2.94**	**2.86**	**2.77**	**2.69**	**2.63**	**2.56**	**2.53**	**2.47**	**2.44**	**2.42**
21	4.32	3.47	3.07	2.84	2.68	2.57	2.49	2.42	2.37	2.32	2.28	2.25	2.20	2.15	2.09	2.05	2.00	1.96	1.93	1.90	1.87	1.84	1.82	1.81
	8.02	**5.78**	**4.87**	**4.37**	**4.04**	**3.81**	**3.65**	**3.51**	**3.40**	**3.31**	**3.24**	**3.17**	**3.07**	**2.99**	**2.88**	**2.80**	**2.72**	**2.63**	**2.58**	**2.51**	**2.47**	**2.42**	**2.38**	**2.36**
22	4.30	3.44	3.05	2.82	2.66	2.55	2.47	2.40	2.35	2.30	2.26	2.23	2.18	2.13	2.07	2.03	1.98	1.93	1.91	1.87	1.84	1.81	1.80	1.78
	7.94	**5.72**	**4.82**	**4.31**	**3.99**	**3.76**	**3.59**	**3.45**	**3.35**	**3.26**	**3.18**	**3.12**	**3.02**	**2.94**	**2.83**	**2.75**	**2.67**	**2.58**	**2.53**	**2.46**	**2.42**	**2.37**	**2.33**	**2.31**
23	4.28	3.42	3.03	2.80	2.64	2.53	2.45	2.38	2.32	2.28	2.24	2.20	2.14	2.10	2.04	2.00	1.96	1.91	1.88	1.84	1.82	1.79	1.77	1.76
	7.88	**5.66**	**4.76**	**4.26**	**3.94**	**3.71**	**3.54**	**3.41**	**3.30**	**3.21**	**3.14**	**3.07**	**2.97**	**2.89**	**2.78**	**2.70**	**2.62**	**2.53**	**2.48**	**2.41**	**2.37**	**2.32**	**2.28**	**2.26**
24	4.26	3.40	3.01	2.78	2.62	2.51	2.43	2.36	2.30	2.26	2.22	2.18	2.13	2.09	2.02	1.98	1.94	1.89	1.86	1.82	1.80	1.76	1.74	1.73
	7.82	**5.61**	**4.72**	**4.22**	**3.90**	**3.67**	**3.50**	**3.36**	**3.25**	**3.17**	**3.09**	**3.03**	**2.93**	**2.85**	**2.74**	**2.66**	**2.58**	**2.49**	**2.44**	**2.36**	**2.33**	**2.27**	**2.23**	**2.21**
25	4.24	3.38	2.99	2.76	2.60	2.49	2.41	2.34	2.28	2.24	2.20	2.16	2.11	2.06	2.00	1.96	1.92	1.87	1.84	1.80	1.77	1.74	1.72	1.71
	7.77	**5.57**	**4.68**	**4.18**	**3.86**	**3.63**	**3.46**	**3.32**	**3.21**	**3.13**	**3.05**	**2.99**	**2.89**	**2.81**	**2.70**	**2.62**	**2.54**	**2.45**	**2.40**	**2.32**	**2.29**	**2.23**	**2.19**	**2.17**
26	4.22	3.37	2.89	2.74	2.59	2.47	2.39	2.32	2.27	2.22	2.18	2.15	2.10	2.05	1.99	1.95	1.90	1.85	1.82	1.78	1.76	1.72	1.70	1.69
	7.72	**5.53**	**4.64**	**4.14**	**3.82**	**3.59**	**3.42**	**3.29**	**3.17**	**3.09**	**3.02**	**2.96**	**2.86**	**2.77**	**2.66**	**2.58**	**2.50**	**2.41**	**2.36**	**2.28**	**2.25**	**2.19**	**2.15**	**2.13**

F^* for $\alpha = .05$ are given in lightface type, and F^* for $\alpha = .01$ are given in boldface type.

If your calculated F is greater than F^*, reject H_0. If your value of *df* is not listed, use the closest listed value of *df* that gives the larger F^*.

Table F^* is reprinted by permission from *Statistical Methods* by George W. Snedecor and William G. Cochran © 1967 by The Iowa State University Press, Ames, Iowa 50010.

TABLE F* *Critical values of F for the analysis of variance—Continued*

Degrees of freedom for denominator	Degrees of freedom for numerator																							
	1	2	3	4	5	6	7	8	9	10	11	12	14	16	20	24	30	40	50	75	100	200	500	∞
27	4.21	3.35	2.96	2.73	2.57	2.46	2.37	2.30	2.25	2.20	2.16	2.13	2.08	2.03	1.97	1.93	1.88	1.84	1.80	1.76	1.74	1.71	1.68	1.67
	7.68	**5.49**	**4.60**	**4.11**	**3.79**	**3.56**	**3.39**	**3.26**	**3.14**	**3.06**	**2.98**	**2.93**	**2.83**	**2.74**	**2.63**	**2.55**	**2.47**	**2.38**	**2.33**	**2.25**	**2.21**	**2.16**	**2.12**	**2.10**
28	4.20	3.34	2.95	2.71	2.56	2.44	2.36	2.29	2.24	2.19	2.15	2.12	2.06	2.02	1.95	1.91	1.87	1.81	1.78	1.75	1.72	1.69	1.67	1.65
	7.64	**5.45**	**4.57**	**4.07**	**3.76**	**3.53**	**3.36**	**3.23**	**3.11**	**3.03**	**2.95**	**2.90**	**2.80**	**2.71**	**2.60**	**2.52**	**2.44**	**2.35**	**2.30**	**2.22**	**2.18**	**2.13**	**2.09**	**2.06**
29	4.18	3.33	2.93	2.70	2.54	2.43	2.35	2.28	2.22	2.18	2.14	2.10	2.05	2.00	1.94	1.90	1.85	1.80	1.77	1.73	1.71	1.68	1.65	1.64
	7.60	**5.52**	**4.54**	**4.04**	**3.73**	**3.50**	**3.32**	**3.20**	**3.08**	**3.00**	**2.92**	**2.87**	**2.77**	**2.68**	**2.57**	**2.49**	**2.41**	**2.32**	**2.27**	**2.19**	**2.15**	**2.10**	**2.06**	**2.03**
30	4.17	3.32	2.92	2.69	2.53	2.42	2.34	2.27	2.21	2.16	2.12	2.09	2.04	1.99	1.93	1.89	1.84	1.79	1.76	1.72	1.69	1.66	1.64	1.62
	7.56	**5.39**	**4.51**	**4.02**	**3.70**	**3.47**	**3.30**	**3.17**	**3.06**	**2.98**	**2.90**	**2.84**	**2.74**	**2.66**	**2.55**	**2.47**	**2.38**	**2.29**	**2.24**	**2.16**	**2.13**	**2.07**	**2.03**	**2.01**
32	4.15	3.30	2.90	2.67	2.51	2.40	2.32	2.25	2.19	2.14	2.10	2.07	2.02	1.97	1.91	1.86	1.82	1.76	1.74	1.69	1.67	1.64	1.61	1.59
	7.50	**5.34**	**4.46**	**3.97**	**3.66**	**3.42**	**3.25**	**3.12**	**3.01**	**2.94**	**2.86**	**2.80**	**2.70**	**2.62**	**2.51**	**2.42**	**2.34**	**2.25**	**2.20**	**2.12**	**2.08**	**2.02**	**1.98**	**1.96**
34	4.13	3.28	2.88	2.65	2.49	2.38	2.30	2.23	2.17	2.12	2.08	2.05	2.00	1.95	1.89	1.84	1.80	1.74	1.71	1.67	1.64	1.61	1.59	1.57
	7.44	**5.29**	**4.42**	**3.93**	**3.61**	**3.38**	**3.21**	**3.08**	**2.97**	**2.89**	**2.82**	**2.76**	**2.66**	**2.58**	**2.47**	**2.38**	**2.30**	**2.21**	**2.15**	**2.08**	**2.04**	**1.98**	**1.94**	**1.91**
36	4.11	3.26	2.86	2.63	2.48	2.36	2.28	2.21	2.15	2.10	2.06	2.03	1.89	1.93	1.87	1.82	1.78	1.72	1.69	1.65	1.62	1.59	1.56	1.55
	7.39	**5.25**	**4.38**	**3.89**	**3.58**	**3.35**	**3.18**	**3.04**	**2.94**	**2.86**	**2.78**	**2.72**	**2.62**	**2.54**	**2.43**	**2.35**	**2.26**	**2.17**	**2.12**	**2.04**	**2.00**	**1.94**	**1.90**	**1.87**
38	4.10	3.25	2.85	2.62	2.46	2.35	2.26	2.19	2.14	2.09	2.05	2.02	1.96	1.92	1.85	1.80	1.76	1.71	1.67	1.63	1.60	1.57	1.54	1.53
	7.35	**5.21**	**4.34**	**3.86**	**3.54**	**3.32**	**3.15**	**3.02**	**2.91**	**2.82**	**2.75**	**2.69**	**2.59**	**2.51**	**2.40**	**2.32**	**2.22**	**2.14**	**2.08**	**2.00**	**1.97**	**1.90**	**1.86**	**1.84**
40	4.08	3.23	2.84	2.61	2.45	2.34	2.25	2.18	2.12	2.07	2.04	2.00	1.95	1.90	1.84	1.79	1.74	1.69	1.66	1.61	1.59	1.55	1.53	1.51
	7.31	**5.18**	**4.31**	**3.83**	**3.51**	**3.29**	**3.12**	**2.99**	**2.88**	**2.80**	**2.73**	**2.66**	**2.56**	**2.49**	**2.37**	**2.29**	**2.20**	**2.11**	**2.05**	**1.97**	**1.94**	**1.88**	**1.84**	**1.81**
42	4.07	3.22	2.83	2.59	2.44	2.32	2.24	2.17	2.11	2.06	2.02	1.90	1.94	1.89	1.82	1.78	1.73	1.68	1.64	1.60	1.57	1.54	1.51	1.49
	7.27	**5.15**	**4.29**	**3.80**	**3.49**	**3.26**	**3.10**	**2.96**	**2.86**	**2.77**	**2.70**	**2.64**	**2.54**	**2.46**	**2.35**	**2.26**	**2.17**	**2.08**	**2.02**	**1.94**	**1.91**	**1.85**	**1.80**	**1.78**
44	4.06	3.21	2.82	2.58	2.43	2.31	2.23	2.16	2.10	2.05	2.01	1.98	1.92	1.88	1.81	1.76	1.72	1.66	1.63	1.58	1.56	1.52	1.50	1.48
	7.24	**5.12**	**4.26**	**3.78**	**3.46**	**3.24**	**3.07**	**2.94**	**2.84**	**2.75**	**2.68**	**2.62**	**2.52**	**2.44**	**2.32**	**2.24**	**2.15**	**2.06**	**2.09**	**1.92**	**1.88**	**1.82**	**1.78**	**1.75**
46	4.05	3.20	2.81	2.57	2.42	2.30	2.22	2.14	2.09	2.04	2.00	1.97	1.91	1.87	1.80	1.75	1.71	1.65	1.62	1.57	1.54	1.51	1.48	1.46
	7.21	**5.10**	**4.24**	**3.76**	**3.44**	**3.22**	**3.05**	**2.92**	**2.82**	**2.73**	**2.66**	**2.60**	**2.50**	**2.42**	**2.30**	**2.22**	**2.13**	**2.04**	**1.98**	**1.90**	**1.86**	**1.80**	**1.76**	**1.72**
48	4.04	3.19	2.80	2.56	2.41	2.30	2.21	2.14	2.08	2.03	1.99	1.96	1.90	1.86	1.79	1.74	1.70	1.64	1.61	1.56	1.53	1.50	1.47	1.45
	7.19	**5.08**	**4.22**	**3.74**	**3.42**	**3.20**	**3.04**	**2.90**	**2.80**	**2.71**	**2.64**	**2.58**	**2.48**	**2.40**	**2.28**	**2.20**	**2.11**	**2.02**	**1.96**	**1.88**	**1.84**	**1.78**	**1.73**	**1.70**

Degrees of freedom for denominator																								
50	4.03	3.18	2.79	2.56	2.40	2.29	2.20	2.13	2.07	2.02	1.98	1.95	1.90	1.85	1.78	1.74	1.69	1.63	1.60	1.55	1.52	1.48	1.46	1.44
	7.17	**5.06**	**4.20**	**3.72**	**3.41**	**3.18**	**3.02**	**2.88**	**2.78**	**2.70**	**2.62**	**2.56**	**2.46**	**2.39**	**2.26**	**2.18**	**2.10**	**2.00**	**1.94**	**1.86**	**1.82**	**1.76**	**1.71**	**1.68**
55	4.02	3.17	2.78	2.54	2.38	2.27	2.18	2.11	2.05	2.00	1.97	1.93	1.88	1.83	1.76	1.72	1.67	1.61	1.58	1.52	1.50	1.46	1.43	1.41
	7.12	**5.01**	**4.16**	**3.68**	**3.37**	**3.15**	**2.98**	**2.85**	**2.75**	**2.66**	**2.59**	**2.53**	**2.43**	**2.35**	**2.23**	**2.15**	**2.06**	**1.96**	**1.90**	**1.82**	**1.78**	**1.71**	**1.66**	**1.64**
60	4.00	3.15	2.76	2.52	2.37	2.25	2.17	2.10	2.04	1.99	1.95	1.92	1.86	1.81	1.75	1.70	1.65	1.59	1.56	1.50	1.48	1.44	1.41	1.39
	7.08	**4.98**	**4.13**	**3.65**	**3.34**	**3.12**	**2.95**	**2.82**	**2.72**	**2.63**	**2.56**	**2.50**	**2.40**	**2.32**	**2.20**	**2.12**	**2.03**	**1.93**	**1.87**	**1.79**	**1.74**	**1.68**	**1.63**	**1.60**
65	3.99	3.14	2.75	2.51	2.36	2.24	2.15	2.08	2.02	1.98	1.94	1.90	1.85	1.80	1.73	1.68	1.63	1.57	1.54	1.49	1.46	1.42	1.39	1.37
	7.04	**4.95**	**4.10**	**3.62**	**3.31**	**3.09**	**2.93**	**2.79**	**2.70**	**2.61**	**2.54**	**2.47**	**2.37**	**2.30**	**2.18**	**2.09**	**2.00**	**1.90**	**1.84**	**1.76**	**1.71**	**1.64**	**1.60**	**1.56**
70	3.98	3.13	2.74	2.50	2.35	2.32	2.14	2.07	2.01	1.97	1.93	1.89	1.84	1.79	1.72	1.67	1.62	1.56	1.53	1.47	1.45	1.40	1.37	1.35
	7.01	**4.92**	**4.08**	**3.60**	**3.29**	**3.07**	**2.91**	**2.77**	**2.67**	**2.59**	**2.51**	**2.45**	**2.35**	**2.28**	**2.15**	**2.07**	**1.98**	**1.88**	**1.82**	**1.74**	**1.69**	**1.62**	**1.56**	**1.53**
80	3.96	3.11	2.72	2.48	2.33	2.21	2.12	2.05	1.99	1.95	1.91	1.88	1.82	1.77	1.70	1.65	1.60	1.54	1.51	1.45	1.42	1.38	1.35	1.32
	6.96	**4.88**	**4.04**	**3.56**	**3.25**	**3.04**	**2.87**	**2.74**	**2.64**	**2.55**	**2.48**	**2.41**	**2.32**	**2.24**	**2.11**	**2.03**	**1.94**	**1.84**	**1.78**	**1.70**	**1.65**	**1.57**	**1.52**	**1.49**
100	3.94	3.09	2.70	2.46	2.30	2.19	2.10	2.03	1.97	1.92	1.88	1.85	1.79	1.75	1.68	1.63	1.57	1.51	1.48	1.42	1.39	1.34	1.30	1.28
	6.90	**4.82**	**3.98**	**3.51**	**3.20**	**2.99**	**2.82**	**2.69**	**2.59**	**2.51**	**2.43**	**2.36**	**2.26**	**2.19**	**2.06**	**1.98**	**1.89**	**1.79**	**1.73**	**1.64**	**1.59**	**1.51**	**1.46**	**1.43**
125	3.92	3.07	2.68	2.44	2.29	2.17	2.08	2.01	1.95	1.90	1.86	1.83	1.77	1.72	1.65	1.60	1.55	1.49	1.45	1.39	1.36	1.31	1.27	1.25
	6.84	**4.78**	**3.94**	**3.47**	**3.17**	**2.95**	**2.79**	**2.65**	**2.56**	**2.47**	**2.40**	**2.33**	**2.23**	**2.15**	**2.03**	**1.94**	**1.85**	**1.75**	**1.68**	**1.59**	**1.54**	**1.46**	**1.40**	**1.37**
150	3.91	3.06	2.67	2.43	2.27	2.16	2.07	2.00	1.94	1.89	1.85	1.82	1.76	1.71	1.64	1.59	1.54	1.47	1.44	1.37	1.34	1.29	1.25	1.22
	6.81	**4.75**	**3.91**	**3.44**	**3.13**	**2.92**	**2.76**	**2.62**	**2.53**	**2.44**	**2.37**	**2.30**	**2.20**	**2.12**	**2.00**	**1.91**	**1.83**	**1.72**	**1.66**	**1.56**	**1.51**	**1.43**	**1.37**	**1.33**
200	3.89	3.04	2.65	2.41	2.26	2.14	2.05	1.98	1.92	1.87	1.83	1.80	1.74	1.69	1.62	1.57	1.52	1.45	1.42	1.35	1.32	1.26	1.22	1.19
	6.76	**4.71**	**3.38**	**3.41**	**3.11**	**2.90**	**2.73**	**2.60**	**2.50**	**2.41**	**2.34**	**2.28**	**2.17**	**2.09**	**1.97**	**1.88**	**1.79**	**1.69**	**1.62**	**1.53**	**1.48**	**1.39**	**1.33**	**1.28**
400	3.86	3.02	2.62	2.39	2.23	2.12	2.03	1.96	1.90	1.85	1.81	1.78	1.72	1.67	1.60	1.54	1.49	1.42	1.38	1.32	1.28	1.22	1.16	1.13
	6.70	**4.66**	**3.83**	**3.36**	**3.06**	**2.85**	**2.69**	**2.55**	**2.46**	**2.37**	**2.29**	**2.23**	**2.12**	**2.04**	**1.92**	**1.84**	**1.74**	**1.64**	**1.57**	**1.47**	**1.42**	**1.32**	**1.24**	**1.19**
1000	3.85	3.00	2.61	2.38	2.22	2.10	2.02	1.95	1.89	1.84	1.80	1.76	1.70	1.65	1.58	1.53	1.47	1.41	1.36	1.30	1.26	1.19	1.13	1.08
	6.66	**4.62**	**3.80**	**3.34**	**3.04**	**2.82**	**2.66**	**2.53**	**2.43**	**2.34**	**2.26**	**2.20**	**2.09**	**2.01**	**1.89**	**1.81**	**1.71**	**1.61**	**1.54**	**1.44**	**1.38**	**1.28**	**1.19**	**1.11**
∞	3.84	2.99	2.60	2.37	2.21	2.09	2.01	1.94	1.88	1.83	1.79	1.75	1.69	1.64	1.57	1.52	1.46	1.40	1.35	1.28	1.24	1.17	1.11	1.00
	6.64	**4.60**	**3.78**	**3.32**	**3.02**	**2.80**	**2.64**	**2.51**	**2.41**	**2.32**	**2.24**	**2.18**	**2.07**	**1.99**	**1.87**	**1.79**	**1.69**	**1.59**	**1.52**	**1.41**	**1.36**	**1.25**	**1.15**	**1.00**

F^* for $\alpha = .05$ are given in lightface type, and F^* for $\alpha = .01$ are given in boldface type.

If your calculated F is greater than F^*, reject H_0. If your value of *df* is not listed, use the closest listed value of *df* that gives the larger F^*.

TABLE r^*

Critical values of r for the Pearson correlation coefficient

	r^* for a two-tailed test		r^* for a one-tailed test	
N	$\alpha = .05$	$\alpha = .01$	$\alpha = .05$	$\alpha = .01$
3	.997	.9999	.988	.9995
4	.950	.990	.900	.980
5	.878	.959	.805	.934
6	.811	.917	.729	.882
7	.754	.874	.669	.833
8	.707	.834	.622	.789
9	.666	.798	.582	.750
10	.632	.765	.549	.716
11	.602	.735	.521	.685
12	.576	.708	.497	.658
13	.553	.684	.476	.634
14	.532	.661	.458	.612
15	.514	.641	.441	.592
16	.497	.623	.426	.574
17	.482	.606	.412	.558
18	.468	.590	.400	.542
19	.456	.575	.389	.528
20	.444	.561	.378	.516
21	.433	.549	.369	.503
22	.423	.537	.360	.492
23	.413	.526	.352	.482
24	.404	.515	.344	.472
25	.396	.505	.337	.462
26	.388	.496	.330	.453
27	.381	.487	.323	.445
28	.374	.479	.317	.437
29	.367	.471	.311	.430
30	.361	.463	.306	.423
31	.355	.456	.301	.416
32	.349	.449	.296	.409
37	.325	.418	.275	.381
42	.304	.393	.257	.358
47	.288	.372	.243	.338
52	.273	.354	.231	.322
62	.250	.325	.211	.295
72	.232	.302	.195	.274
82	.217	.283	.183	.256
92	.205	.267	.173	.242
102	.195	.254	.164	.230

If your calculated r is greater than r^*, reject H_0. If your value of N is not listed, use r^* for the next smaller value of N.

From Table VI of Fisher and Yates: *Statistical Tables for Biological, Agricultural and Medical Research,* 6th edition, 1974, published by Longman Group Ltd., London. (previously published by Oliver and Boyd, Edinburgh), and by permission of the authors and publishers.

TABLE t^*

Critical values of t for the t tests

	t^* for a two-tailed test		t^* for a one-tailed test	
df	α = .05	α = .01	α = .05	α = .01
1	12.706	63.657	6.314	31.821
2	4.303	9.925	2.920	6.965
3	3.182	5.841	2.353	4.541
4	2.776	4.604	2.132	3.747
5	2.571	4.032	2.015	3.365
6	2.447	3.707	1.943	3.143
7	2.365	3.499	1.895	2.998
8	2.306	3.355	1.860	2.896
9	2.262	3.250	1.833	2.821
10	2.228	3.169	1.812	2.764
11	2.201	3.106	1.796	2.718
12	2.179	3.055	1.782	2.681
13	2.160	3.012	1.771	2.650
14	2.145	2.977	1.761	2.624
15	2.131	2.947	1.753	2.602
16	2.120	2.921	1.746	2.583
17	2.110	2.898	1.740	2.567
18	2.101	2.878	1.734	2.552
19	2.093	2.861	1.729	2.539
20	2.086	2.845	1.725	2.528
21	2.080	2.831	1.721	2.518
22	2.074	2.819	1.717	2.508
23	2.069	2.807	1.714	2.500
24	2.064	2.797	1.711	2.492
25	2.060	2.787	1.708	2.485
26	2.056	2.779	1.706	2.479
27	2.052	2.771	1.703	2.473
28	2.048	2.763	1.701	2.467
29	2.045	2.756	1.699	2.462
30	2.042	2.750	1.697	2.457
40	2.021	2.704	1.684	2.423
60	2.000	2.660	1.671	2.390
120	1.980	2.617	1.658	2.358
∞	1.960	2.576	1.645	2.326

If your calculated t is greater than t^*, reject H_0. If your value of *df* is not listed, use t^* for the next smaller *df*.

From Tables III of Fisher and Yates: *Statistical Tables for Biological, Agricultural and Medical Research*, 6th edition, 1974, published by Longman Group Ltd., London. (previously published by Oliver and Boyd, Edinburgh), and by permission of the authors and publishers.

TABLE U^*

Two-tailed test for $\alpha = .05$

Critical values of U for the Mann-Whitney U test, U for a two-tailed test for* $\alpha = .05$.

	N_1																			
N_2	1	2	3	4	5	6	7	8	9	10	11	12	13	14	15	16	17	18	19	20
1	—	—	—	—	—	—	—	—	—	—	—	—	—	—	—	—	—	—	—	—
2	—	—	—	—	—	—	—	0 16	0 18	0 20	0 22	1 23	1 25	1 27	1 29	1 31	2 32	2 34	2 36	2 38
3	—	—	—	—	0 15	1 17	1 20	2 22	2 25	3 27	3 30	4 32	4 35	5 37	5 40	6 42	6 45	7 47	7 50	8 52
4	—	—	—	0 16	1 19	2 22	3 25	4 28	4 32	5 35	6 38	7 41	8 44	9 47	10 50	11 53	11 57	12 60	13 63	13 67
5	—	—	0 15	1 19	2 23	3 27	5 30	6 34	7 38	8 42	9 46	11 49	12 53	13 57	14 61	15 65	17 68	18 72	19 76	20 80
6	—	—	1 17	2 22	3 27	5 31	6 36	8 40	10 44	11 49	13 53	14 58	16 62	17 67	19 71	21 75	22 80	24 84	25 89	27 93
7	—	—	1 20	3 25	5 30	6 36	8 41	10 46	12 51	14 56	16 61	18 66	20 71	22 76	24 81	26 86	28 91	30 96	32 101	34 106
8	—	0 16	2 22	4 28	6 34	8 40	10 46	13 51	15 57	17 63	19 69	22 74	24 80	26 86	29 91	31 97	34 102	36 108	38 114	41 119
9	—	0 18	2 25	4 32	7 38	10 44	12 51	15 57	17 64	20 70	23 76	26 82	28 89	31 95	34 101	37 107	39 114	42 120	45 126	48 132
10	—	0 20	3 27	5 35	8 42	11 49	14 56	17 63	20 70	23 77	26 84	29 91	33 97	36 104	39 111	42 118	45 125	48 132	52 138	55 145
11	—	0 22	3 30	6 38	9 46	13 53	16 61	19 69	23 76	26 84	30 91	33 99	37 106	40 114	44 121	47 129	51 136	55 143	58 151	62 158
12	—	1 23	4 32	7 41	11 49	14 58	18 66	22 74	26 82	29 91	33 99	37 107	41 115	45 123	49 131	53 139	57 147	61 155	65 163	69 171
13	—	1 25	4 35	8 44	12 53	16 62	20 71	24 80	28 89	33 97	37 106	41 115	45 124	50 132	54 141	59 149	63 158	67 167	72 175	76 184
14	—	1 27	5 37	9 47	13 51	17 67	22 76	26 86	31 95	36 104	40 114	45 123	50 132	55 141	59 151	64 160	67 171	74 178	78 188	83 197
15	—	1 29	5 40	10 50	14 61	19 71	24 81	29 91	34 101	39 111	44 121	49 131	54 141	59 151	64 161	70 170	75 180	80 190	85 200	90 210
16	—	1 31	6 42	11 53	15 65	21 75	26 86	31 97	37 107	42 118	47 129	53 139	59 149	64 160	70 170	75 181	81 191	86 202	92 212	98 222
17	—	2 32	6 45	11 57	17 68	22 80	28 91	34 102	39 114	45 125	51 136	57 147	63 158	67 171	75 180	81 191	87 202	93 213	99 224	105 235
18	—	2 34	7 47	12 60	18 72	24 84	30 96	36 108	42 120	48 132	55 143	61 155	67 167	74 178	80 190	86 202	93 213	99 225	106 236	112 248
19	—	2 36	7 50	13 63	19 76	25 89	32 101	38 114	45 126	52 138	58 151	65 163	72 175	78 188	85 200	92 212	99 224	106 236	113 248	119 261
20	—	2 38	8 52	13 67	20 80	27 93	34 106	41 119	48 132	55 145	62 158	69 171	76 184	83 197	90 210	98 222	105 235	112 248	119 261	127 273

If your calculated U is greater than or equal to the larger U^* in the pair, or if your calculated U is less than or equal to the smaller U^* in the pair, reject H_0.

Dashes in the body of the table indicate that no decision is possible.

TABLE U^*

Critical values of U for the Mann-Whitney U test, U^* for a two-tailed test for $\alpha = .01$—Continued

Two-tailed test for $\alpha = .01$

N_2 \ N_1	1	2	3	4	5	6	7	8	9	10	11	12	13	14	15	16	17	18		
1	—	—	—	—	—	—	—	—	—	—	—	—	—	—	—	—	—	—	—	—
2	—	—	—	—	—	—	—	—	—	—	—	—	—	—	—	—	—	—	0 38	0 40
3	—	—	—	—	—	—	—	—	0 27	0 30	0 33	1 35	1 38	1 41	2 43	2 46	2 49	2 52	3 54	3 57
4	—	—	—	—	—	0 24	0 28	1 31	1 35	2 38	2 42	3 45	3 49	4 52	5 55	5 59	6 62	6 66	7 69	8 72
5	—	—	—	—	0 25	1 29	1 34	2 38	3 42	4 46	5 50	6 54	7 58	7 63	8 67	9 71	10 75	11 79	12 83	13 87
6	—	—	—	0 24	1 29	2 34	3 39	4 44	5 49	6 54	7 59	9 63	10 68	11 73	12 78	13 83	15 87	16 92	17 97	18 102
7	—	—	—	0 28	1 34	3 39	4 45	6 50	7 56	9 61	10 67	12 72	13 78	15 83	16 89	18 94	19 100	21 105	22 111	24 116
8	—	—	—	1 31	2 38	4 44	6 50	7 57	9 63	11 69	13 75	15 81	17 87	18 94	20 100	22 106	24 112	26 118	28 124	30 130
9	—	—	0 27	1 35	3 42	5 49	7 56	9 63	11 70	13 77	16 83	18 90	20 97	22 104	24 111	27 117	29 124	31 131	33 138	36 144
10	—	—	0 30	2 38	4 46	6 54	9 61	11 69	13 77	16 84	18 92	21 99	24 106	26 114	29 121	31 129	34 136	37 143	39 151	42 158
11	—	—	0 33	2 42	5 50	7 59	10 67	13 75	16 83	18 92	21 100	24 108	27 116	30 124	33 132	36 140	39 148	42 156	45 164	48 172
12	—	—	1 35	3 45	6 54	9 63	12 72	15 81	18 90	21 99	24 108	27 117	31 125	34 134	37 143	41 151	44 160	47 169	51 177	54 186
13	—	—	1 38	3 49	7 58	10 68	13 78	17 87	20 97	24 106	27 116	31 125	34 135	38 144	42 153	45 163	49 172	53 181	56 191	60 200
14	—	—	1 41	4 52	7 63	11 73	15 83	18 94	22 104	26 114	30 124	34 134	38 144	42 154	46 164	50 174	54 184	58 194	63 203	67 213
15	—	—	2 43	5 55	8 67	12 78	16 89	20 100	24 111	29 121	33 132	37 143	42 153	46 164	51 174	55 185	60 195	64 206	69 216	73 227
16	—	—	2 46	5 59	9 71	13 83	18 94	22 106	27 117	31 129	36 140	41 151	45 163	50 174	55 185	60 196	65 207	70 218	74 230	79 241
17	—	—	2 49	6 62	10 75	15 87	19 100	24 112	29 124	34 136	39 148	44 160	49 172	54 184	60 195	65 297	70 219	75 231	81 242	86 254
18	—	—	2 52	6 66	11 79	16 92	21 105	26 118	31 131	37 143	42 156	47 169	53 181	58 194	64 206	70 218	75 231	81 243	87 255	92 268
19	—	0 38	3 54	7 69	12 83	17 97	22 111	28 124	33 138	39 151	45 164	51 177	56 191	63 203	69 216	74 230	81 242	87 255	93 268	99 281
20	—	0 40	3 57	8 72	13 87	18 102	24 116	30 130	36 144	42 158	48 172	54 186	60 200	67 213	73 227	79 241	86 254	92 268	99 281	105 295

If your calculated U is greater than or equal to the larger U^* in the pair, or if your calculated U is less than or equal to the smaller U^* in the pair, reject H_0.

Dashes in the body of the table indicate that no decision is possible.

TABLE U*

Critical values of U for the Mann-Whitney U test, U for a one-tailed test for α = .05—Continued*

One-tailed test for α = .05

	N_1																			
N_2	1	2	3	4	5	6	7	8	9	10	11	12	13	14	15	16	17	18	19	20
1	—	—	—	—	—	—	—	—	—	—	—	—	—	—	—	—	—	—	0 19	0 20
2	—	—	—	—	0 10	0 12	0 14	1 15	1 17	1 19	1 21	2 22	2 24	2 26	3 27	3 29	3 31	4 32	4 34	4 36
3	—	—	0 9	0 12	1 14	2 16	2 19	3 21	3 24	4 26	5 28	5 31	6 33	7 35	7 38	8 40	9 42	9 45	10 47	11 49
4	—	—	0 12	1 15	2 18	3 21	4 24	5 27	6 30	7 33	8 36	9 39	10 42	11 45	12 48	14 50	15 53	16 56	17 59	18 62
5	—	0 10	1 14	2 18	4 21	5 25	6 29	8 32	9 36	11 39	12 43	13 47	15 50	16 54	18 57	19 61	20 65	22 68	23 72	25 75
6	—	0 12	2 16	3 21	5 25	7 29	8 34	10 38	12 42	14 46	16 50	17 55	19 59	21 63	23 67	25 71	26 76	28 80	30 84	32 88
7	—	0 14	2 19	4 24	6 29	8 34	11 38	13 43	15 48	17 53	19 58	21 63	24 67	26 72	28 77	30 82	33 86	35 91	37 96	39 101
8	—	1 15	3 21	5 27	8 32	10 38	13 43	15 49	18 54	20 60	23 65	26 70	28 76	31 81	33 87	36 92	39 97	41 103	44 108	47 113
9	—	1 17	3 24	6 30	9 36	12 42	15 48	18 54	21 60	24 66	27 72	30 78	33 84	36 90	39 96	42 102	45 108	48 114	51 120	54 126
10	—	1 19	4 26	7 33	11 39	14 46	17 53	20 60	24 66	27 73	31 79	34 86	37 93	41 99	44 106	48 112	51 119	55 125	58 132	62 138
11	—	1 21	5 28	8 36	12 43	16 50	19 58	23 65	27 72	31 79	34 87	38 94	42 101	46 108	50 115	54 122	57 130	61 137	65 144	69 151
12	—	2 22	5 31	9 39	13 47	17 55	21 63	26 70	30 78	34 86	38 94	42 102	47 109	51 117	55 125	60 132	64 140	68 148	72 156	77 163
13	—	2 24	6 33	10 42	15 50	19 59	24 67	28 76	33 84	37 93	42 101	47 109	51 118	56 126	61 134	65 143	70 151	75 159	80 167	84 176
14	—	2 26	7 35	11 45	16 54	21 63	26 72	31 81	36 90	41 99	46 108	51 117	56 126	61 135	66 144	71 153	77 161	82 170	87 179	92 188
15	—	3 27	7 38	12 48	18 57	23 67	28 77	33 87	39 96	44 106	50 115	55 125	61 134	66 144	72 153	77 163	83 172	88 182	94 191	100 200
16	—	3 29	8 40	14 50	19 61	25 71	30 82	36 92	42 102	48 112	54 122	60 132	65 143	71 153	77 163	83 173	89 183	95 193	101 203	107 213
17	—	3 31	9 42	15 53	20 65	26 76	33 86	39 97	45 108	51 119	57 130	64 140	70 151	77 161	83 172	89 183	96 193	102 204	109 214	115 225
18	—	4 32	9 45	16 56	22 68	28 80	35 91	41 103	48 114	55 123	61 137	68 148	75 159	82 170	88 182	95 193	102 204	109 215	116 226	123 237
19	0 19	4 34	10 47	17 59	23 72	30 84	37 96	44 108	51 120	58 132	65 144	72 156	80 167	87 179	94 191	101 203	109 214	116 226	123 238	130 250
20	0 20	4 36	11 49	18 62	25 75	32 88	39 101	47 113	54 126	62 138	69 151	77 163	84 176	92 188	100 200	107 213	115 225	123 237	130 250	138 262

If you predict that the Group 1 population distribution lies above that of the Group 2 population, then if your calculated U is greater than or equal to the larger U^*, reject H_0; if you predict that the Group 1 population distribution lies below that of the Group 2 population, then if your calculated U is less than or equal to the smaller U^*, reject H_0.

Dashes in the body of the table indicate that no decision is possible.

TABLE U^*

Critical values of U for the Mann-Whitney U test, U^* for a one-tailed test for $\alpha = .01$—Continued

One-tailed test for $\alpha = .01$

N_2 \ N_1	1	2	3	4	5	6	7	8	9	10	11	12	13	14	15	16	17	18	19	20
1	—	—	—	—	—	—	—	—	—	—	—	—	—	—	—	—	—	—	—	—
2	—	—	—	—	—	—	—	—	—	—	—	—	0 26	0 28	0 30	0 32	0 34	0 36	1 37	1 39
3	—	—	—	—	—	—	0 21	0 24	1 26	1 29	1 32	2 34	2 37	2 40	3 42	3 45	4 47	4 50	4 52	5 55
4	—	—	—	—	0 20	1 23	1 27	2 30	3 33	3 37	4 40	5 43	5 47	6 50	7 53	7 57	8 60	9 63	9 67	10 70
5	—	—	—	0 20	1 24	2 28	3 32	4 36	5 40	6 44	7 48	8 52	9 56	10 60	11 64	12 68	13 72	14 76	15 80	16 84
6	—	—	—	1 23	2 28	3 33	4 38	6 42	7 47	8 52	9 57	11 61	12 66	13 71	15 75	16 80	18 84	19 89	20 94	22 98
7	—	—	0 21	1 27	3 32	4 38	6 43	7 49	9 54	11 59	12 65	14 70	16 75	17 81	19 86	21 91	23 96	24 102	26 107	28 112
8	—	—	0 24	2 30	4 36	6 42	7 49	9 55	11 61	13 67	15 73	17 79	20 84	22 90	24 96	26 102	28 108	30 114	32 120	34 126
9	—	—	1 26	3 33	5 40	7 47	9 54	11 61	14 67	16 74	18 81	21 87	23 94	26 100	28 107	31 113	33 120	36 126	38 133	40 140
10	—	—	1 29	3 37	6 44	8 52	11 59	13 67	16 74	19 81	22 88	24 96	27 103	30 110	33 117	36 124	38 132	41 139	44 146	47 153
11	—	—	1 32	4 40	7 48	9 57	12 65	15 73	18 81	22 88	25 96	28 104	31 112	34 120	37 128	41 135	44 143	47 151	50 159	53 167
12	—	—	2 34	5 43	8 52	11 61	14 70	17 79	21 87	24 96	28 104	31 113	35 121	38 130	42 138	46 146	49 155	53 163	56 172	60 180
13	—	0 26	2 37	5 47	9 56	12 66	16 75	20 84	23 94	27 103	31 112	35 121	39 130	43 139	47 148	51 157	55 166	59 175	63 184	67 193
14	—	0 28	2 40	6 50	10 60	13 71	17 81	22 90	26 100	30 110	34 120	38 130	43 139	47 149	51 159	56 168	60 178	65 187	69 197	73 207
15	—	0 30	3 42	7 53	11 64	15 75	19 86	24 96	28 107	33 117	37 128	42 138	47 148	51 159	56 169	61 179	66 189	70 200	75 210	80 220
16	—	0 32	3 45	7 57	12 68	16 80	21 91	26 102	31 113	36 124	41 135	46 146	51 157	56 168	61 179	66 190	71 201	76 212	82 222	87 233
17	—	0 34	4 47	8 60	13 72	18 84	23 96	28 108	33 120	38 132	44 143	49 155	55 166	60 178	66 189	71 201	77 212	82 224	88 234	93 247
18	—	0 36	4 50	9 63	14 76	19 89	24 102	30 114	36 126	41 139	47 151	53 163	59 175	65 187	70 200	76 212	82 224	88 236	94 248	100 260
19	—	1 37	4 53	9 67	15 80	20 94	26 107	32 120	38 133	44 146	50 159	56 172	63 184	69 197	75 210	82 222	88 235	94 248	101 260	107 273
20	—	1 39	5 55	10 70	16 84	22 98	28 112	34 126	40 140	47 153	53 167	60 180	67 193	73 207	80 220	87 233	93 247	100 260	107 273	114 286

If you predict that the Group 1 population distribution lies above that of the Group 2 population, then if your calculated U is greater than or equal to the larger U^*, reject H_0; if you predict that the Group 1 population distribution lies below that of the Group 2 population, then if your calculated U is less than or equal to the smaller U^*, reject H_0.

Dashes in the body of the table indicate that no decision is possible.

From D. Auble, Extended Tables for the Mann-Whitney statistic, Tables 2, 3, 5, and 7. *Bulletin of the Institute of Educational Research,* Vol. 1, No. 2, 1953, Indiana University.

TABLE W^*

Critical values of W for the Wilcoxon test

	W^* for a two-tailed test		W^* for a one-tailed test	
N	$\alpha = .05$	$\alpha = .01$	$\alpha = .05$	$\alpha = .01$
5	—	—	1	—
6	1	—	2	—
7	2	—	4	0
8	4	0	6	2
9	6	2	8	3
10	8	3	11	5
11	11	5	14	7
12	14	7	17	10
13	17	10	21	13
14	21	13	26	16
15	25	16	30	20
16	30	19	36	24
17	35	23	41	28
18	40	28	47	33
19	46	32	54	38
20	52	37	60	43
21	59	43	68	49
22	66	49	75	56
23	73	55	83	62
24	81	61	92	69
25	90	68	101	77
26	98	76	110	85
27	107	84	120	93
28	117	92	130	102
29	127	100	141	111
30	137	109	152	120
31	148	118	163	130
32	159	128	175	141
33	171	138	188	151
34	183	149	201	162
35	195	160	214	174
36	208	171	228	186
37	222	183	242	198
38	235	195	256	211
39	250	208	271	224
40	264	221	287	238
41	279	234	303	252
42	295	248	319	267
43	311	262	336	281
44	327	277	353	297
45	344	292	371	313
46	361	307	389	329
47	379	323	408	345
48	397	339	427	362
49	415	356	446	380
50	434	373	466	398

If your calculated W is less than W^*, reject H_0.

Dashes in the body of the table indicate that no decision is possible.

From F. Wilcoxon and R. A. Wilcox. *Some rapid approximate statistical procedures,* 1964 revision, published by the Lederle Laboratories Division of American Cyanamid Company, Pearl River, New York. Reprinted by permission of American Cyanamid Company.

TABLE z^*

Critical values of z for the z test

z^* for a two-tailed test		z^* for a one-tailed test	
$\alpha = .05$	$\alpha = .01$	$\alpha = .05$	$\alpha = .01$
1.96	2.58	1.65	2.33

If your calculated z is greater than z^*, reject H_0.

TABLE z

Percentage of cases under the normal distribution

Digits in the ones and tenths places of z score	Digit in the hundredths place of z score: .00	.01	.02	.03	.04	.05	.06	.07	.08	.09
0.0	00.00	00.40	00.80	01.20	01.60	01.99	02.39	02.79	03.19	03.59
0.1	03.98	04.38	04.78	05.17	05.57	05.96	06.36	06.75	07.14	07.53
0.2	07.93	08.32	08.71	09.10	09.48	09.87	10.26	10.64	11.03	11.41
0.3	11.79	12.17	12.55	12.93	13.31	13.68	14.06	14.43	14.80	15.17
0.4	15.54	15.91	16.28	16.64	17.00	17.36	17.72	18.08	18.44	18.79
0.5	19.15	19.50	19.85	20.19	20.54	20.88	21.23	21.57	21.90	22.24
0.6	22.57	22.91	23.24	23.57	23.89	24.22	24.54	24.86	25.17	25.49
0.7	25.80	26.11	26.42	26.73	27.04	27.34	27.64	27.94	28.23	28.52
0.8	28.81	29.10	29.39	29.67	29.95	30.23	30.51	30.78	31.06	31.33
0.9	31.59	31.86	32.12	32.38	32.64	32.90	33.15	33.40	33.65	33.89
1.0	34.13	34.38	34.61	34.85	35.08	35.31	35.54	35.77	35.99	36.21
1.1	36.43	36.65	36.86	37.08	37.29	37.49	37.70	37.90	38.10	38.30
1.2	38.49	38.69	38.88	39.07	39.25	39.44	39.62	39.80	39.97	40.15
1.3	40.32	40.49	40.66	40.82	40.99	41.15	41.31	41.47	41.62	41.77
1.4	41.92	42.07	42.22	42.36	42.51	42.65	42.79	42.92	43.06	43.19
1.5	43.32	43.45	43.57	43.70	43.83	43.94	44.06	44.18	44.29	44.41
1.6	44.52	44.63	44.74	44.84	44.95	45.05	45.15	45.25	45.35	45.45
1.7	45.54	45.64	45.73	45.82	45.91	45.99	46.08	46.16	46.25	46.33
1.8	46.41	46.49	46.56	46.64	46.71	46.78	46.86	46.93	46.99	47.06
1.9	47.13	47.19	47.26	47.32	47.38	47.44	47.50	47.56	47.61	47.67
2.0	47.72	47.78	47.83	47.88	47.93	47.98	48.03	48.08	48.12	48.17
2.1	48.21	48.26	48.30	48.34	48.38	48.42	48.46	48.50	48.54	48.57
2.2	48.61	48.64	48.68	48.71	48.75	48.78	48.81	48.84	48.87	48.90
2.3	48.93	48.96	48.98	49.01	49.04	49.06	49.09	49.11	49.13	49.16
2.4	49.18	49.20	49.22	49.25	49.27	49.29	49.31	49.32	49.34	49.36
2.5	49.38	49.40	49.41	49.43	49.45	49.46	49.48	49.49	49.51	49.52
2.6	49.53	49.55	49.56	49.57	49.59	49.60	49.61	49.62	49.63	49.64
2.7	49.65	49.66	49.67	49.68	49.69	49.70	49.71	49.72	49.73	49.74
2.8	49.74	49.75	49.76	49.77	49.77	49.78	49.79	49.79	49.80	49.81
2.9	49.81	49.82	49.82	49.83	49.84	49.84	49.85	49.85	49.86	49.86
3.0	49.87									
3.5	49.98									
4.0	49.997									
5.0	49.99997									

Values in the body of the table indicate the percentage of scores between the mean and a given z score.

From *A First Course in Statistics* by E. F. Lindquist. ©1942 by E. F. Lindquist. Reprinted by permission of Houghton Mifflin.

ANSWERS TO EXERCISES

ANSWERS TO EXERCISES, CHAPTER 1

NOTE ON COMPUTATIONAL ACCURACY

You may find minor discrepancies between the answers you obtain for these problems and the answers shown here. We have rounded answers to three digits to the right of the decimal place (the thousands place). We have used the convention of rounding to the nearest even digit when the rounded number is a "5." Where the discrepancies are small, they are probably due to differences in rounding off. Such discrepancies are quite common due to varying capacities of calculators (and hand calculations).

1 nominal; discrete

2 interval/ratio; continuous

3 ordinal; discrete

4 interval/ratio; discrete

5 interval/ratio; continuous

6 nominal; discrete

7 ordinal; discrete

8 interval/ratio; discrete

9 ordinal; discrete

10 ordinal; discrete

11 nominal; discrete

12 The population consists of all the students enrolled at the university. The sample consists of the 50 students actually interviewed.

13 The population is all widows over 65 years of age living in Miami, Florida. The sample is the group of 200 widows actually questioned.

14 The population is comprised of all students enrolled in introductory psychology courses. The sample consists of the 20 students who were administered the memory test.

15 The population consists of all male tower operators in the United States between the ages of 26 and 30. The sample is the 80 male tower operators examined. This is an example of inferential statistics because you made a conclusion about *all* male tower operators between the ages of 26 and 30 on the basis of your sample of 80 tower operators.

16 The population consists of the ten rats that were administered amphetamine. There is no sample. This is an example of descriptive statistics because the average lever presses of 187 summarizes the results of the session.

17 The population consists of the 674 sixth-graders in the school district. The sample consists of the 135 sixth-graders whose reading achievement was actually tested. This is an example of inferential statistics because you inferred that all 674 sixth-graders, on the average, read at the eighth-grade level.

ANSWERS TO EXERCISES, CHAPTER 2

1

	Interval width	*Midpoint*
a.	3	1
b.	10	15.5
c.	10	69.5
d.	10	−.5
e.	5	−8
f.	3	4
g.	5	13
h.	25	37

2 Note: Data for this problem are nominal, and hence can be listed in any order in the frequency distribution or histogram.

a.

X	f
Car pool	2
Bus	7
Private auto	6
Motorcycle	3
Bicycle	2

b.

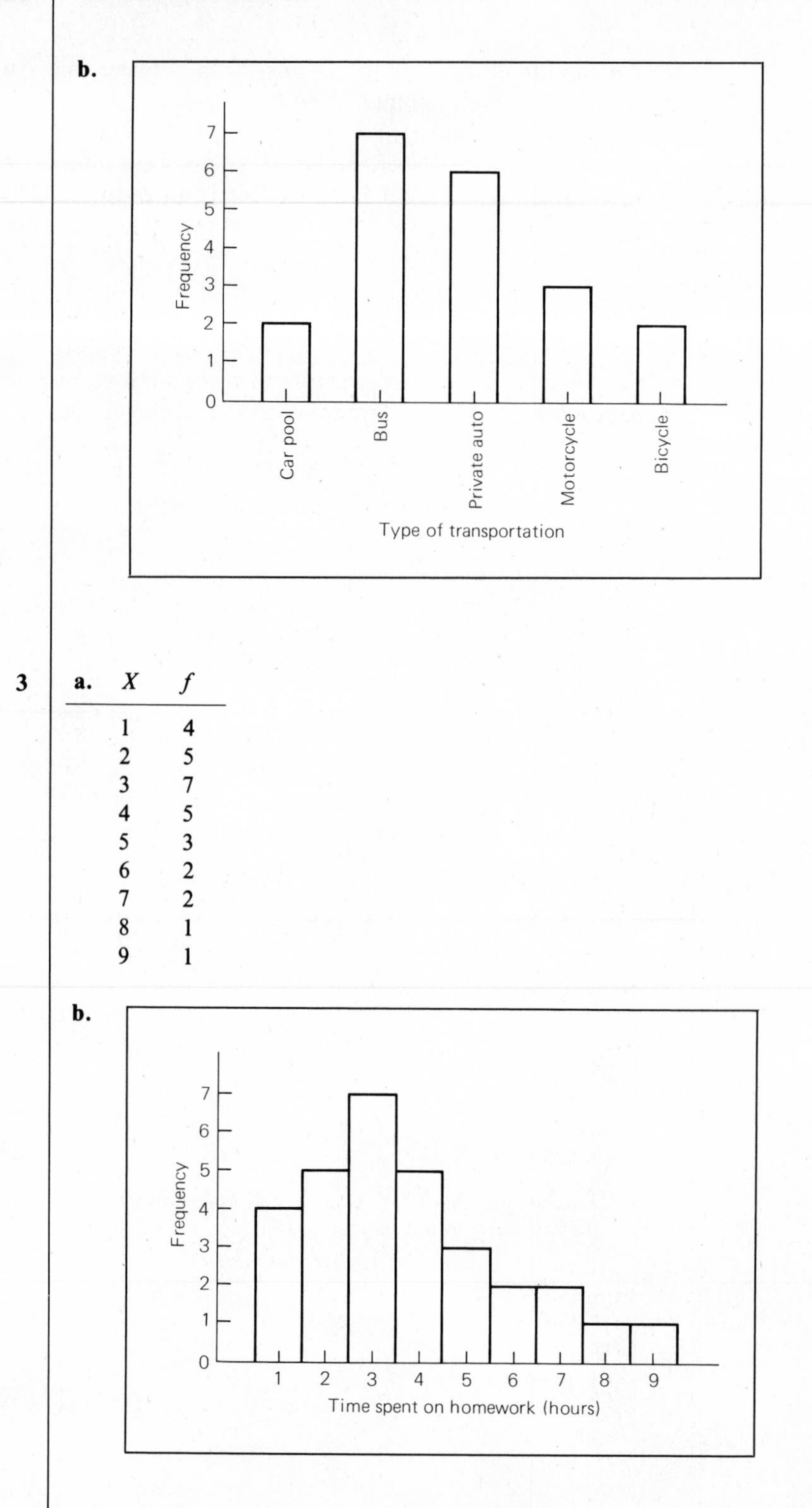

3 **a.**

X	f
1	4
2	5
3	7
4	5
5	3
6	2
7	2
8	1
9	1

b.

c. Positively skewed

d. X	f	cf
1	4	4
2	5	9
3	7	16
4	5	21
5	3	24
6	2	26
7	2	28
8	1	29
9	1	30

e.

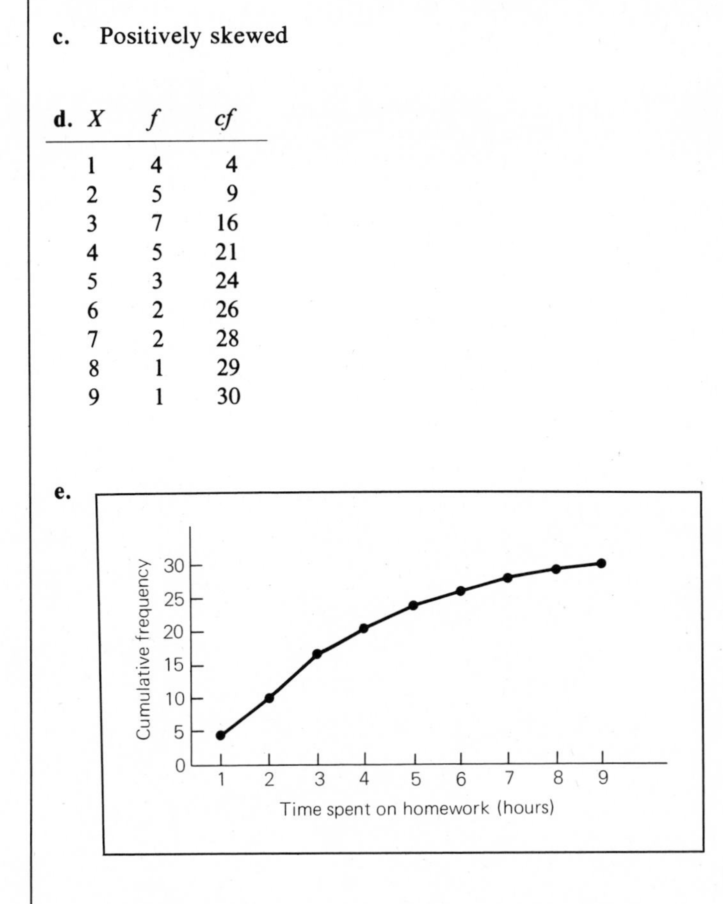

4 a.

Pacific High X	f	*Atlantic High* X	f
68–72	1	68–72	1
73–77	1	73–77	0
78–82	1	78–82	1
83–87	1	83–87	0
88–92	2	88–92	0
93–97	2	93–97	0
98–102	4	98–102	0
103–107	2	103–107	1
108–112	2	108–112	2
113–117	1	113–117	3
118–122	1	118–122	2
123–127	1	123–127	3
128–132	1	128–132	3
		133–137	2

b.

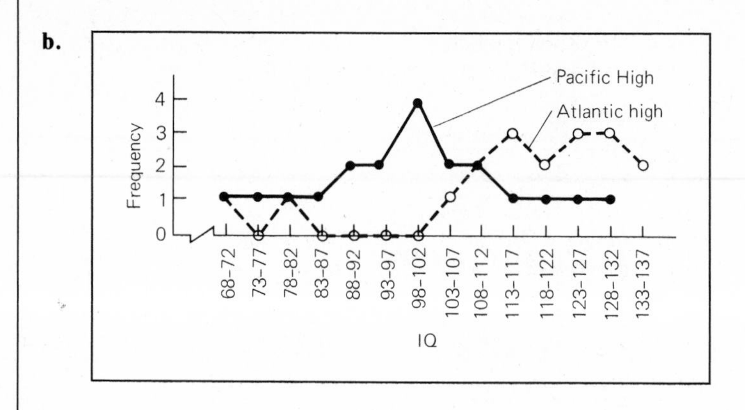

c. The distribution of Pacific High scores is symmetrical. The distribution of Atlantic High scores is negatively skewed.

ANSWERS TO EXERCISES, CHAPTER 3

1

	Mode	*Median*	*Mean*
a.	3	3.5	4.125
b.	3 and 4.	3.5	17.875
c.	0	2	4
d.	15	13	14
e.	14	14	14
f.	2.5 and 4.2	3.6	3.454

2 mode = 70, median = 68, mean = 67.125

3

a. interval/ratio; the median is preferred since these data are very skewed; median = 19
b. ordinal; median = medium
c. nominal; mode = conduct disorder
d. interval/ratio; mean = 171.15
e. ordinal; median = 1.5

4 mode = 1000; median = 850; mean = 854.167; mode was reported; either median or mean would be more accurate.

5
- **a.** range = 7
- **b.** range = 13
- **c.** range = 41
- **d.** range = 20
- **e.** range = 22

6

	ΣX	$(\Sigma X)^2$	ΣX^2	$\frac{(\Sigma X)^2}{N}$	σ^2	σ	μ
a.	30	900	140	112.5	3.438	1.854	3.75
b.	56	3136	492	313.6	17.84	4.224	5.6
c.	270	72900	9140	9112.5	3.438	1.854	33.75
d.	856	732736	73452	73273.6	17.84	4.224	85.6
e.	77	5929	569	539	2.727	1.651	7
f.	77	5929	1031	539	44.727	6.688	7
g.	737	543169	49409	49379	2.727	1.651	67
h.	517	267289	24791	24299	44.727	6.688	47

7 Groups—(*e*) and (*f*)—have the same mean but different standard deviations because one group of scores is more spread out from the mean than the other. Some pairs of groups have different means but the same standard deviations—because the scores within each group of the pair are spread out equivalently above and below their respective means: (*a*) and (*c*); (*b*) and (*d*); (*e*) and (*g*); and (*f*) and (*h*).

8

	ΣX	μ	$\Sigma(X - \mu)^2$	σ^2	σ
a.	30	3.75	27.5	3.438	1.854
b.	270	33.75	27.5	3.438	1.854
c.	77	7	492	44.727	6.688
d.	46	4.6	23.9	2.39	1.546

9
- **a.** $\mu = 4$ $\sigma = 0$
- **b.** $\mu = 34.262$ $\sigma = 7.425$
- **c.** $\mu = 334.333$ $\sigma = 99.369$
- **d.** $\mu = 4$ $\sigma = 2.204$

10
- **a.** 43 and 57
- **b.** 36 and 64
- **c.** 29 and 71

ANSWERS TO EXERCISES, CHAPTER 4

1 First, add a column of *cf* values to the given frequency distribution, making a cumulative frequency distribution:

X	f	cf
0	1	1
1	0	1
2	1	2
3	2	4
4	1	5
5	6	11
6	7	18
7	12	30
8	10	40
9	5	45
10	5	50

a. Using Formula 4.1, percentile rank $= \frac{5}{50}(100) = 10.$

b. Using Formula 4.1, percentile rank $= \frac{45}{50}(100) = 90.$

c. Using Formula 4.2, $\frac{25}{100}(50) = 12.5$, which corresponds to a raw score of 6.

d. Using Formula 4.2, $\frac{50}{100}(50) = 25$, which corresponds to a raw score of 7.

e. Using Formula 4.2, $\frac{75}{100}(50) = 37.5$, which corresponds to a raw score of 8.

2 First, make a cumulative frequency distribution:

X	f	cf
7	1	1
8	2	3
9	3	6
10	4	10
11	3	13
12	2	15
13	0	15
14	2	17
15	1	18

a. Using Formula 4.1, percentile rank $= \frac{3}{18}(100) = 16.667.$

b. Using Formula 4.1, percentile rank $= \frac{15}{18}(100) = 83.333$.

c. Using Formula 4.2, $\frac{50}{100}(18) = 9$, which corresponds to a raw score of 10.

d. Using Formula 4.2, $\frac{90}{100}(18) = 16.2$, which corresponds to a raw score of 14.

3 Using Formula 4.3:

a. $z = \frac{57 - 50}{6.5} = 1.077$

b. $z = \frac{43.5 - 50}{6.5} = -1$

c. $z = \frac{75 - 50}{6.5} = 3.846$

d. $z = \frac{64.1 - 50}{6.5} = 2.169$

e. $z = \frac{57.5 - 50}{6.5} = 1.154$

4 Using Formula 4.4:

a. $X = 18.5 + (1.5)(3) = 23$
b. $X = 18.5 + (3.25)(3) = 28.25$
c. $X = 18.5 + (-1.5)(3) = 14$
d. $X = 18.5 + (-.67)(3) = 16.49$
e. $X = 18.5 + (2.33)(3) = 25.49$

5 **a.** The strategy is to use Formula 4.3 to transform your raw score to a z score, and then use Table z to derive the percentage of scores between the mean and the raw score.

For 95: $z = \frac{95 - 75}{20} = 1$, and hence 34.13%.

For 45: $z = \frac{45 - 75}{20} = -1.5$, and hence 43.32%.

For 84: $z = \frac{84 - 75}{20} = .45$, and hence 17.36%.

For 37: $z = \frac{37 - 75}{20} = -1.9$, and hence 47.13%.

For 77.6 $z = \dfrac{77.6 - 75}{20} = .13$, and hence 5.17%.

For 28.25: $z = \dfrac{28.25 - 75}{20} = -2.338$, and hence 49.04%.

b. For problems concerning the percentage of scores falling above a given raw score, the strategy is to proceed as in (a). Then, if z is positive, subtract the percentage found in Table z from 50%. If z is negative, add the percentage found in Table z to 50%. Draw diagrams to make things clearer.

For 95: Since $z = 1$, then subtract: 50% − 34.13% = 15.87%

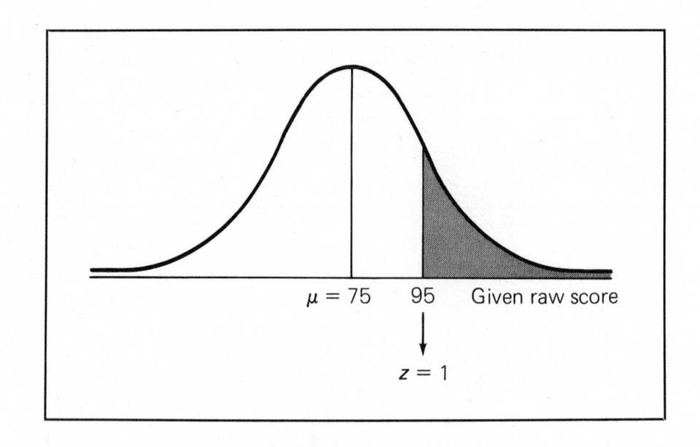

For 45: Since $z = -1.5$, then add: 50% + 43.32% = 93.32%

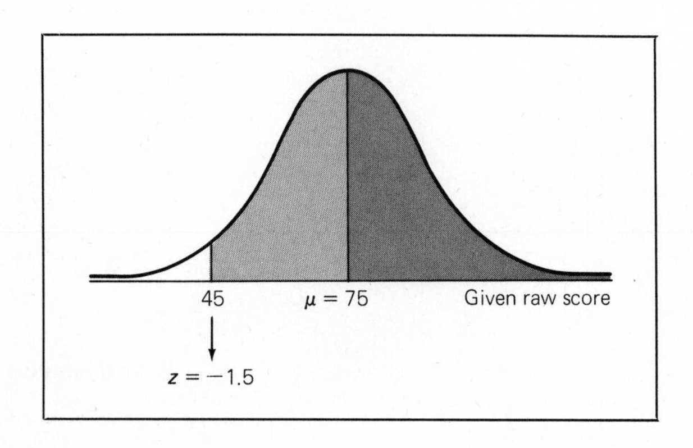

For 84: Since z = .45, then subtract: 50% − 17.36% = 32.64%

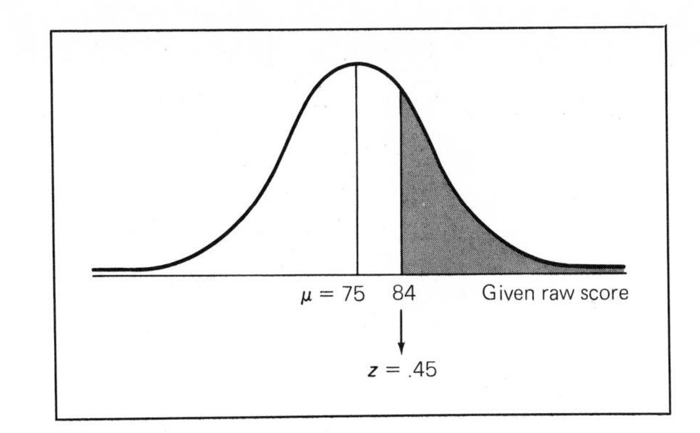

For 37: Since z = −1.9, then add: 50% + 47.13% = 97.13%.

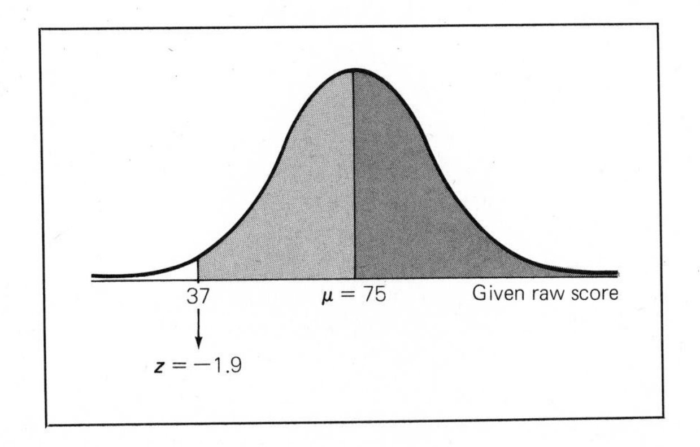

For 77.6: Since z = .13, then subtract: 50% − 5.17% = 44.83%.

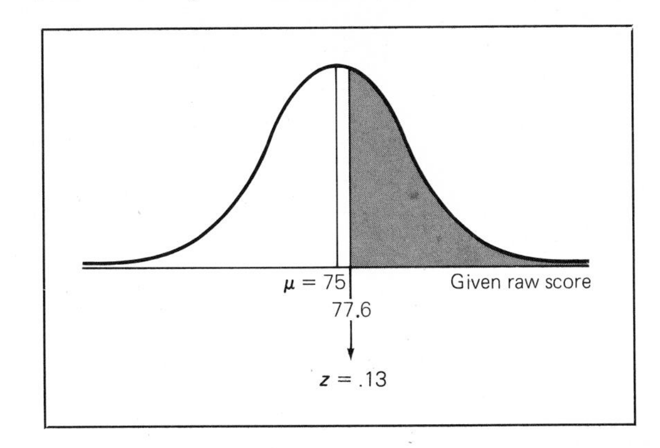

For 28.25: Since $z = -2.338$, then add: 50% + 49.04% = 99.04%.

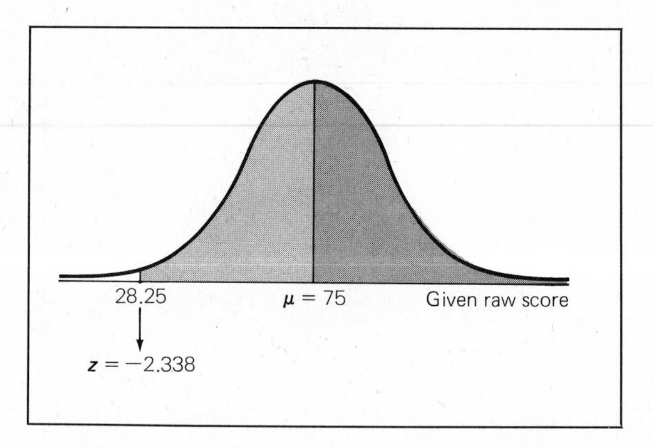

c. For problems concerning the percentage of scores falling below a given raw score, the strategy is to proceed as in (a). Then, if z is positive, add the percentage found in Table z to 50%. If z is negative, subtract the percentage found in Table z from 50%. Draw diagrams!

For 95: Since $z = 1$, then add: 50% + 34.13% = 84.13%.
For 45: Since $z = -1.5$, then subtract: 50% − 43.32% = 6.68%.
For 84: Since $z = .45$, then add: 50% + 17.36% = 67.36%.
For 37: Since $z = -1.9$, then subtract: 50% − 47.13% = 2.87%.
For 77.6: Since $z = .13$, then add: 50% + 5.17% = 55.17%.
For 28.25: Since $z = -2.338$, then subtract: 50% − 49.04% = .96%.

d. For problems concerning the percentage of scores falling between the given raw scores, the strategy is to proceed as in (a). Then, if the two z scores are on opposite sides of the mean, add the percentages found in Table z. If the two z scores are on the same side of the mean, subtract the smaller percentage found in Table z from the larger percentage. Draw diagrams!

For 28.25 to 37: Since the z scores of −2.338 and −1.9 lie on the same side of the mean, then subtract: 49.04% − 47.13% = 1.91%.

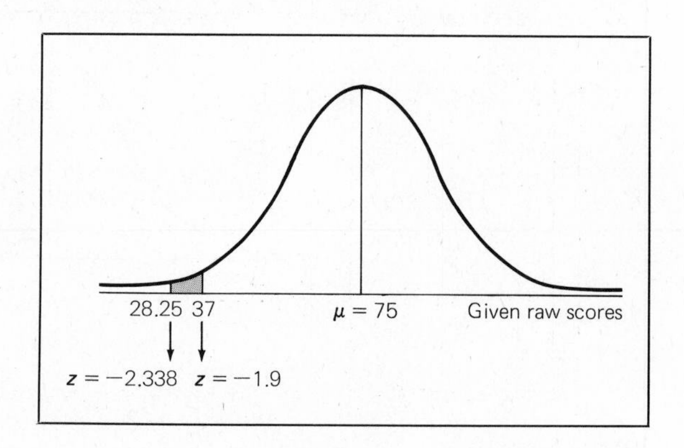

For 45 to 84: Since the z scores of -1.5 and .45 lie on opposite sides of the mean, then add: 43.32% + 17.36% = 60.68%.

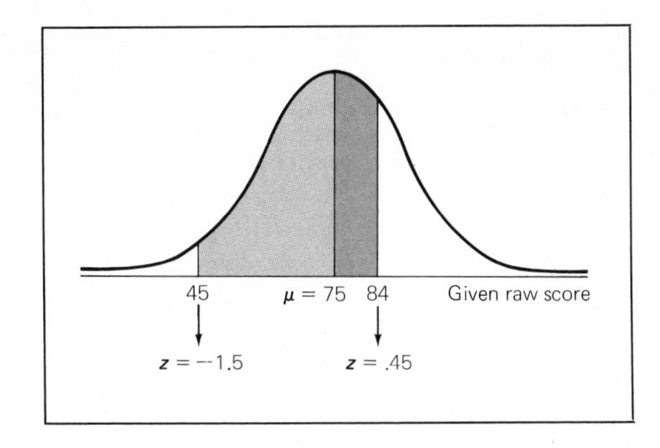

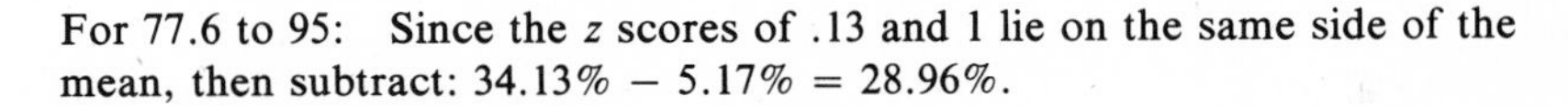

For 77.6 to 95: Since the z scores of .13 and 1 lie on the same side of the mean, then subtract: 34.13% − 5.17% = 28.96%.

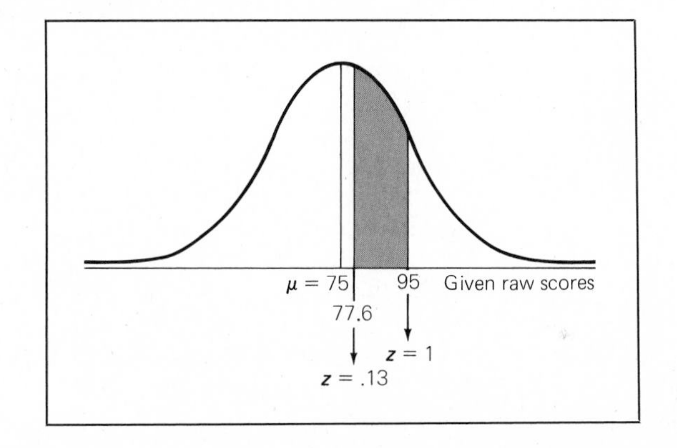

6 The strategy is to proceed as in Exercise 5. Since Exercise 6 asks for the frequencies, convert your ultimate percentage to a proportion (by moving the decimal two places to the left) and then multiply this proportion by N.

For 380: $z = \frac{380 - 500}{50} = -2.4$, and hence 49.18% from Table z.

For 590: $z = \frac{590 - 500}{50} = 1.8$, and hence 46.41% from Table z.

For 415: $z = \frac{415 - 500}{50} = -1.7$, and hence 45.54% from Table z.

For 515: $z = \frac{515 - 500}{50} = .3$, and hence 11.79% from Table z.

For 646: $z = \dfrac{646 - 500}{50} = 2.92$, and hence 49.82% from Table z.

a. For 380: (.4918)(400) = 196.72 scores
For 590: (.4641)(400) = 185.64 scores
For 415: (.4554)(400) = 182.16 scores
For 515: (.1179)(400) = 47.16 scores
For 646: (.4982)(400) = 199.28 scores

b. For 380: 50% + 49.18% = 99.18%, hence (.9918)(400) = 396.72 scores.
For 590: 50% − 46.41% = 3.59%, hence (.0359)(400) = 14.36 scores.
For 415: 50% + 45.54% = 95.54%, hence (.9554)(400) = 382.16 scores.
For 515: 50% − 11.79% = 38.21%, hence (.3821)(400) = 152.84 scores.
For 646: 50% − 49.82% = .18%, hence (.0018)(400) = .72 scores.

c. For 380: 50% − 49.18% = .82%, hence (.0082)(400) = 3.28 scores.
For 590: 50% + 46.41% = 96.41%, hence (.9641)(400) = 385.64 scores.
For 415: 50% − 45.54% = 4.46%, hence (.0446)(400) = 17.84 scores.
For 515: 50% + 11.79% = 61.79%, hence (.6179)(400) = 247.16 scores.
For 646: 50% + 49.82% = 99.82%, hence (.9982)(400) = 399.28 scores.

d. For 380 to 415: 49.18% − 45.54% = 3.64%, hence (.0364)(400) = 14.56 scores.
For 415 to 590: 45.54% + 46.41% = 91.95%, hence (.9195)(400) = 367.8 scores.
For 515 to 646: 49.82% − 11.79% = 38.03%, hence (.3803)(400) = 152.12 scores.

7 **a.** Draw a diagram to avoid confusion. Look in the body of Table z for 40% (which will leave the top 10%) and find its associated value of z. The percentage in the table closest to 40% is 39.97%, and its associated z is 1.28. Then, using Formula 4.4, the raw score that cuts off the top 10% is 694:

$$X = 530 + (1.28)(128) = 693.84 \text{ or } 694$$

b. Yes, since his score of 720 is above the cutoff score of 694.

c. Yes, since her z score of 2.13 is above the z score of 1.28, which cuts off the top 10%. You might also determine that her raw score is 803:

$$X = 530 + (2.13)(128) = 802.64 \text{ or } 803$$

d. Draw a diagram. Determine that 600 is 8.333% of 7200 by:

$$\frac{600}{7200}(100) = 8.333\%.$$

Look in the body of Table z for 41.667% (which will leave the top 8.333%) and find its associated value of z. The percentage in the table closest to 41.667% is 41.62%, and its associated z is 1.38. Then, using Formula 4.4, the quantitative ability score that cuts off the top 600 people is 707:

$$X = 530 + (1.38)(128) = 706.64 \text{ or } 707$$

e. College Joe is better in quantitative ability.

For verbal ability, $z = \frac{740 - 510}{157} = 1.465.$

For quantitative ability, $z = \frac{720 - 530}{128} = 1.484.$

ANSWERS TO EXERCISES, CHAPTER 5

1 Naturalistic

2 Naturalistic

3 Experimental

4 Experimental

5

a. Independent variable: the location of electrode placement.
Dependent variable: the number of minutes it takes to remember their name.

b. Hypothesis: The location of electrode placement affects amount of memory loss following ECT.

c. Extraneous variables: The intensity and duration of ECT are controlled extraneous variables because they are kept constant for all three groups; the individual histories and duration of depressive episodes, and the length of time spent in the mental health institutions are two extraneous variables that have not been controlled.

d. Experimental design: multiple-group design.

6

a. Independent variable: the amount of time spent watching TV.
Dependent variable: the number of arguments.

b. Hypothesis: A reduction in the time spent watching TV will decrease the number of arguments between husbands and wives.

c. Extraneous variables: The length of time the couples have been married and the number of children (none) are controlled extraneous variables; the effect of the passage of time is one extraneous variable that is not controlled (that is, regardless of the amount of time spent watching TV, the couples may alter their frequency of arguing because several weeks have elapsed and many other factors can enter the situation during this time to produce changes in arguing).

d. Experimental design: two-sample dependent design.

7

a. Independent variable: the consumption of alcohol.
Dependent variable: RT.

b. Hypothesis: The consumption of alcohol will affect RT.

c. Extraneous variables: The context of the situation calling for braking is a controlled extraneous variable because the simulated motorcycle ride is the same for subjects in both groups; the subjects' ages and the consumption of 12 ounces of liquid are also extraneous variables that have been controlled.

d. Experimental design: two-sample independent design.

8

a. Independent variable: the instructions to use mnemonic imagery.
Dependent variable: the number of digits recalled in correct order.

b. Hypothesis: The instruction to use mnemonic imagery affects recall.

c. Extraneous variables: Two uncontrolled extraneous variables are the amount of previous experience using imagery as a memory aid, and the motivation to do well.

d. Experimental design: one-sample design.

9

a. Independent variable: watching violence on TV.
Dependent variable: the amount of aggressiveness.

b. Watching violence on TV: The number of hours spent per week watching violent TV programs. (Violent TV programs are identified from a list of "violent TV programs" designated by the researcher or other media researchers.)

Amount of aggressiveness: the number of times the child strikes a "boxing doll" suspended from the ceiling in a play session.

c. Take a random sample of 3- to 4-year-old children and measure the amount of time spent watching violence on TV and their amount of aggressiveness.

Extraneous variables: the TV viewing habits of parents, behavior patterns of parents (particularly their child-rearing practices), socioeconomic status, and so on.

d. Take a random sample of 3- to 4-year-old children and make them watch, say, four hours of violent programs on TV each day for a certain number of days. Then measure the amount of aggressiveness displayed in a one-hour play session. You assume that you know the level of aggressiveness in the population of 3- to 4-year-olds, and you assume that this level does not change during the course of your experiment. You also assume that the level of violent TV viewing you selected (four hours daily) is above the level in the general population of 3- to 4-year-olds.
Extraneous variables: The effects of making a child watch TV may be frustrating regardless of its content; and the mere passage of time over the duration of the experiment may affect changes.

e. Randomly select a large number of children from the population of 3 to 4 year olds and then randomly assign each to one of two groups such that there is an equal number of subjects in both groups. In the control group, have the subjects watch a large amount of randomly chosen nonviolent TV programs, say 4 hours per day. In the experimental group, have the subjects watch 4 hours per day of violent TV programs. Afterward, measure the amount of aggressiveness in each group.

f. There are several ways to do this. First, randomly select a sample of 3- to 4-year-old children. Then, you could proceed in either of two ways. One way is to pretest the subjects' initial level of aggressiveness, then require a large amount of time spent viewing violent TV programs per day, and then retest the same subjects' level of aggressiveness. A second way is to pretest the subjects' initial level of aggressiveness and then match subjects by pairs on the basis of similar pretest results. Then randomly assign one member of each pair to the control group; the other member of the pair is placed in the experimental group. Have subjects in the control group watch a large amount of randomly chosen nonviolent TV programs, say four hours per day. Have the subjects in the experimental group watch four hours per day of violent TV programs. Afterward, measure the amount of aggressiveness in each group.

g. Suppose your hypothesis is that increasing levels of viewing violence on TV produce increasing levels of aggressiveness in children aged 3 and 4. Take a random sample of subjects from the population of 3- to 4-year-old children and then randomly assign each to one of five groups such that there is an equal number of subjects in each group. One group is a control group whose subjects watch a large amount of randomly selected nonviolent TV programs, say four hours per day. A second group watches random nonviolent TV for three hours and violent TV for one hour per day; a third group watches nonviolent TV for two hours and violent TV for two hours per day; a fourth group watches nonviolent TV for one hour and violent TV for three hours per day; and the fifth group watches violent TV for four hours per day. Afterward, measure the amount of aggressiveness in each group.

h. Chance sampling error.

ANSWERS TO EXERCISES, CHAPTER 6

1

a. $p\text{ (one)} = \frac{1}{6} = .167$

b. $p\text{ (three ones in a row)} = \left(\frac{1}{6}\right)\left(\frac{1}{6}\right)\left(\frac{1}{6}\right) = .005$

c. $p\text{ (one, then two, then three)} = \left(\frac{1}{6}\right)\left(\frac{1}{6}\right)\left(\frac{1}{6}\right) = .005$

2

a. $p\text{ (heart)} = \left(\frac{13}{52}\right) = .25$

b. $p\text{ (two hearts in a row)} = \left(\frac{13}{52}\right)\left(\frac{13}{52}\right) = .062$

c. $p\text{ (heart and then a spade)} = \left(\frac{13}{52}\right)\left(\frac{13}{52}\right) = .062$

d. p (heart and then an ace) $= \left(\frac{13}{52}\right)\left(\frac{4}{52}\right) = .019$

3

H_0: The sample comes from the general population having $\mu = 104$. (The stress has no effect on recognition memory.)
H_A: The sample comes from a different population having $\mu \neq 104$. (Stress has an effect on recognition memory.)

4

H_0: The sample comes from the general population having $\mu = 1.6$. (Passive observation of an instructional film has no effect on free-throw shooting.)
H_A: The sample comes from a different population having $\mu \neq 1.6$. (Passive observation of an instructional film has an effect on free-throw shooting.]

5

H_0: The sample comes from the general population having $\mu = 1$. (Coffee has no effect on the speed with which the liver metabolizes alcohol.)
H_A: The sample comes from a different population having $\mu \neq 1$. (Coffee alters the speed with which the liver metabolizes alcohol.)

6

H_0: The sample comes from the general population of having $\mu = 20$. (The prescribed labor room conditions have no effect on developmental behavior patterns and strength.)
H_A: The sample comes from a different population having $\mu \neq 20$. (The prescribed labor room conditions have an effect on developmental behavior patterns and strength.)

7

Step 1. H_0: p (club) $= .25$ (The deck is fair)
H_A: p (club) $\neq .25$ (The deck is biased)
Step 2. Let $\alpha = .05$.
Step 3. Assuming H_0 is true, the probability of getting your result by chance is:

p (four clubs in a row) $= (.25)(.25)(.25)(.25) = .004$

Step 4. Since the probability of getting this result by chance ($p = .004$) is less than alpha, you should reject H_0 and conclude that the deck is biased.

8

The friend has not "assumed the null hypothesis." The two competing hypotheses are: The probability of a head (or tail) is .5 (this is to say that the coin is fair); and the probability of a head (or tail) is not .5 (this is to say that the coin is biased). If H_0 is that the coin is biased, then you could not complete Step 3 of the hypothesis test because you cannot calculate the probability of getting some result when the probability of a head (or tail) is not precisely specified. The proper H_0 is that the coin is fair:

H_0: probability of heads (or tails) $= .5$

9

Step 1. H_0: probability of being correct $= \frac{5}{25} = .2$ (Your friend is performing at chance and has no ESP.)

H_A: probability of being correct $\neq .2$. (Your friend is performing extraordinarily.)

Step 2. Let $\alpha = .01$.
Step 3. Assuming H_0 is true, the probability of getting your result by chance is:

p (three correct guesses in a row) = (.2)(.2)(.2) = .008

Step 4. Since the probability of getting this result by chance ($p = .008$) is less than alpha, you should reject H_0 and conclude that your friend is extraordinary.

10 By "accepting" H_0, you imply that you have shown it to be true. In fact, however, all you do in the hypothesis-testing process is gather evidence against it.

11 Alpha is the *probability, p,* of rejecting a true H_0, whereas an alpha error is an actual *instance* of rejecting a true H_0. Beta is the probability of retaining a false H_0. Power is the probability of rejecting a false H_0.

ANSWERS TO EXERCISES, CHAPTER 7

1

a. $\mu = 7$
b. As the size of the sample you select increases, the variability among sample means decreases.

2

a. H_0: $\mu = 79$ (Giving valentines has no effect on final scores)
H_A: $\mu \neq 79$ (Giving valentines has an effect on final scores)

This is a two-tailed test since no direction is specified.

b. H_0: $\mu \leq 79$ (Giving valentines does not increase final scores)
H_A: $\mu > 79$ (Giving valentines increases final scores)

This is a one-tailed test since an increase in scores is expected.

3

a. The four steps of the hypothesis testing procedure are:
(1) H_0: $\mu = 45$ (Working jigsaw puzzles has no effect on the time to match with blocks)
H_A: $\mu \neq 45$ (Working jigsaw puzzles has an effect on the time to match with blocks)
(2) Setting $\alpha = .05$, since this is a two-tailed test, then $z^* = 1.96$.
(3) Using Formula 7.1, the calculated z for your sample mean of 37.5 is:

$$z = \frac{\bar{X} - \mu}{\frac{\sigma}{\sqrt{N}}} = \frac{37.5 - 45}{\frac{16}{\sqrt{20}}} = -2.096$$

(4) Since $|-2.096| > 1.96$, reject H_0 and conclude that working with jigsaw puzzles has an effect on the time to match with blocks.

b. $z = 2.096$, $p < .05$

4

a. $\bar{X} = 3.667$, $s = 4.619$
b. $\bar{X} = 7$, $s = 4.195$
c. $\bar{X} = 6.4$, $s = 3.534$
d. $\bar{X} = 7.267$, $s = 3.535$
e. In general, as the size of the sample you select increases, s comes closer to σ.

5

a. The z test, since σ is known.
b. The four steps of the hypothesis-testing procedure are:

(1) H_0: $\mu \geq 1100$ (The applicants are not inferior)
H_A: $\mu < 1100$ (The applicants are inferior)

(2) Setting $\alpha = .01$, since this is a one-tailed test, then $z^* = 2.33$. This is a one-tailed test, so you check to be sure that your sample mean is in the direction expected with H_A. Since $\bar{X} = 1073$ is in the expected direction, you proceed with Step (3).
(3) Using Formula 7.1, the calculated z for your sample mean of 960 is:

$$z = \frac{\bar{X} - \mu}{\frac{\sigma}{\sqrt{N}}} = \frac{1073 - 1100}{\frac{80}{\sqrt{45}}} = -2.264$$

(4) Since $|-2.264| < 2.33$, retain H_0 and conclude that there is not sufficient evidence to say the applicants are inferior.

c. $z = 2.264$, N.S.

6

a. The one-sample t test, since σ is unknown.
b. The four steps of the hypothesis-testing procedure are:

(1) H_0 $\mu = 8$ (Vitamin C has no effect on absenteeism due to colds)
H_A: $\mu \neq 8$ (Vitamin C has an effect on absenteeism due to colds)

(2) Setting $\alpha = .01$, since this is a two-tailed test and $df = N - 1 = 69$, then $t^* = 2.660$.
(3) Using Formula 7.3, the calculated t score for your sample mean of 6.55 is:

$$t = \frac{\bar{X} - \mu}{\frac{s}{\sqrt{N}}} = \frac{6.55 - 8}{\frac{3.21}{\sqrt{70}}} = 3.779$$

(4) Since $|-3.779| > 2.660$, reject H_0 and conclude that vitamin C has an effect on absenteeism.

c. $t(69) = 3.779$, $p < .01$

7

a. The z test, since σ is known.
b. The steps of the hypothesis-testing procedure are:

(1) H_0: $\mu \leq 70$ (The extra time did not increase the scores)
H_A: $\mu > 70$ (The extra time increased the scores)

(2) Setting $\alpha = .05$. Since this is a one-tailed test, $z^* = 1.65$. Because this is a one-tailed test, you check to be sure your sample mean is in the direction expected with H_A. Since $\bar{X} = 69.867$ is not in the direction specified in H_A, you terminate the hypothesis-testing procedure at this stage and retain H_0. You may conclude that the extra time did not increase the scores.

8 The correct procedure is **d**.

9

a. The region of rejection is the same in terms of percentage of scores in the sampling distribution; it is just divided differently in the two cases.

b. The region of rejection is larger for $\alpha = .05$ because you are more willing to risk an alpha error in this case.

10

a. $\alpha = .05$

b. The steps of the hypothesis-testing procedure are:

(1) H_0: $\mu \geq 60$ (The drug does not decrease dreaming time in monkeys)
H_A: $\mu < 60$ (The drug decreases dreaming time in monkeys)

(2) Setting $\alpha = .05$, since this is a one-tailed test and $df = N - 1 = 9$, then $t^* = 1.833$. This is a one-tailed test, so you check to be sure that your sample mean is in the direction specified with H_A. Since $\bar{X} = 43.2$ is in the expected direction, you proceed with Step (3).

(3) Using Formula 7.3, the calculated t for your sample mean is:

$$t = \frac{\bar{X} - \mu}{\frac{s}{\sqrt{N}}} = \frac{43.2 - 60}{\frac{18.949}{\sqrt{10}}} = -2.804$$

where, using Formula 7.2,

$$s = \sqrt{\frac{\Sigma(X - \bar{X})^2}{N - 1}} = \sqrt{\frac{3231.6}{9}} = 18.949$$

(4) Since $|-2.804| > 1.833$, reject H_0 and conclude that the drug decreases dreaming time in monkeys. You thus would decide not to pursue marketing the drug for human consumption in its present chemical form.

c. $t\,(9) = 2.804,\ p < .05$

d.
$$t = \frac{\bar{X} - \mu}{\sqrt{\frac{\Sigma X^2 - \frac{(\Sigma X)^2}{N}}{N(N-1)}}} = \frac{43.2 - 60}{\sqrt{\frac{21894 - \frac{186{,}624}{10}}{10(9)}}} = -2.804$$

11

a. The one-sample t test, since σ is unknown.

b. The steps of the hypothesis-testing procedure are:

(1) H_0: $\mu \leq 36$ (Disruptive adolescents do not have high ergic tension)
H_A: $\mu > 36$ (Disruptive adolescents have high ergic tension)

(2) Setting $\alpha = .05$, since this is a one-tailed test and $df = N - 1 = 19$, then $t^* = 1.729$. This is a one-tailed test, so you check to be sure that your sample mean is in the direction specified with H_A. Since $\bar{X} = 40$ is in the expected direction, you proceed with Step (3).

(3) Using Formula 7.5, the calculated t is:

$$t = \frac{\bar{X} - \mu}{\sqrt{\dfrac{\Sigma X^2 - \dfrac{(\Sigma X)^2}{N}}{N(N-1)}}} = \frac{40 - 36}{\sqrt{\dfrac{33{,}176 - \dfrac{640{,}000}{20}}{20(19)}}} = 2.274$$

(4) Since $|2.274| > 1.729$, reject H_0 and conclude that disruptive adolescents have high ergic tension.

c. $t(19) = 2.274,\ p < .05$

ANSWERS TO EXERCISES, CHAPTER 8

1

a. The four steps of the hypothesis-testing procedure are:

(1) H_0: $\mu_1 \geq \mu_2$ (Coaching does not increase LSAT scores)
H_A: $\mu_1 < \mu_2$ (Coaching increases LSAT scores)

(2) Setting $\alpha = .05$, since this is a one-tailed test and $df = N_1 + N_2 - 2 = 48$, then $t^* = 1.684$. This is a one-tailed test where you predict $\mu_1 < \mu_2$ in H_A, so you expect a negative difference when you subtract $\bar{X}_1 - \bar{X}_2$. Since $\bar{X}_1 - \bar{X}_2 = 619 - 625 = -6$ is indeed a negative difference, you proceed with Step (3).

(3) Using definitional Formula 8.5, the calculated t score for the difference between your sample means is:

$$t = \frac{(\bar{X}_1 - \bar{X}_2) - 0}{\sqrt{\dfrac{(N_1 - 1)\,s_1^2 + (N_2 - 1)\,s_2^2}{N_1 + N_2 - 2}\left(\dfrac{1}{N_1} + \dfrac{1}{N_2}\right)}}$$

$$= \frac{619 - 625}{\sqrt{\dfrac{(24)56 + (24)52}{48}(.04 + .04)}} = -2.887$$

(4) Since $|-2.887| > 1.684$, reject H_0 and conclude that coaching increases LSAT scores.

b. $t(48) = 2.887,\ p < .05$

2 s^2_{pooled} is the weighted average of s_1^2 and s_2^2. It is the "pooled" estimate of σ^2.

3 **a.** The steps of the hypothesis-testing procedure are:

(1) H_0: $\mu_1 = \mu_2$ (Encouragement has no effect on the accuracy of reproductions)

H_A: $\mu_1 \neq \mu_2$ (Encouragement has an effect on the accuracy of reproductions)

(2) Setting $\alpha = .05$, since this is a two-tailed test and $df = N - 1 = 4$, then $t^* = 2.776$. You create a set of D scores and calculate $\bar{D} = \Sigma D/N = 20/5 = 4$.

Subject	X_1	X_2	D $(X_1 - X_2)$	D^2
S_1	3	2	1	1
S_2	5	0	5	25
S_3	5	1	4	16
S_4	7	0	7	49
S_5	4	1	3	9
			$\Sigma D = 20$	$\Sigma D^2 = 100$
			$(\Sigma D)^2 = (20)^2 = 400$	

(3) Using definitional Formula 8.8, the calculated t score for your $\bar{D} = 4$ is:

$$t = \frac{\bar{D} - 0}{\frac{s_D}{\sqrt{N}}} = \frac{4}{\frac{2.236}{\sqrt{5}}} = 4$$

where, using Formula 8.7,

$$s_D = \sqrt{\frac{\Sigma D^2 - \frac{(\Sigma D)^2}{N}}{N - 1}} = \sqrt{\frac{100 - \frac{400}{5}}{4}} = 2.236$$

(4) Since $|4| > 2.776$, reject H_0 and conclude that encouragement has an effect on accuracy of reproductions.

b. $t(4) = 4, p < .05$

4 **a.** The steps of the hypothesis-testing procedure are:

(1) H_0: $\mu_1 \leq \mu_2$ (Encouragement does not increase the accuracy of reproductions)

H_A: $\mu_1 > \mu_2$ (Encouragement increases the accuracy of reproductions)

(2) Setting $\alpha = .05$, since this is a one-tailed test and $df = N_1 + N_2 - 2 = 8$, then $t^* = 1.860$. This is a one-tailed test where you predict $\mu_1 > \mu_2$ in H_A, so you expect a positive difference when you subtract $\bar{X}_1 - \bar{X}_2$. Since $\bar{X}_1 - \bar{X}_2 = 4 - 1 = 3$ is indeed a positive difference, you proceed with Step (3).

(3) Using computational Formula 8.9, the calculated t score for the difference between your sample means is:

$$t = \frac{(\bar{X}_1 - \bar{X}_2)}{\sqrt{\dfrac{\Sigma X_1^2 - \dfrac{(\Sigma X_1)^2}{N_1} + \Sigma X_2^2 - \dfrac{(\Sigma X_2)^2}{N_2}}{N_1 + N_2 - 2}\left(\dfrac{1}{N_1} + \dfrac{1}{N_2}\right)}}$$

$$= \frac{4 - 1}{\sqrt{\dfrac{82 - \dfrac{400}{5} + 9 - \dfrac{25}{5}}{8}(.2 + .2)}}$$

$$= 5.477$$

(4) Since $|5.477| > 1.860$, reject H_0 and conclude that encouragement increases the accuracy of reproductions.

b. $t(8) = 5.477, p < .05$

5 **a.** The steps of the hypothesis-testing procedure are:

(1) H_0: $\mu_1 \geq \mu_2$ (The math instructor is not effective)
H_A: $\mu_1 < \mu_2$ (The math instructor is effective)

(2) Setting $\alpha = .01$, since this is a one-tailed test and $df = N - 1 = 19$, then $t^* = 2.539$. This is a one-tailed test where you predict $\mu_1 < \mu_2$ in H_A and, since $\bar{D} = -8$, you found the negative mean difference you expected. You proceed with Step (3).

(3) Using definitional Formula 8.8, the calculated t score for your $\bar{D} = -8$ is:

$$t = \frac{\bar{D} - 0}{\dfrac{s_D}{\sqrt{N}}} = \frac{-8}{\dfrac{16}{\sqrt{20}}} = -2.236$$

(4) Since $|-2.236| < 2.539$, retain H_0 and conclude that the math instructor is not effective.

b. $t(19) = 2.236$, N.S.

6 **a.** The dependent t test, since the same subjects are tested and retested.

b. The steps of the hypothesis testing-procedure are:

(1) H_0: $\mu_1 \leq \mu_2$ (The campaign does not reduce cigarette smoking at work)
H_A: $\mu_1 > \mu_2$ (The campaign reduces cigarette smoking at work)

(2) Setting $\alpha = .05$, since this is a one-tailed test and $df = N - 1 = 7$, then $t^* = 1.895$. This is a one-tailed test where you predict $\mu_1 > \mu_2$ in H_A, and expect $\bar{D}$ to be positive. Since $\bar{D} = 3.25$ is positive, you proceed with Step (3).

(3) Using computational Formula 8.10, the calculated t score for your $\bar{D} =$ 3.25 is:

$$t = \frac{\bar{D}}{\sqrt{\dfrac{\Sigma D^2 - \dfrac{(\Sigma D)^2}{N}}{N(N-1)}}} = \frac{3.25}{\sqrt{\dfrac{184 - \dfrac{676}{8}}{56}}} = 2.438$$

(4) Since $|2.438| > 1.895$, reject H_0 and conclude that the campaign reduces cigarette smoking at work.

c. $t(7) = 2.438,\ p < .05$

7 The steps of the hypothesis-testing procedure are:

(1) H_0: $\mu_1 \le \mu_2$ (The passage of time does not reduce cigarette smoking at work)
H_A: $\mu_1 > \mu_2$ (The passage of time reduces cigarette smoking at work)

(2) Setting $\alpha = .05$, since this is a one-tailed test and $df = N - 1 = 7$, then $t^* = 1.895$. This is a one-tailed test where you predict $\mu_1 > \mu_2$ in H_A, and expect $\bar{D}$ to be positive. Since

$$\bar{D} = \frac{\Sigma D}{N} = \frac{-1}{8} = -.125$$

is not in the direction expected under H_A, you terminate the hypothesis-testing procedure and retain H_0. You cannot conclude that the passage of time reduces smoking. Therefore you cannot conclude that the passage of time is a viable explanation for the reduction in smoking observed in the study conducted in Exercise 6.

8 **a.** The independent t test, since different subjects are randomly assigned to one of the two conditions, and there is no attempt to match up subjects into pairs.

b. The steps of the hypothesis-testing procedure are:

(1) H_0: $\mu_1 = \mu_2$ (Noise does not affect recall)
H_A: $\mu_1 \ne \mu_2$ (Noise affects recall)

(2) Setting $\alpha = .05$, since this is a two-tailed test and $df = N_1 + N_2 - 2 =$ 14, then $t^* = 2.145$.

(3) Using definitional Formula 8.5, the calculated t score for the observed difference between your sample means is:

$$t = \frac{(\bar{X}_1 - \bar{X}_2) - 0}{\sqrt{\dfrac{(N_1 - 1)s_1^2 + (N_2 - 1)s_2^2}{N_1 + N_2 - 2}\left(\dfrac{1}{N_1} + \dfrac{1}{N_2}\right)}}$$

$$= \frac{17.5 - 18.5}{\sqrt{\frac{(7)6.857 + (7)11.714}{14}(.125 + .125)}}$$

$$= -.656$$

where

$$s_1^2 = \frac{\Sigma X_1^2 - \frac{(\Sigma X_1)^2}{N_1}}{N_1 - 1} = \frac{2498 - \frac{19,600}{8}}{7} = 6.875$$

where

$$s_2^2 = \frac{\Sigma X_2^2 - \frac{(\Sigma X_2)^2}{N_2}}{N_2 - 1} = \frac{2820 - \frac{21,904}{8}}{7} = 11.714$$

(4) Since $|-.656| < 2.145$, retain H_0 and conclude that there is not sufficient evidence to say that noise affects recall.

c. $t(14) = .656, p > .05$

9 **a.** The dependent t test, since pairs of subjects are matched on several characteristics that could affect job satisfaction.

b. The steps of the hypothesis-testing procedure are:

(1) H_0: $\mu_1 = \mu_2$ (A city's general unemployment rate does not affect job satisfaction)

H_A: $\mu_1 \neq \mu_2$ (A city's general unemployment rate affects job satisfaction)

(2) Setting $\alpha = .05$, since this is a two-tailed test and $df = N - 1 = 9$, then $t^* = 2.262$. You create a set of D scores and calculate

$$\bar{D} = \frac{\Sigma D}{N} = \frac{-19}{10} = -1.9$$

(3) Using computational Formula 8.10, the t score for your $\bar{D} = -1.9$ is:

$$t = \frac{\bar{D}}{\sqrt{\frac{\Sigma D^2 - \frac{(\Sigma D)^2}{N}}{N(N-1)}}} = \frac{-1.9}{\sqrt{\frac{85 - \frac{361}{10}}{90}}} = -2.578$$

(4) Since $|-2.578| > 2.262$, reject H_0 and conclude that a city's general unemployment rate affects job satisfaction.

c. $t(9) = 2.578, p < .05$.

10

a. The independent t test, because different subjects are randomly assigned to either of the two conditions, and there was no attempt to pair subjects through matching.

b. The steps of the hypothesis-testing procedure are:

(1) H_0: $\mu_1 \leq \mu_2$ (Auditory feedback does not facilitate self-control)
H_A: $\mu_1 > \mu_2$ (Auditory feedback facilitates self-control)

(2) Setting $\alpha = .01$, since this is a one-tailed test and $df = N_1 + N_2 - 2 = 12 + 11 - 2 = 21$, then $t^* = 2.518$. This is a one-tailed test where you predict $\mu_1 > \mu_2$ in H_A, so you expect a positive difference when you subtract $\bar{X}_1 - \bar{X}_2$. Since $\bar{X}_1 - \bar{X}_2 = 2.45 - 2.136 = .314$ is indeed a positive difference, you proceed with Step (3).

(3) Using computational Formula 8.9, the calculated t score for the observed difference between your sample means is:

$$t = \frac{(\bar{X}_1 - \bar{X}_2)}{\sqrt{\dfrac{\Sigma X_1^2 - \dfrac{(\Sigma X_1)^2}{N_1} + \Sigma X_2^2 - \dfrac{(\Sigma X_2)^2}{N_2}}{N_1 + N_2 - 2}\left(\dfrac{1}{N_1} + \dfrac{1}{N_2}\right)}}$$

$$= \frac{2.450 - 2.136}{\sqrt{\dfrac{79.62 - \dfrac{864.36}{12} + 55.79 - \dfrac{552.25}{11}}{21}(.083 + .091)}}$$

$$= .950$$

(4) Since $|.950| < 2.518$, retain H_0 and conclude that auditory feedback does not facilitate self-control.

c. $t(21) = .950$, N.S.

ANSWERS TO EXERCISES, CHAPTER 9

1

a. The Mann-Whitney U test, since the samples are independent and there is no attempt to match up babies into pairs.

b. The radically different ranges for male and female babies suggests that the variance of the population of male babies may not be equal to the variance of the population of female babies. The independent t test assumes that the two populations have equal variances.

c. The steps of the hypothesis-testing procedure are:

(1) H_0: Baby gender does not affect fathers' perceptions of cuddliness. That is, the distribution of scores for the population of male babies is identical to the distribution of scores for the population of female babies.

H_A: Baby gender makes a difference in fathers' perceptions of cuddliness. That is, the distribution of scores for the population of male babies is not identical to the distribution of scores for the population of female babies.

(2) Setting $\alpha = .05$, since this is a two-tailed test and $N_1 = 7$ and $N_2 = 8$, then $U^* = 10$ and 46. You then organize the data, assign ranks, and calculate $R_1 = 38$ for male babies.

(3) Using Formula 9.1, you calculate U:

$$U = R_1 - \frac{(N_1)(N_1 + 1)}{2}$$

$$= 38 - \frac{(7)(7 + 1)}{2} = 10$$

Since this is a two-tailed test you proceed to Step (4a).

(4) Since the calculated U of 10 is equal to the smaller critical value of 10, reject H_0 and conclude that baby gender influences father perceptions of cuddliness.

d. Mann-Whitney $U = 10$, $p < .05$.

2

a. The Wilcoxon test, because the same subjects are tested and retested.

b. Both distributions of reading scores are very negatively skewed, suggesting that the populations of "before" and "after" scores are not normally distributed. The dependent t test assumes that the two population distributions are normal.

c. The steps of the hypothesis-testing procedure are:

(1) H_0: The home reading program does not increase reading achievement. That is, the distribution of the population of "before" reading equivalence scores lies above or is identical to the distribution of the population of "after" scores.

H_A: The home reading program increases reading achievement. That is, the distribution of the population of "before" scores lies below the distribution of the population of "after" scores.

(2) Although you collected 15 pairs of scores, since two pairs of X_1 and X_2 scores are equal (for S_3 and S_{13}), you ignore these pairs, hence $N = 13$. Setting $\alpha = .05$, since this is a one-tailed test and $N = 13$, then $W^* = 21$

(3) You create a set of D scores, assign ranks, and calculate W:

Subject	D $(X_1 - X_2)$	$\lvert D \rvert$	*Rank of* $\lvert D \rvert$	*Ranks for positive* D*s*	*Ranks for negative* D*s*
S_1	−.5	.5	2.5		2.5
S_2	−.8	.8	7		7
S_3	0				
S_4	−.5	.5	2.5		2.5
S_5	−.7	.7	5		5

Subject	D $(X_1 - X_2)$	$\|D\|$	Rank of $\|D\|$	Ranks for positive Ds	Ranks for negative Ds
S_6	−.9	.9	9		9
S_7	−1.8	1.8	13		13
S_8	−.7	.7	5		5
S_9	−.7	.7	5		5
S_{10}	−.9	.9	9		9
S_{11}	−1.2	1.2	12		12
S_{12}	.3	.3	1	1	
S_{13}	0				
S_{14}	−.9	.9	9		9
S_{15}	−1.1	1.1	11		11
				Sum = 1 = W	Sum = 90

The sum of the ranks for the positive D scores is smallest, thus $W = 1$. Since this is a one-tailed test, where you predict that the "before" population lies below the "after" population, then you expect W will be associated with the sum of the ranks for the positive D scores. You proceed with Step (4).

(4) Since $1 < 21$, reject H_0 and conclude that the home reading program increases reading achievement.

d. Wilcoxon $W = 1, p < .05$

3 **a.** The steps of the hypothesis-testing procedure are:

(1) H_0: Gender stereotyping of occupations does not change. That is, the distribution of the population of last year's ratings is identical to the distribution of the population of this year's ratings.

H_A: Gender stereotyping of occupations has changed. That is, the distribution of the population of last year's ratings is not identical to the distribution of this year's ratings.

(2) Although you collected 10 pairs of scores, since two pairs of X_1 and X_2 scores are equal (for Nurse and Taxi Driver) you ignore these pairs and hence $N = 8$. Setting $\alpha = .05$, since this is a two-tailed test and $N = 8$, then $W^* = 4$.

(3) You create a set of D scores, assign ranks, and calculate W. The sum of the positive ranks is 16 and the sum of the negative ranks is 20. Thus, $W = 16$.

(4) Since $16 > 4$, retain H_0 and conclude that there is not sufficient evidence to say that gender stereotyping of occupations has changed.

b. Wilcoxon $W = 16$, N.S.

c. The Wilcoxon test might be preferred because there is no reason to believe that the populations of last year's scores and this year's scores are normally distributed, as required by the dependent t test.

4

a. The Mann-Whitney U test, since the samples are independent.

b. The ranges are considerably different for the two groups, suggesting that the variances of the populations may not be identical. Furthermore, the distribution of errors for the amphetamine group is very negatively skewed, suggesting that the population of amphetamine errors is not normally distributed. The independent t test requires equal population variances and that both populations be normally distributed.

c. Since the sample sizes are greater than 20, U must be converted to a z score. The steps of the hypothesis-testing procedure using the Mann-Whitney U for larger samples are:

(1) H_0: Amphetamine does not reduce the number of errors. That is, the distribution of errors for the population of rats given amphetamine lies above or is identical to the distribution of errors for the population of rats given a placebo.

H_A: Amphetamine reduces the number of errors. That is, the distribution of errors for the population given amphetamine lies below the distribution of errors for the population of rats given a placebo.

(2) Setting $\alpha = .01$, since this is a one-tailed test, then $z^* = 2.33$. You then organize the data, assign ranks, and calculate $R_1 = 459.5$ for the amphetamine group.

(3) Using Formula 9.1, you calculate U:

$$U = R_1 - \frac{(N_1)(N_1 + 1)}{2}$$

$$= 459.5 - \frac{(25)(25 + 1)}{2} = 134.5$$

Using Formula 9.2, you transform your U to a z score:

$$z = \frac{U - \dfrac{(N_1)(N_2)}{2}}{\sqrt{\dfrac{(N_1)(N_2)(N_1 + N_2 + 1)}{12}}}$$

$$= \frac{134.5 - \dfrac{(25)(25)}{2}}{\sqrt{\dfrac{(25)(25)(25 + 25 + 1)}{12}}} = -3.454$$

Since this is a one-tailed test, you proceed with step (4b).

(4b) Because you predict that the scores in Group 1 (Amphetamine) lie below those in Group 2 (Placebo), you expect z to be negative. Since $z = -3.454$ and $|-3.454| > 2.33$, then reject H_0 and conclude that amphetamine facilitates memory insofar as it reduces the number of errors.

d. Mann-Whitney $U = 134.5$, $p < .01$

5 **a.** The steps of the hypothesis-testing procedure are:

(1) H_0: Auditory feedback does not facilitate self-control. That is, the distribution of scores for the population receiving no feedback lies below or is identical to the distribution of scores for the population receiving auditory feedback.

H_A: Auditory feedback facilitates self-control. That is, the distribution of scores for the population receiving no feedback lies above the distribution of scores for the population receiving auditory feedback.

(2) Setting $\alpha = .01$, since this is a one-tailed test and $N_1 = 12$ and $N_2 = 11$, then $U^* = 28$ and 104. You then calculate $R_1 = 156.5$ for the No-Feedback group.

(3) Using Formula 9.1, you calculate U:

$$U = R_1 - \frac{(N_1)(N_1 + 1)}{2}$$

$$= 156.5 - \frac{(12)(12 + 1)}{2} = 78.5$$

Since this is a one-tailed test, you proceed with Step (4b).

(4b) Because you predict that the scores in Group 1 (no feedback) lie above those in Group 2 (auditory feedback), U must be greater than or equal to the larger U^* to reject H_0. Since the calculated U of 78.5 falls below the larger critical value of 104, retain H_0 and conclude that there is not sufficient evidence to say that auditory feedback facilitates self-control of the frontalis muscle.

b. Mann-Whitney $U = 78.5$, $p > .01$.

6 **a.** The steps of the hypothesis-testing procedure are:

(1) H_0: The campaign does not reduce cigarette smoking at work. That is, the distribution of the population of "before" scores lies below or is identical to the distribution of the population of "after" scores.

H_A: The campaign reduces cigarette smoking at work. That is, the distribution of the population of "before" scores lies above the distribution of the population of "after" scores.

(2) Although you collected eight pairs of scores, since one pair of X_1 and X_2 scores is equal (S_8), you ignore this pair and hence $N = 7$. Setting $\alpha = .05$, since this is a one-tailed test and $N = 7$, then $W^* = 4$.

(3) You create a set of D scores, assign ranks, and calculate W. The sum of the positive ranks is 26.5 and the sum of the negative ranks is 1.5. Thus, $W = 1.5$.

(4) Since $1.5 < 4$, reject H_0 and conclude that the campaign reduces cigarette smoking at work.

b. Wilcoxon $W = 1.5$, $p < .05$.

7 **a.** The steps of the hypothesis-testing procedure are:

(1) H_0: A city's general unemployment rate does not affect job satisfaction. That is, the distribution of the population of scores from City 1 is identical to the distribution of the population of scores from City 2.

H_A: A city's general employment rate affects job satisfaction. That is, the distribution of the population of scores from City 1 is not identical to the distribution of the population of scores from City 2.

(2) Although you collected ten pairs of scores, since one pair of X_1 and X_2 scores is equal (pair number 9) you ignore this pair and hence $N = 9$. Setting $\alpha = .05$, since this is a two-tailed test and $N = 9$, then $W^* = 6$.

(3) You create a set of D scores, assign ranks, and calculate W. The sum of the positive ranks is 5 and the sum of the negative ranks is 40. Thus, $W = 5$.

(4) Since $5 < 6$, reject H_0 and conclude that a city's general unemployment rate affects job satisfaction.

b. Wilcoxon $W = 5$, $p < .05$

ANSWERS TO EXERCISES, CHAPTER 10

1 The four steps of the hypothesis-testing procedure are:

(1) H_0: The coins are fair. That is, your observed frequencies occurred by chance from a population of coins where heads and tails are, equally likely to occur.

H_A: The coins are biased. That is, your observed frequencies came from a population of coins in which heads and tails are not equally likely to occur.

(2) Setting $\alpha = .01$, since you have two categories of tosses (heads and tails), $df = K_c - 1 = 2 - 1 = 1$, then $\chi^{2*} = 6.635$.

(3) You create tables of the observed and expected frequencies.

Observed Frequencies

heads	*tails*
66	34

Under H_0, in a sample of 100 tosses, you expect to find 50 heads and 50 tails.

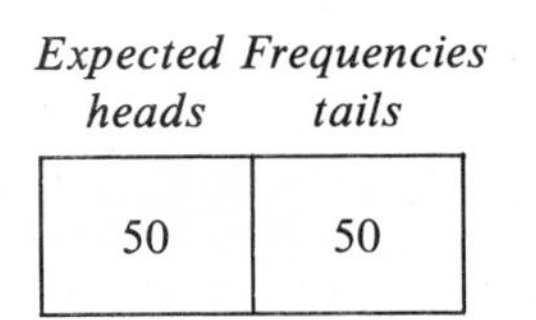

Expected Frequencies

heads	*tails*
50	50

Using Formula 10.1, χ^2 for your observed frequencies is:

$$\chi^2 = \Sigma \frac{(O - E)^2}{E} = \frac{(66 - 50)^2}{50} + \frac{(34 - 50)^2}{50} = 10.24$$

(4) Since $10.24 > 6.635$, reject H_0 and conclude that the new minting process produces biased coins.

2 **a.** Setting $\alpha = .05$, since you have three categories of precincts ($K_r = 3$) and five categories of crime ($K_c = 5$), $df = (K_r - 1)(K_c - 1) = (3 - 1)(5 - 1) = 8$, then $\chi^{2*} = 15.507$. Since the calculated χ^2 of $28.656 > 15.507$, your result is significant. Yes, the frequency of the type of crime committed depends on the given precinct.

b. Setting $\alpha = .05$, since you have six precincts ($K_r = 6$) and nine types of crime ($K_c = 9$), $df = (6 - 1)(9 - 1) = 40$, then $\chi^{2*} = 55.759$. Since $49.148 < 55.759$, you should retain H_0 and conclude that there is not sufficient evidence to say that the type of crime depends on the particular precinct.

3 **a.** The steps of the hypothesis-testing procedure are:

(1) H_0: Suicide-related phone calls are equally likely to occur in fall, winter, spring, and summer. That is, your observed frequencies occurred by chance from a population with equal numbers of suicide-related calls for each season.

H_A: Suicide-related calls are not equally likely to occur in fall, winter, spring, and summer. That is, your observed frequencies arose from a population with unequal numbers of suicide-related calls for each season.

(2) Setting $\alpha = .05$, since you have four categories of season, $df = K_c - 1 = 4 - 1 = 3$, then $\chi^{2*} = 7.815$.

(3) You create tables of the observed and expected frequencies.

Observed Frequencies

fall	*winter*	*spring*	*summer*
27	58	71	44

Under H_0, in a sample of 200 suicide-related calls you expect to find 50 calls in each season:

Expected Frequencies

fall	*winter*	*spring*	*summer*
50	50	50	50

Using Formula 10.1, χ^2 for your observed frequencies is:

$$\chi^2 = \Sigma \frac{(O - E)^2}{E}$$

$$= \frac{(27 - 50)^2}{50} + \frac{(58 - 50)^2}{50} + \frac{(71 - 50)^2}{50} + \frac{(44 - 50)^2}{50}$$

$$= 21.4$$

(4) Since $21.4 > 7.815$, reject H_0 and conclude that suicide-related calls are not equally likely to occur in each season.

b. $\chi^2(3) = 21.4, p < .05$

c. The observed frequencies in each season may not be independent if an individual with suicidal tendencies called more than once over the year.

4 **a.** Using the χ^2 goodness-of-fit test, the steps of the hypothesis-testing procedure are:

(1) H_0: The distribution of occupations of female state senators does not differ from that of state senators in general. That is, your observed frequencies occurred by chance from the general population of state senators whose pattern is represented by the expected frequencies.

H_A: The distribution of occupations of female state senators differs from that of state senators in general. That is, your observed frequencies came from a population whose pattern is not represented by the expected frequencies.

(2) Setting $\alpha = .01$, since you have five categories of occupation, $df = 5 - 1 = 4$, then $\chi^{2*} = 13.277$.

(3) You create tables of the observed and expected frequencies:

Observed Frequencies

teachers, social workers	*lawyers*	*executives*	*real estate sales*	*major government employee*
14	45	30	8	3

Expected Frequencies

18	61	10	5	6

Using Formula 10.1, you calculate χ^2:

$$\chi^2 = \Sigma \frac{(O - E)^2}{E}$$

$$= \frac{(14 - 18)^2}{18} + \frac{(45 - 61)^2}{61} + \frac{(30 - 10)^2}{10} + \frac{(8 - 5)^2}{5} + \frac{(3 - 6)^2}{6}$$

$$= 48.386$$

(4) Since $48.386 > 13.277$, reject H_0 and conclude that the distribution of occupations of female state senators differs from that of state senators in general.

b. $\chi^2(4) = 48.368$, $p < .01$

c. The expected frequency for each occupation must be at least 5. An $E = 4$ for real estate sales violates this assumption.

5 **a.** Using the χ^2 goodness-of-fit test, the steps of the hypothesis-testing procedure are:

(1) H_0: Students are equally knowledgeable about the two water pollutants.
H_A: Students are not equally knowledgeable about the two water pollutants.

(2) Setting $\alpha = .05$, since you have two categories of water pollutants, $df = 2 - 1 = 1$, then $\chi^{2*} = 3.841$.

(3) You create tables of the observed and expected frequencies:

Observed Frequencies

know more about mercury	*know more about nitrates/phosphates*
17	8

Under H_0, in a sample of 25 students you expect to find 12.5 knowing more about each pollutant:

Expected Frequencies

know more about mercury	*know more about nitrates/phosphates*
12.5	12.5

You calculate χ^2:

$$\chi^2 = \Sigma \frac{(O - E)^2}{E} = \frac{(17 - 12.5)^2}{12.5} + \frac{(8 - 12.5)^2}{12.5} = 3.24$$

(4) Since $3.24 < 3.841$, retain H_0 and conclude that there is not sufficient evidence to say that students know more about one type of water pollutant than the other.

b. $\chi^2(1) = 3.24$, N.S.

c. The data must be frequency data. If you used such a rating scale, you would have ordinal measurement and could not employ the χ^2 test.

6

a. The χ^2 test of independence, because there are two variables: (1) the sex of the self-disclosing traveler; and (2) the sex of the approacher.

b. The steps of the hypothesis-testing procedure are:

(1) H_0: The sex of the self-disclosing traveler and the sex of the approacher are independent.

H_A: The sex of the self-disclosing traveler and the sex of the approacher are dependent.

(2) Setting $\alpha = .05$, since you have two categories of approachers ($K_r = 2$) and two categories of self-disclosing travelers ($K_c = 2$), $df = (K_r - 1)(K_c - 1) = (2 - 1)(2 - 1) = 1$, then $\chi^2 = 3.841$.

(3) You create tables of the observed and expected frequencies.

Observed frequencies

	male travelers	*female travelers*	*Total*
male approachers	5	9	14
female approachers	41	27	68
Total	46	36	82

Using Formula 10.2, you calculate the expected frequencies for each cell:

Expected Frequencies

	male travelers	*female travelers*
male approachers	$E = \dfrac{(14)(46)}{82} = 7.854$	$E = \dfrac{(14)(36)}{82} = 6.146$
female approachers	$E = \dfrac{(68)(46)}{82} = 38.146$	$E = \dfrac{(68)(36)}{82} = 29.854$

Using Formula 10.1, the calculated χ^2 is:

$$\chi^2 = \Sigma \frac{(O - E)^2}{E}$$

$$= \frac{(5 - 7.854)^2}{7.854} + \frac{(9 - 6.146)^2}{6.146} + \frac{(41 - 38.146)^2}{38.146} + \frac{(27 - 29.854)^2}{29.854}$$

$$= 2.848$$

(4) Since $2.848 < 3.841$, retain H_0 and conclude that there is not sufficient evidence to say that sex of self-disclosing traveler depends on sex of approacher.

c. $\chi^2(1) = 2.848$, $p > .05$.

7 **a.** The four steps of the hypothesis-testing procedure are:

(1) H_0: There is no relationship between drug use and region. This is to say that the pattern of drug use and region are independent.
H_A: Drug use and region are related. This is to say that the pattern of drug use depends on the region in which the high school seniors live.

(2) Setting $\alpha = .01$, since you have three categories of region ($K_r = 3$) and six categories of drugs ($K_c = 6$), $df = (K_r - 1)(K_c - 1) = (3 - 1)(6 - 1) = 10$, then $\chi^{2*} = 23.209$

(3) You create tables of the observed and expected frequencies.

Observed Frequencies

	alcohol	*amphetamines*	*barbiturates*	*cocaine*	*hallucinogens*	*marijuana*	*Total*
Northeast	25	13	12	6	6	13	75
South	29	5	18	2	4	10	68
West	21	12	6	10	12	16	77
Total	75	30	36	18	22	39	220

Using Formula 10.2, you calculate the expected frequencies for each cell.

Expected frequencies

	alcohol	*amphetamines*	*barbiturates*	*cocaine*	*hallucinogens*	*marijuana*
Northeast	25.568	10.227	12.273	6.136	7.5	13.295
South	23.182	9.273	11.127	5.564	6.8	12.054
West	26.25	10.5	12.6	6.3	7.7	13.65

Using Formula 10.1, the calculated χ^2 is:

$$\chi^2 = \Sigma \frac{(O - E)^2}{E}$$

$$= \frac{(25 - 25.568)^2}{25.568} + \frac{(13 - 10.227)^2}{10.227} + \frac{(12 - 12.273)^2}{12.273} + \frac{(6 - 6.136)^2}{6.136}$$

$$+ \frac{(6 - 7.5)^2}{7.5} + \frac{(13 - 13.295)^2}{13.295} + \frac{(29 - 23.182)^2}{23.182} + \frac{(5 - 9.273)^2}{9.273}$$

$$+ \frac{(18 - 11.127)^2}{11.127} + \frac{(2 - 5.564)^2}{5.564} + \frac{(4 - 6.8)^2}{6.8} + \frac{(10 - 12.054)^2}{12.054}$$

$$+ \frac{(21 - 26.25)^2}{26.25} + \frac{(12 - 10.5)^2}{10.5} + \frac{(6 - 12.6)^2}{12.6} + \frac{(10 - 6.3)^2}{6.3}$$

$$+ \frac{(12 - 7.7)^2}{7.7} + \frac{(16 - 13.65)^2}{13.65}$$

$= .013 + .752 + .006 + .003$

$+ .3 + .012 + 1.46 + 1.969$

$+ 4.245 + 2.283 + 1.153 + .35$

$+ 1.05 + .214 + 3.457 + 2.173$

$+ 2.401 + .404$

$= 24.245$

(4) Since $24.245 < 23.209$, retain H_0 and conclude that there is not sufficient evidence to say that the pattern of drug use depends on the region in which the high school seniors live.

b. $\chi^2(10) = 22.245$ N.S.

ANSWERS TO EXERCISES, CHAPTER 11

1

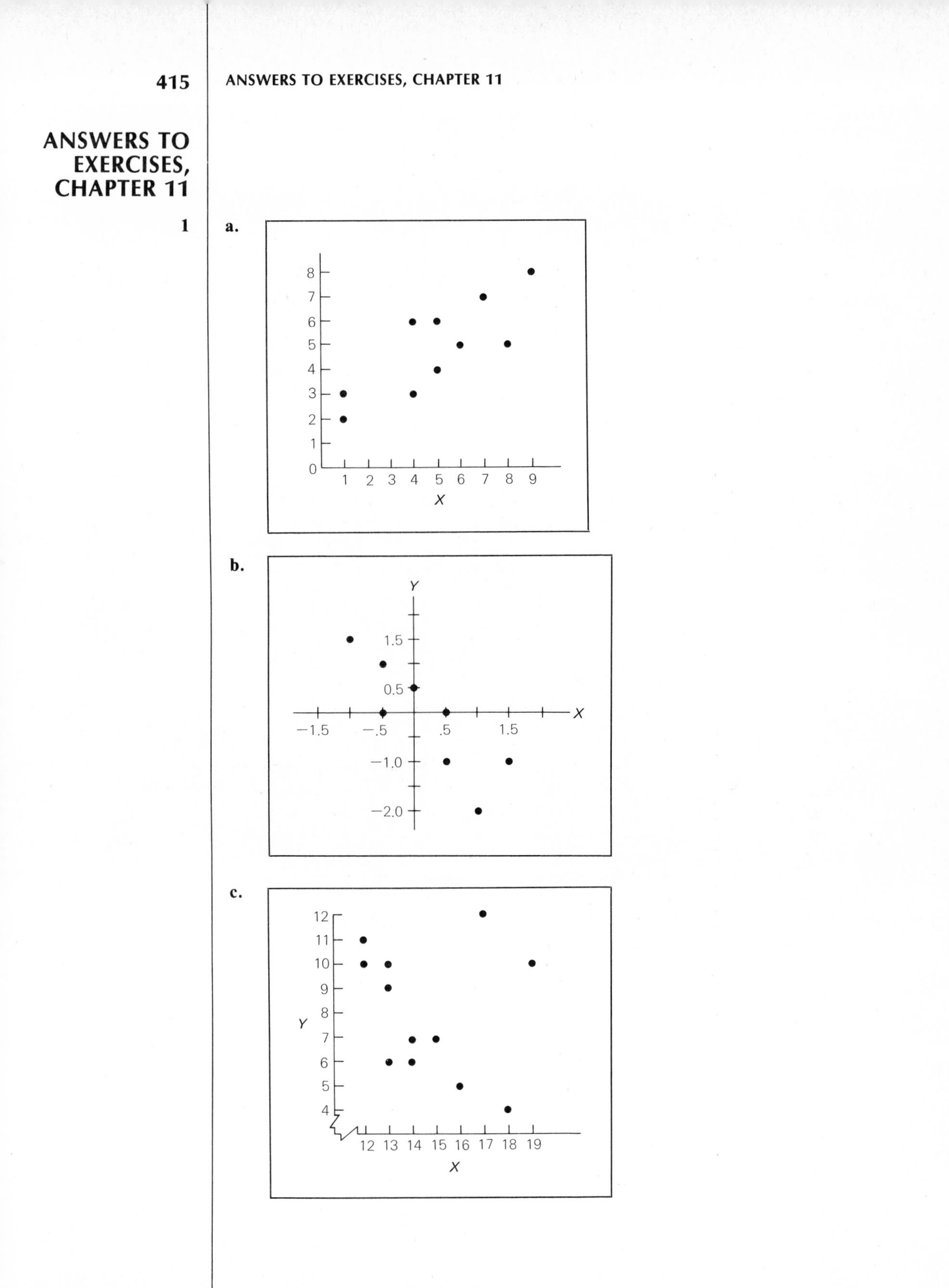

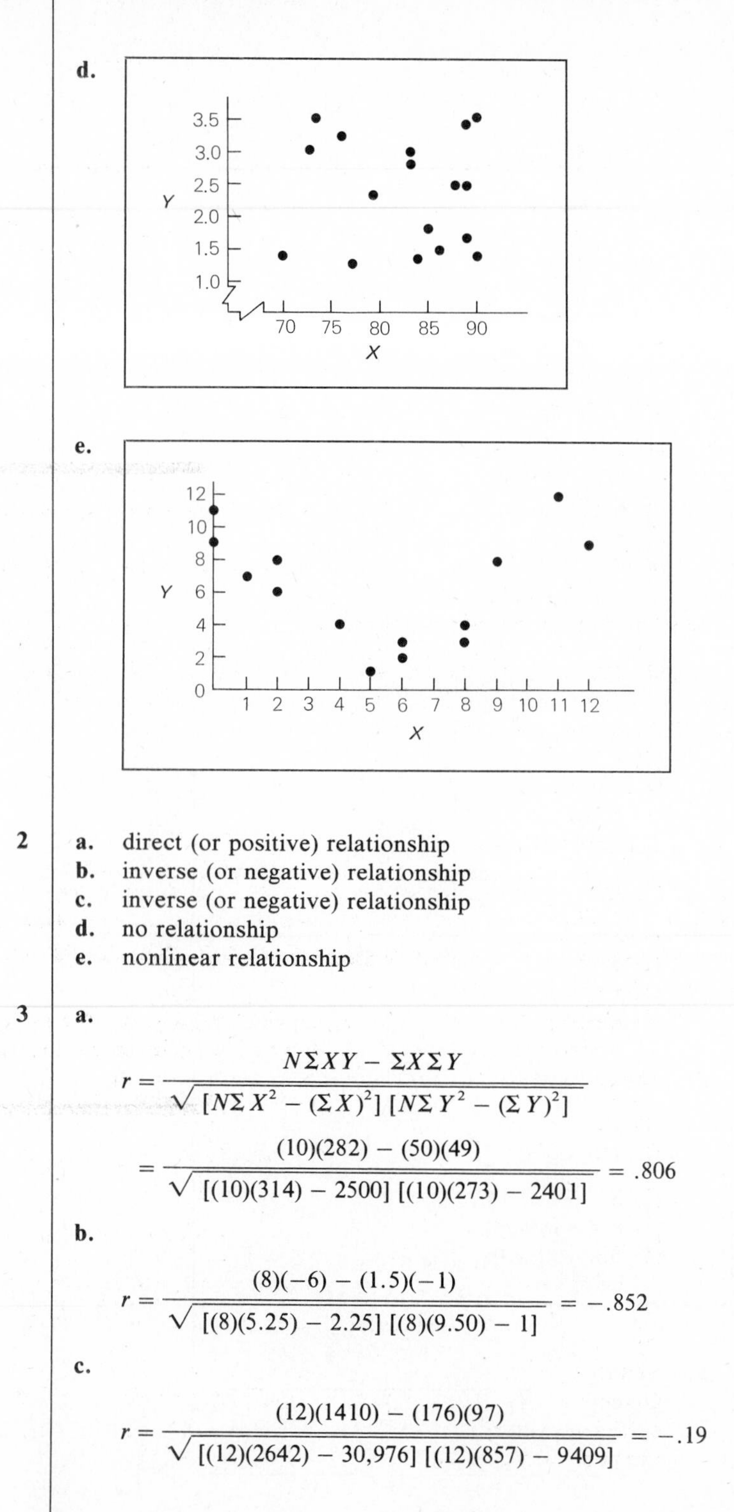

2 **a.** direct (or positive) relationship
b. inverse (or negative) relationship
c. inverse (or negative) relationship
d. no relationship
e. nonlinear relationship

3 **a.**

$$r = \frac{N\Sigma XY - \Sigma X \Sigma Y}{\sqrt{[N\Sigma X^2 - (\Sigma X)^2]\,[N\Sigma Y^2 - (\Sigma Y)^2]}}$$

$$= \frac{(10)(282) - (50)(49)}{\sqrt{[(10)(314) - 2500]\,[(10)(273) - 2401]}} = .806$$

b.

$$r = \frac{(8)(-6) - (1.5)(-1)}{\sqrt{[(8)(5.25) - 2.25]\,[(8)(9.50) - 1]}} = -.852$$

c.

$$r = \frac{(12)(1410) - (176)(97)}{\sqrt{[(12)(2642) - 30{,}976]\,[(12)(857) - 9409]}} = -.19$$

d.

$$r = \frac{(17)(3291.2) - (1404)(39.9)}{\sqrt{[(17)(116{,}638) - 1{,}971{,}216]\,[(17)(104.65) - (1592.01)]}}$$

$$= -.047$$

e.

$$r = \frac{(14)(454) - (74)(87)}{\sqrt{[(14)(596) - 5476]\,[(14)(695) - 7569]}} = -.033$$

4 **a.** The steps of the hypothesis-testing procedure for data set (a) are:

(1) H_0: $\rho = 0$ (The population correlation coefficient is 0)
H_A: $\rho \neq 0$ (The population correlation coefficient is different from 0)
(2) Setting $\alpha = .05$, since this is a two-tailed test and $N = 10$, then $r^* = .632$.
(3) $r = .806$
(4) Since $|.806| > .632$, reject H_0 and conclude that the population correlation coefficient is different from 0.

b. The steps of the hypothesis-testing procedure for data set (b) are:

(1) H_0: $\rho = 0$ (The population correlation coefficient is 0)
H_A: $\rho \neq 0$ (The population correlation coefficient is different from 0)
(2) Setting $\alpha = .05$, since this is a two-tailed test and $N = 8$, then $r^* = .707$.
(3) $r = -.852$
(4) Since $|-.852| > .707$, reject H_0 and conclude that the population correlation coefficient is different from 0.

c. The steps of the hypothesis-testing procedure for data set (c) are:

(1) H_0: $\rho = 0$ (The population correlation coefficient is 0)
H_A: $\rho \neq 0$ (The population correlation coefficient is different from 0)

(2) Setting $\alpha = .05$, since this is a two-tailed test and $N = 12$, then $r^* = .576$.
(3) $r = -.19$
(4) Since $|-.19| < .576$, retain H_0 and conclude that there is not sufficient evidence to say the population correlation coefficient is different from 0.

d. The steps of the hypothesis-testing procedure for data set (d) are:

(1) H_0: $\rho = 0$ (The population correlation coefficient is 0)
H_A: $\rho \neq 0$ (The population correlation coefficient is different from 0)

(2) Setting $\alpha = .05$, since this is a two-tailed test and $N = 17$, then $r^* = .482$.
(3) $r = -.047$
(4) Since $|-.047| > .482$, retain H_0 and conclude that there is not sufficient evidence to say the population correlation coefficient is different from 0.

5 **a.** Setting $\alpha = .01$, since this is a two-tailed test and $N = 30$, then $r^* = .463$. Given $r = .46$, since $|.46| < .463$, you can conclude that there is not sufficient evidence to say a relationship exists between the two variables in the population.

b. Setting $\alpha = .01$ with a two-tailed test, a larger value of N would have meant your $r = .46$ was significant and that a relationship exists between the two variables. In general, it is easier to reject H_0 and achieve significance with larger sample sizes.

6 The friend has made an error in computing the correlation coefficient. Values of r can only range between -1 and $+1$.

7 **a.**

$$r = 1 - \left(\frac{1}{2}\right)\left(\frac{\Sigma(z_x - z_y)^2}{N}\right)$$

$$= 1 - (.5)\left(\frac{31.2}{40}\right) = .61$$

b.

$$r = \frac{\Sigma z_x z_y}{N} = \frac{24.4}{40} = .61$$

c. The steps of the hypothesis-testing procedure are:

(1) H_0: $\rho \leq 0$ (College entrance examination scores and GPA are not positively related)
H_A: $\rho > 0$ (College entrance examination scores and GPA are positively related)

(2) Setting $\alpha = .05$, since this is a one-tailed test and $N = 40$, then $r^* = .275$. (Note, if your value of N is not listed in Table r^*, use the critical values of r for the closest value of N that is smaller than your N.)
(3) This is a one-tailed test where you predict a positive population correlation coefficient in H_A. Since $r = .61$ is positive, as expected, you proceed to Step (4).
(4) Since $|.61| > .275$, reject H_0 and conclude that college entrance examination scores and GPA are positively related.

d. $r = .61, p < .05$

8 **a.**

$$r = 1 - \left(\frac{1}{2}\right)\left(\frac{\Sigma(z_x - z_y)^2}{N}\right).$$

$$= 1 - (.5)\left(\frac{3.664}{14}\right) = .869$$

b.

$$r = \frac{\Sigma z_x z_y}{N} = \frac{12.166}{14} = .869$$

c. The steps of the hypothesis-testing procedure are:

(1) H_0: $\rho \leq 0$ (The two halves of the test are not directly related)
H_A: $\rho > 0$ (The two halves of the test are directly related)

(2) Setting $\alpha = .01$, since this is a one-tailed test and $N = 14$, then $r^* = .612$.
(3) This is a one-tailed test where you predict a positive population correlation coefficient in H_A. Since $r = .869$ is positive, as expected, you proceed to Step (4).
(4) Since $|.869| > .612$, you reject H_0 and conclude that the two halves of the test are directly related to each other.

d. $r = .869, p < .01$

9 **a.** The relationship can be assumed to be approximately linear since there is no pronounced nonlinearity.

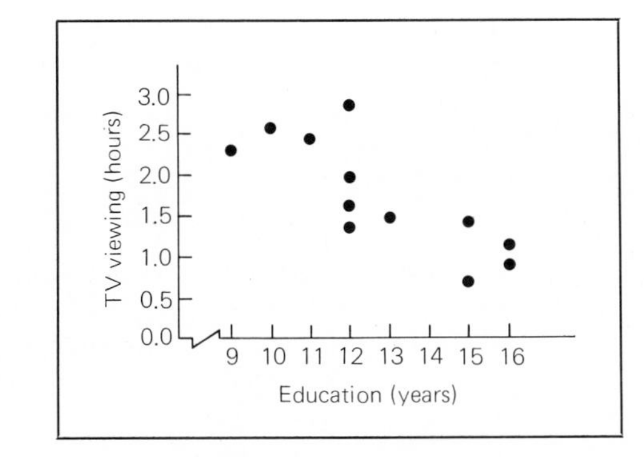

b.

$$r = \frac{N\Sigma XY - \Sigma X \Sigma Y}{\sqrt{[N\Sigma X^2 - (\Sigma X)^2][N\Sigma Y^2 - (\Sigma Y)^2]}}$$

$$= \frac{(12)(259.1) - (153)(21.4)}{\sqrt{[(12)(2009) - 23{,}409][(12)(43.28) - 457.96]}} = -.796$$

c. The steps of the hypothesis-testing procedure are:

(1) H_0: $\rho \geq 0$ (TV viewing and educational level are not inversely related)
H_A: $\rho < 0$ (TV viewing and educational level are inversely related)

(2) Setting $\alpha = .05$, since this is a one-tailed test and $N = 12$, then $r^* = .497$.
(3) This is a one-tailed test where you predict a negative population correlation coefficient in H_A. Since $r = -.796$ is negative as expected, you proceed with Step (4).
(4) Since $|-.796] > .497$, reject H_0 and conclude that the time spent viewing TV is inversely related to educational level.

d. $r = -.796, p < .05$

11 a. No. The correlation coefficient only describes observed relationships between two variables, it does not imply a cause-effect relationship. Therefore you cannot infer that increases in cigarette smoking cause increases in respiratory disease. Both cigarette smoking and respiratory disease may be caused by one or more as yet unidentified variables. For example, people who are vulnerable to respiratory disease may for some unknown reason be more prone to smoke cigarettes.

b. No. An $r = .8$ describes a stronger positive relationship than an $r = .4$, but you cannot say it is twice as strong. The strength of a relationship is not directly proportional to the magnitude of r.

ANSWERS TO EXERCISES, CHAPTER 12

1 a. Two chosen values of X lying at opposite ends of the abscissa are 0 and 10. Using the regression equation the $\hat{Y}$ predictions for these values of X are:

For X = 0	*For X = 10*
$\hat{Y} = r\frac{\sigma_y}{\sigma_x}(X - \bar{X}) + \bar{Y}$	$\hat{Y} = r\frac{\sigma_y}{\sigma_x}(X - \bar{X}) + \bar{Y}$
$= .934\left(\frac{2.071}{2.182}\right)(0 - 3.8) + 4.1$	$= .934\left(\frac{2.071}{2.182}\right)(10 - 3.8) + 4.1$
$= .731$	$= 9.596$

The points corresponding to $X = 0$, $Y = .731$, and $X = 10$, $Y = 9.596$ are then connected with a straight line in the scatter plot.

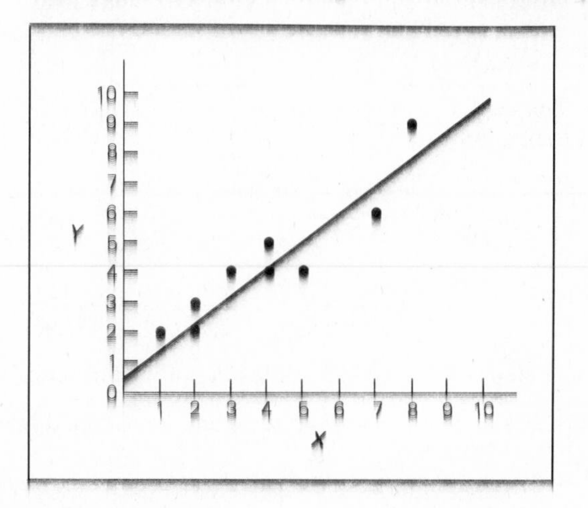

$$\hat{Y} = -.009\left(\frac{3.068}{2.182}\right)(0 - 3.8) + 4.7 \qquad \hat{Y} = -.009\left(\frac{3.068}{2.182}\right)(10 - 3.8) + 4.7$$

$$= 4.748 \qquad\qquad = 4.622$$

The points corresponding to $X = 0$, $Y = 4.748$, and $X = 10$, $Y = 4.622$ are then connected with a straight line in the scatter plot.

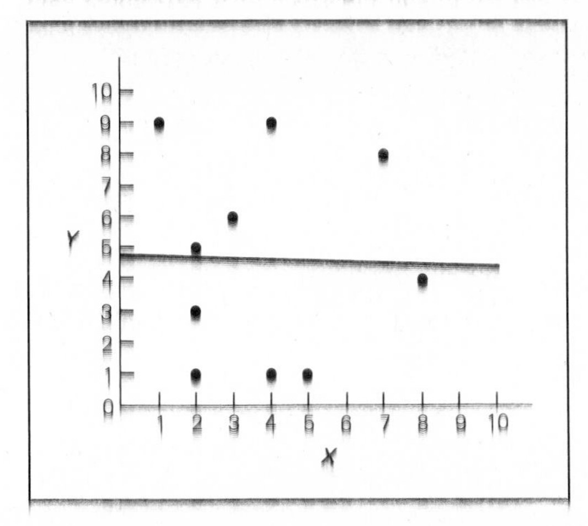

c. Two chosen values of X lying at opposite ends of the abscissa are 0 and 10. The $\hat{Y}$ predictions for these scores are:

For $X = 0$ *For* $X = 10$

$$\hat{Y} = -.925\left(\frac{2.1}{2.182}\right)(0 - 3.8) + 4.3 \qquad \hat{Y} = -.925\left(\frac{2.1}{2.182}\right)(10 - 3.8) + 4.3$$

$$= 7.683 \qquad\qquad = -1.219$$

The points corresponding to $X = 0$, $Y = 7.683$, and $X = 10$, $Y = -1.219$ are then connected with a straight line in the scatter plot.

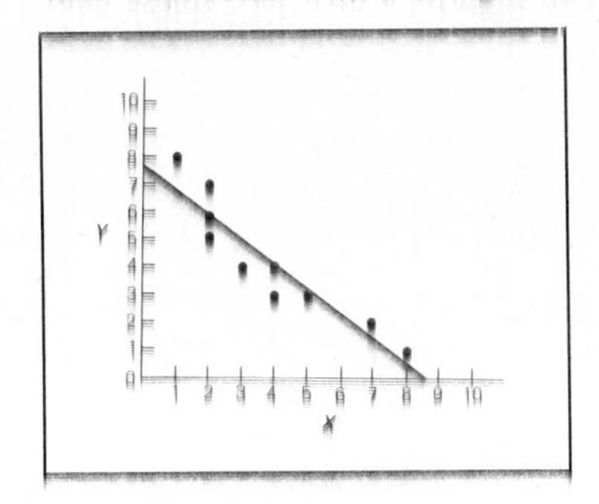

d. Two chosen values of X lying at opposite ends of the abscissa are 0 and 10. The $\hat{Y}$ predictions for these scores are:

For $X = 0$	*For* $X = 10$
$\hat{Y} = .619\left(\frac{1.628}{2.182}\right)(0 - 3.8) + 4.5$	$\hat{Y} = .619\left(\frac{1.628}{2.182}\right)(10 - 3.8) + 4.5$
$= 2.745$	$= 7.363$

The points corresponding to $X = 0$, $Y = 2.745$, and $X = 10$, $Y = 7.363$ are then connected with a straight line in the scatter plot.

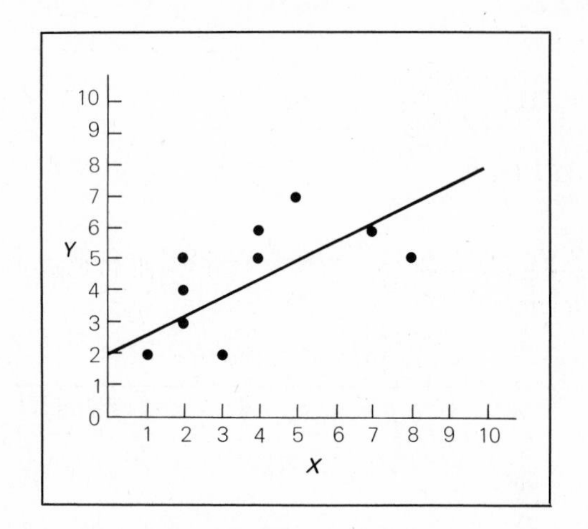

2 The standard error of the estimate is the standard deviation of the errors in prediction.

3 **a.** For data set (a): $\sigma_{\hat{y}} = \sigma_y \sqrt{1 - r^2}$

$$= 2.071\sqrt{1 - (.934)^2} = .740$$

For data set (b): $\sigma_{\hat{y}} = 3.068\sqrt{1 - (-.009)^2} = 3.068$

For data set (c): $\sigma_{\hat{y}} = 2.1\sqrt{1 - (-.925)^2} = .798$

For data set (d): $\sigma_{\hat{y}} = 1.628\sqrt{1 - (.619)^2} = 1.279$

b. Large errors in prediction indicate greater variability of the actual values of Y from the predicted values of $\hat{Y}$, and hence a larger standard error of the estimate is obtained.

c. For data set (a): $\hat{Y} = .934\left(\frac{2.071}{2.182}\right)(6 - 3.8) + 4.1 = 6.050$

For data set (b): $\hat{Y} = -.009\left(\frac{3.068}{2.182}\right)(6 - 3.8) + 4.7 = 4.672$

For data set (c): $\hat{Y} = -.925\left(\frac{2.1}{2.182}\right)(6 - 3.8) + 4.3 = 2.341$

For data set (d): $\hat{Y} = .619\left(\frac{1.628}{2.182}\right)(6 - 3.8) + 4.5 = 5.516$

4 **a.** Two chosen values of X lying on opposite sides of the abscissa are 0 and 5. Using the regression equation, the $\hat{Y}$ predictions for these X scores are:

For X = 0	*For X* = 5
$\hat{Y} = .986\left(\frac{9.466}{1.414}\right)(0 - 3) + 51$	$\hat{Y} = .986\left(\frac{9.466}{1.414}\right)(5 - 3) + 51$
$= 31.198$	$= 64.202$

The points corresponding to $X = 0$, $Y = 31.198$ and $X = 5$, $Y = 64.202$ are then connected with a straight line in the scatter plot.

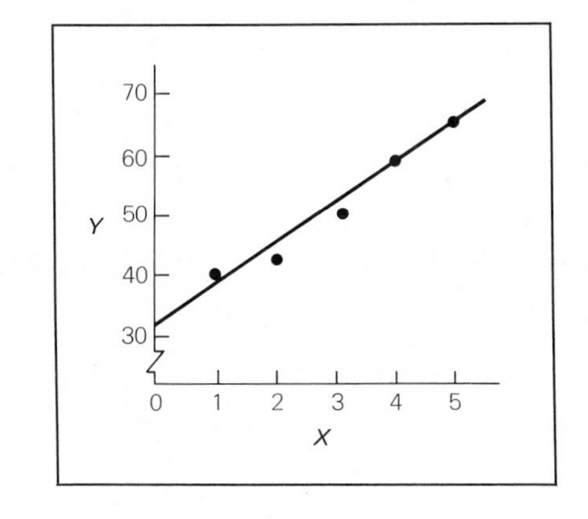

b. Two chosen values of X lying on opposite ends of the abscissa are 40 and 100. The $\hat{Y}$ predictions for these X values are:

For X = 40	*For X* = 100
$\hat{Y} = -.397\left(\frac{18.188}{18.112}\right)(40 - 62.6) + 65.7$	$\hat{Y} = -.397\left(\frac{18.188}{18.112}\right)(100 - 62.6) + 65.7$
$= 74.710$	$= 50.790$

The points corresponding to $X = 40$, $Y = 74.710$, and $X = 100$, $Y = 50.790$ are then connected with a straight line in the scatter plot.

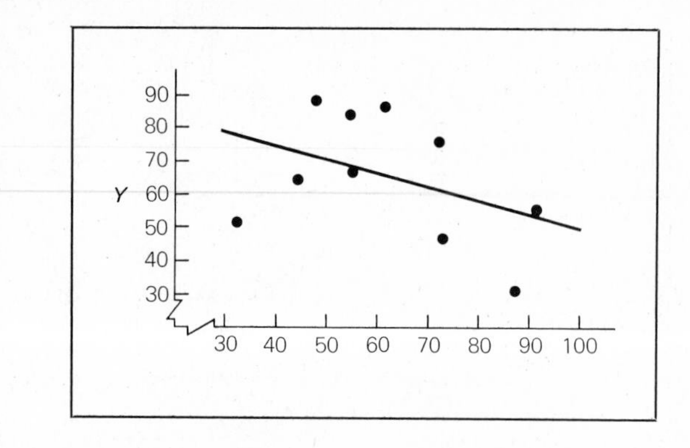

c. Two chosen values of X lying at opposite ends of the abscissa are 9.9 and 11. The $\hat{Y}$ predictions for these values of X are:

For $X = 9.9$	*For* $X = 11$
$\hat{Y} = .549\left(\frac{.744}{.465}\right)(9.9 - 10.429) + 9.143$	$\hat{Y} = .549\left(\frac{.744}{.465}\right)(11 - 10.429) + 9.143$
$= 8.678$	$= 9.645$

The points corresponding to $X = 9.9$, $Y = 8.678$, and $X = 11$, $Y = 9.645$ are then connected with a straight line in the scatter plot.

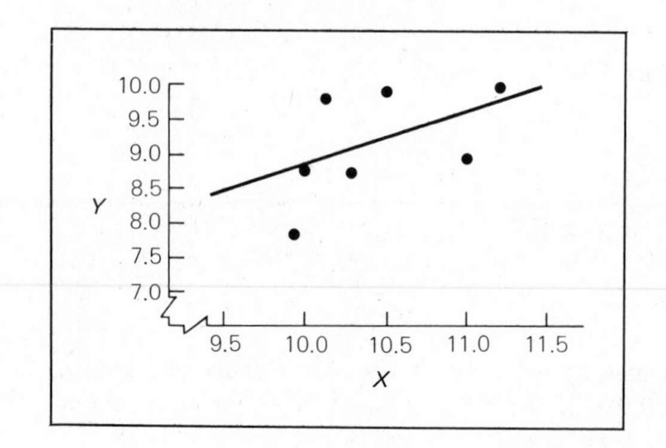

d. Two chosen values of X lying at the opposite ends of the abscissa are 5 and 10. The $\hat{Y}$ predictions for these X values are:

For $X = 5$	*For* $X = 10$
$\hat{Y} = -.919\left(\frac{2.514}{1.492}\right)(5 - 8.077) + 12.855$	$\hat{Y} = -.919\left(\frac{2.514}{1.492}\right)(10 - 8.077) + 12.855$
$= 17.620$	$= 9.877$

The points corresponding to $X = 5$, $Y = 17.620$, and $X = 10$, $Y = 9.877$ are then connected with a straight line in the scatter plot.

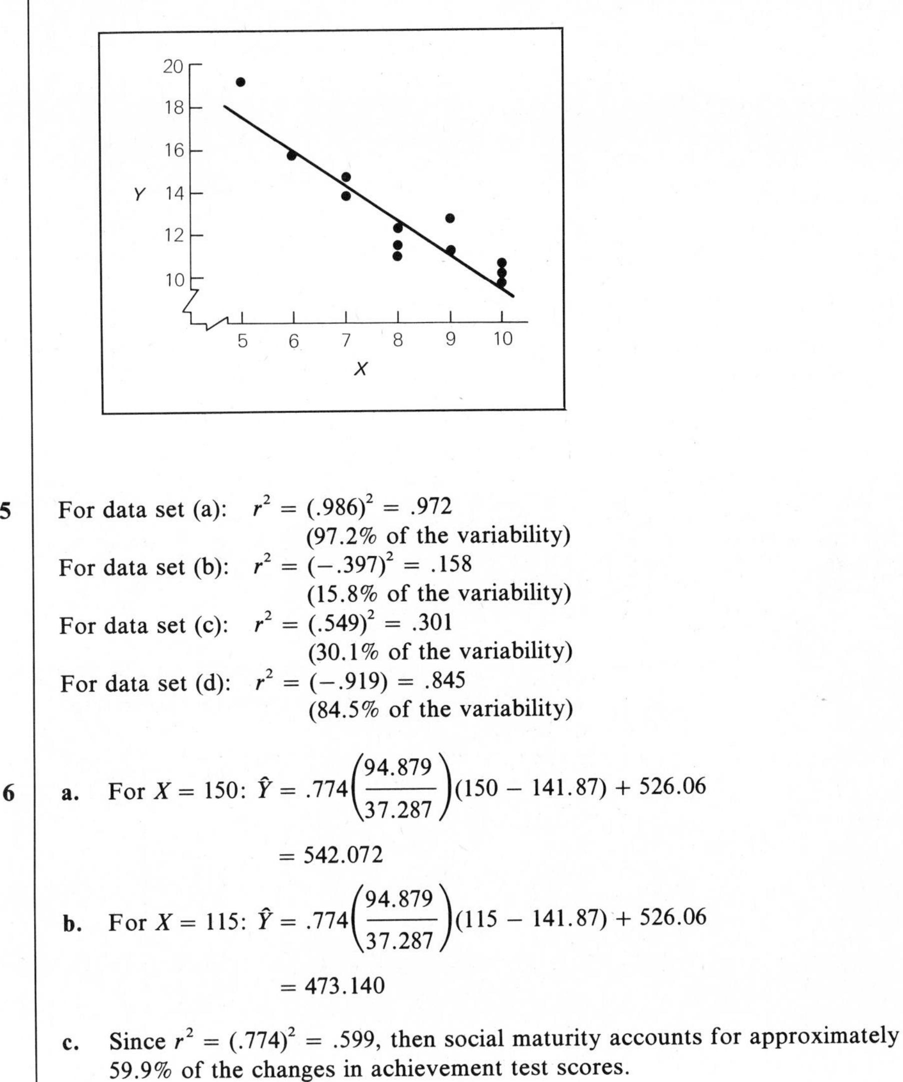

5 For data set (a): $r^2 = (.986)^2 = .972$
(97.2% of the variability)

For data set (b): $r^2 = (-.397)^2 = .158$
(15.8% of the variability)

For data set (c): $r^2 = (.549)^2 = .301$
(30.1% of the variability)

For data set (d): $r^2 = (-.919) = .845$
(84.5% of the variability)

6 **a.** For $X = 150$: $\hat{Y} = .774\left(\dfrac{94.879}{37.287}\right)(150 - 141.87) + 526.06$

$= 542.072$

b. For $X = 115$: $\hat{Y} = .774\left(\dfrac{94.879}{37.287}\right)(115 - 141.87) + 526.06$

$= 473.140$

c. Since $r^2 = (.774)^2 = .599$, then social maturity accounts for approximately 59.9% of the changes in achievement test scores.

d. $z_{\hat{y}} = rz_x = (.774)(-1) = -.774$

7 Two values of X lying at opposite ends of the abscissa are 100 and 400. The $\hat{Y}$ predictions for these X values are:

For $X = 100$

$$\hat{Y} = .933\left(\frac{2.63}{112.965}\right)(100 - 264.7) + 5.87$$

$= 2.291$

For $X = 400$

$$\hat{Y} = .933\left(\frac{2.63}{112.965}\right)(400 - 264.7) + 5.87$$

$= 8.810$

The points corresponding to $X = 100$, $Y = 2.291$ and $X = 400$, $Y = 8.810$ are then connected with a straight line in the graph:

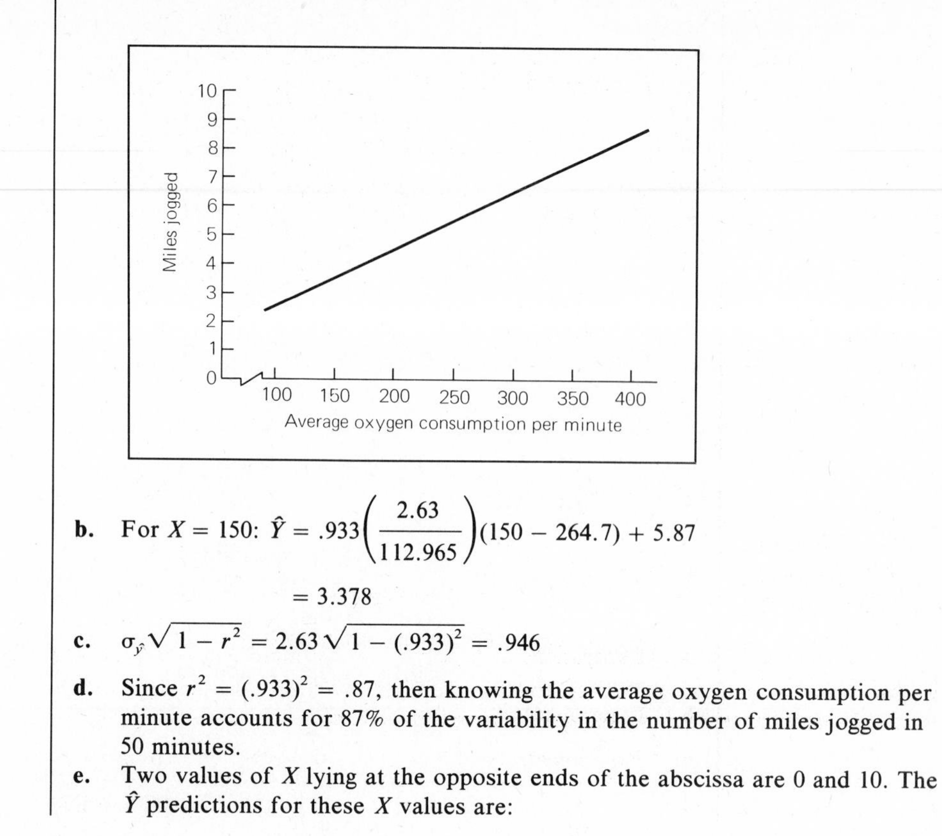

b. For $X = 150$: $\hat{Y} = .933\left(\dfrac{2.63}{112.965}\right)(150 - 264.7) + 5.87$

$$= 3.378$$

c. $\sigma_{\hat{y}}\sqrt{1 - r^2} = 2.63\sqrt{1 - (.933)^2} = .946$

d. Since $r^2 = (.933)^2 = .87$, then knowing the average oxygen consumption per minute accounts for 87% of the variability in the number of miles jogged in 50 minutes.

e. Two values of X lying at the opposite ends of the abscissa are 0 and 10. The $\hat{Y}$ predictions for these X values are:

For X = 0	*For X* = 10
$\hat{Y} = .933\left(\dfrac{112.965}{2.63}\right)(0 - 5.87) + 264.7$	$\hat{Y} = .933\left(\dfrac{112.965}{2.63}\right)(10 - 5.87) + 264.7$
$= 29.462$	$= 430.208$

The points corresponding to $X = 0$, $Y = 29.462$, and $X = 10$, $Y = 430.208$ are then connected with a straight line in the graph.

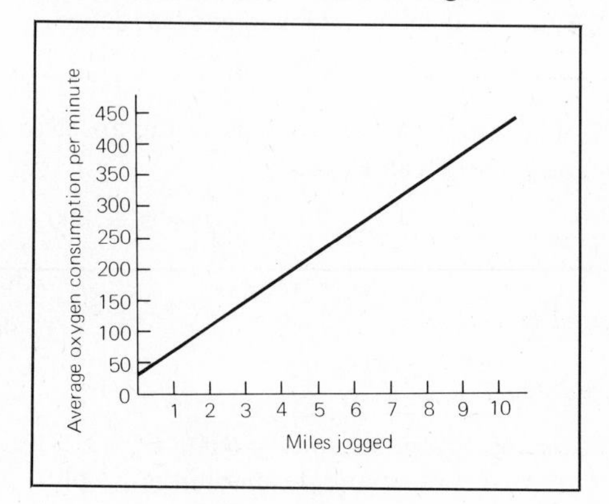

f. For $X = 4.5$: $\hat{Y} = .933\left(\frac{112.965}{2.63}\right)(4.5 - 5.87) + 264.7$

$= 209.798$

g. For $X = 8$: $\hat{Y} = .933\left(\frac{112.965}{2.63}\right)(8 - 5.87) + 264.7$

$= 350.059$

h. $\sigma_{y\hat{y}} = \sigma_y \sqrt{1 - r^2} = 112.965\sqrt{1 - (.933)^2} = 40.653$

ANSWERS TO EXERCISES, CHAPTER 13

1

a. Experiment *E*. Since *df* for the numerator is $K - 1 = 4$, where K is the number of groups, then K must equal 5.

b. Experiment B.

c.

Experiment	*df*	$\alpha = .05$	$\alpha = .01$
A	2 and 33	$F^* = 3.30$	$F^* = 5.34$
B	3 and 62	$F^* = 2.76$	$F^* = 4.13$
C	2 and 24	$F^* = 3.40$	$F^* = 5.61$
D	2 and 15	$F^* = 3.68$	$F^* = 6.36$
E	4 and 18	$F^* = 2.93$	$F^* = 4.58$
F	3 and 36	$F^* = 2.86$	$F^* = 4.38$

d. Experiments B, E, and F.

2

a. For no rewards:

$$s_1^2 = \frac{\Sigma X_1^2 - \frac{(\Sigma X_1)^2}{N_1}}{N_1 - 1}$$

$$= \frac{126{,}746 - \frac{1{,}258{,}884}{10}}{10 - 1} = 95.289$$

For social rewards:

$$s_2^2 = \frac{158{,}588 - \frac{1{,}577{,}536}{10}}{10 - 1} = 92.711$$

For candy rewards:

$$s_3^2 = \frac{152{,}714 - \dfrac{1{,}517{,}824}{10}}{10 - 1} = 103.511$$

For token rewards:

$$s_4^2 = \frac{180{,}364 - \dfrac{1{,}795{,}600}{10}}{10 - 1} = 89.333$$

b. Within-groups estimate of σ^2:

$$s^2_{\text{WG}} = \frac{s_1^2 + s_2^2 + s_3^2 + s_4^2}{4}$$

$$= \frac{95.289 + 92.711 + 103.511 + 89.333}{4}$$

$$= 95.211$$

c. Within-groups estimate of $\sigma_{\bar{X}}^2$:

$$\frac{s^2_{\text{WG}}}{N} = \frac{95.211}{10} = 9.521$$

d. Between groups estimate of $\sigma_{\bar{X}}^2$:

$$s_{\bar{X}}^2 = \frac{\Sigma \bar{X}^2 - \dfrac{(\Sigma \bar{X})^2}{K}}{K - 1} = \frac{61{,}498.44 - \dfrac{245{,}025}{4}}{4 - 1} = 80.73$$

e. The four steps of the hypothesis-testing procedure are:

(1) H_0: $\mu_1 = \mu_2 = \mu_3 = \mu_4$ (Type of reward does not affect mastery of multiplication)

H_A: $\mu_1 \neq \mu_2 \neq \mu_3 \neq \mu_4$ (Type of reward affects mastery of multiplication)

(2) Since you have four groups ($K = 4$), the numerator $df = K - 1 = 4 - 1 = 3$, and the denominator $df = K(N - 1) = 4(10 - 1) = 36$. Setting $\alpha = .05$, then $F^* = 2.86$.

(3) Using definitional Formula 13.1, the F ratio for your data is:

$$F = \frac{\text{between-groups estimate of } \sigma_{\bar{X}}^2}{\text{within-groups estimate of } \sigma_{\bar{X}}^2}$$

(4) Since 8.479 > 2.86, reject H_0 and conclude that the type of reward affects mastery of multiplication tables.

f. $F(3, 36) = 8.479, p < .05.$

3 a.

$$F = \frac{\text{between groups estimate of } \sigma^2}{\text{within groups estimate of } \sigma^2}$$

$$= \frac{Ns_{\bar{x}}^2}{s_{WG}^2} = \frac{10(80.73)}{95.211} = \frac{807.3}{95.211} = 8.479$$

b. Calculate the following values to substitute in the computational formulas.

$$\Sigma\Sigma X = 1122 + 1256 + 1232 + 1340 = 4950$$

$$(\Sigma\Sigma X)^2 = (4950)^2 = 24{,}502{,}500$$

$$\Sigma(\Sigma X)^2 = 1{,}258{,}884 + 1{,}577{,}536 + 1{,}517{,}824 + 1{,}795{,}600 = 6{,}149{,}844$$

$$\Sigma\Sigma X^2 = 126{,}746 + 158{,}588 + 152{,}714 + 180{,}364 = 618{,}412$$

Formula 13.3

$$SS_{WG} = \Sigma\Sigma X^2 - \frac{\Sigma(\Sigma X)^2}{N} = 618{,}412 - \frac{6{,}149{,}844}{10}$$

$$= 3427.6$$

Formula 13.4

$$SS_{BG} = \frac{\Sigma(\Sigma X)^2}{N} - \frac{(\Sigma\Sigma X)^2}{KN}$$

$$= \frac{6{,}149{,}844}{10} - \frac{24{,}502{,}500}{4(10)}$$

$$= 2421.9$$

c. Between-groups estimate of σ^2:

$$Ns_{\bar{x}}^2 = \frac{SS_{BG}}{df} = \frac{SS_{BG}}{K - 1} = \frac{2421.9}{3} = 807.3$$

d.

Source	SS	*df*	MS	*F*
BG	2421.9	3	807.3	8.479
WG	3427.6	36	95.211	

4 **a.** The steps of the hypothesis-testing procedure are:

(1) H_0: $\mu_1 = \mu_2 = \mu_3$ (The retention interval has no effect on recognition memory for comparative adjectives)

H_A: $\mu_1 \neq \mu_2 \neq \mu_3$ (The retention interval affects recognition memory for comparative adjectives)

(2) Since you have 3 groups ($K = 3$), the numerator $df = K - 1 = 3 - 1 = 2$, and the denominator $df = K(N - 1) = 3(10 - 1) = 27$. Setting $\alpha = .05$, then $F^* = 3.35$.

(3) Calculate the following values to substitute in the computational formulas:

	Group 1	*Group 2*	*Group 3*
$\Sigma X =$	146	121	107
$(\Sigma X)^2 =$	21,316	14,641	11,449
$\Sigma X^2 =$	2152	1497	1173

$$\Sigma\Sigma X = 146 + 121 + 107 = 374$$
$$(\Sigma\Sigma X)^2 = (374)^2 = 139{,}876$$
$$\Sigma(\Sigma X)^2 = 21{,}316 + 14{,}641 + 11{,}449 = 47{,}406$$
$$\Sigma\Sigma X^2 = 2152 + 1497 + 1173 = 4822$$

Formula 13.3

$$SS_{WG} = \Sigma\Sigma X^2 - \frac{\Sigma(\Sigma X)^2}{N}$$
$$= 4822 - \frac{47{,}406}{10}$$
$$= 81.4$$

Formula 13.4

$$SS_{BG} = \frac{\Sigma(\Sigma X)^2}{N} - \frac{(\Sigma\Sigma X)^2}{KN}$$
$$= \frac{47{,}406}{10} - \frac{139{,}876}{3(10)}$$
$$= 78.067$$

Adapting Formula 13.2:

$$F = \frac{Ns_{\bar{x}}^2}{s_{WG}^2} = \frac{\frac{SS_{BG}}{df}}{\frac{SS_{WG}}{df}}$$

$$= \frac{\frac{78.067}{2}}{\frac{81.4}{27}}$$

$$= \frac{39.034}{3.015} = 12.947$$

(4) Since $12.947 > 3.35$, reject H_0 and conclude that the retention interval affects recognition memory for comparative adjectives.

b.

Source	SS	*df*	MS	*F*
BG	78.067	2	39.034	12.947
WG	81.4	27	3.015	

c. $F(2, 27) = 12.947$, $p < .05$.

d. For equal-size groups, $df = K(N - 1) = 3(10 - 1) = 27$.
For both equal- and unequal-size groups, $df = N_{\text{total}} - K = 30 - 3 = 27$.

5 **a.** The steps of the hypothesis-testing procedure are:

(1) H_0: $\mu_1 = \mu_2 = \mu_3 = \mu_4 = \mu_5$ (Mode of presentation has no effect on the reduction of math anxiety)

H_A: $\mu_1 \neq \mu_2 \neq \mu_3 \neq \mu_4 \neq \mu_5$ (Mode of presentation has an effect on the reduction of math anxiety)

(2) Since you have five groups ($K = 5$), the numerator $df = K - 1 = 5 - 1 = 4$. Since you have a total of 32 scores ($N_{\text{total}} = 32$), the denominator $df = N_{\text{total}} - K = 32 - 5 = 27$. Setting $\alpha = .05$, then $F^* = 2.73$.

(3) Calculate the following values to substitute in the computational formulas:

	Group 1	*Group 2*	*Group 3*	*Group 4*	*Group 5*
$\Sigma X =$	67	56	29	35	40
$(\Sigma X)^2 =$	4489	3136	841	1225	1600
$\frac{(\Sigma X)^2}{n_g} =$	$\frac{4489}{7}$	$\frac{3136}{7}$	$\frac{841}{6}$	$\frac{1225}{6}$	$\frac{1600}{6}$
$=$	641.286	448	140.167	204.167	266.667

	Group 1	*Group 2*	*Group 3*	*Group 4*	*Group 5*
$\Sigma X^2 =$	679	476	167	241	306

$$\Sigma\frac{(\Sigma X)^2}{n_g} = 641.286 + 448 + 140.167 + 204.167 + 266.667 = 1700.287$$

$$\Sigma\Sigma X = 67 + 56 + 29 + 35 + 40 = 227$$

$$(\Sigma\Sigma X)^2 = (227)^2 = 51{,}529$$

$$N_{\text{total}} = 7 + 7 + 6 + 6 + 6 = 32$$

$$\frac{(\Sigma\Sigma X)^2}{N_{\text{total}}} = \frac{51{,}529}{32} = 1610.281$$

$$\Sigma\Sigma X^2 = 679 + 476 + 167 + 241 + 306 = 1869$$

Formula 13.5

$$SS_{BG} = \Sigma\frac{(\Sigma X)^2}{n_g} - \frac{(\Sigma\Sigma X)^2}{N_{\text{total}}}$$

$$= 1770.287 - 1610.281 = 160.006$$

Formula 13.6

$$SS_{WG} = \Sigma\Sigma X^2 - \Sigma\frac{(\Sigma X)^2}{n_g}$$

$$= 1869 - 1700.287 = 168.713$$

Adapting

Formula 13.2

$$F = \frac{Ns_{\bar{x}}^2}{s_{WG}^2} = \frac{\frac{SS_{BG}}{df}}{\frac{SS_{WG}}{df}}$$

$$= \frac{\frac{160.006}{4}}{\frac{168.713}{2}}$$

$$= \frac{40.002}{6.249} = 6.401$$

(4) Since $6.401 > 2.73$, reject H_0 and conclude that the mode of presentation has an effect on the reduction of math anxiety.

b.

Source	*SS*	*df*	*MS*	*F*
BG	160.006	4	40.002	6.401
WG	168.713	27	6.249	

c. $F(4, 27) = 6.401, p < .05$

6 **a.** The steps of the hypothesis-testing procedure are:

(1) H_0: $\mu_1 = \mu_2 = \mu_3 = \mu_4$ (Method of relaxation has no effect on subsequent anxiety scores)

H_A: $\mu_1 \neq \mu_2 \neq \mu_3 \neq \mu_4$ (Method of relaxation has an effect on subsequent anxiety scores)

(2) Since you have four groups ($K = 4$), the numerator $df = K - 1 = 4 - 1 = 3$. Since you have equal-size groups the denominator $df = K(N - 1) = 4(10 - 1) = 36$. Setting $\alpha = .01$, then $F^* = 4.38$.

(3) Since you have equal-size groups, you use the computational Formulas 13.3 and 13.4. Calculate the following values to substitute in the formulas:

	Group 1	*Group 2*	*Group 3*	*Group 4*
$\Sigma X =$	350	355	348	344
$(\Sigma X)^2 =$	122,500	126,025	121,104	118,336
$\Sigma X^2 =$	12,316	12,675	12,190	11,874
$\Sigma\Sigma X =$	1397			
$(\Sigma\Sigma X)^2 =$	1,951,609			
$\Sigma(\Sigma X)^2 =$	487,965			
$\Sigma\Sigma X^2 =$	49,055			

Formula 13.4

$$SS_{BG} = \frac{487{,}965}{10} - \frac{1{,}951{,}609}{4(10)} = 6.275$$

Formula 13.3

$$SS_{WG} = 49{,}055 - \frac{487{,}965}{10} = 258.5$$

Adapting Formula 13.2

$$F = \frac{\dfrac{SS_{BG}}{df}}{\dfrac{SS_{WG}}{df}} = \frac{\dfrac{6.275}{3}}{\dfrac{258.5}{36}}$$

$$= \frac{2.092}{7.181}$$

$$= .291$$

(4) Since $.291 < 4.38$, retain H_0 and conclude that there is not sufficient evidence to say that the method of relaxation has an effect on subsequent anxiety scores.

b.

Source	*SS*	*df*	*MS*	*F*
BG	6.275	3	2.092	.291
WG	258.5	36	7.181	

c. $F(3, 36) = .291$, N.S.

INDEX

(See also Glossaries beginning at p. 318.)

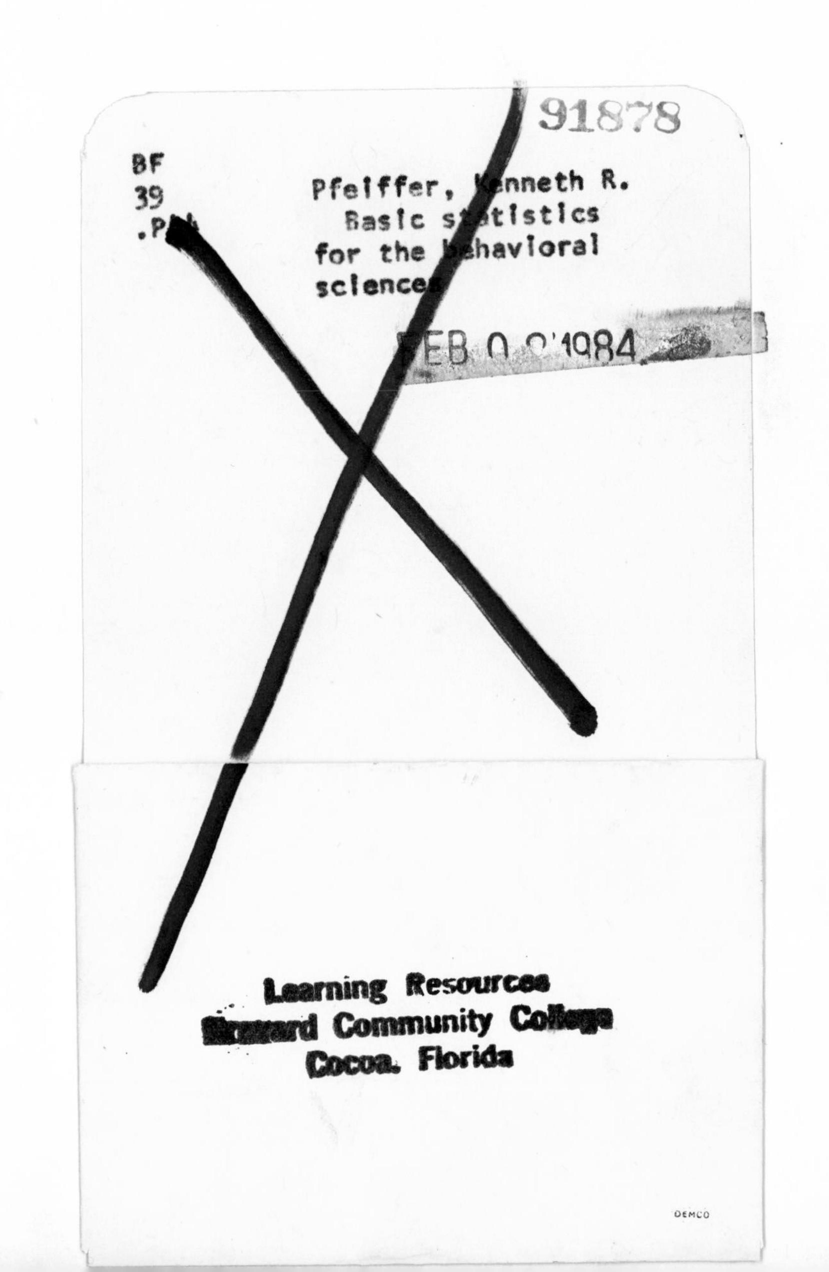
91878
BF
39
.P
Pfeiffer, Kenneth R.
Basic statistics
for the behavioral
sciences
FEB 0 9 1984
Learning Resources
Brevard Community College
Cocoa. Florida
DEMCO

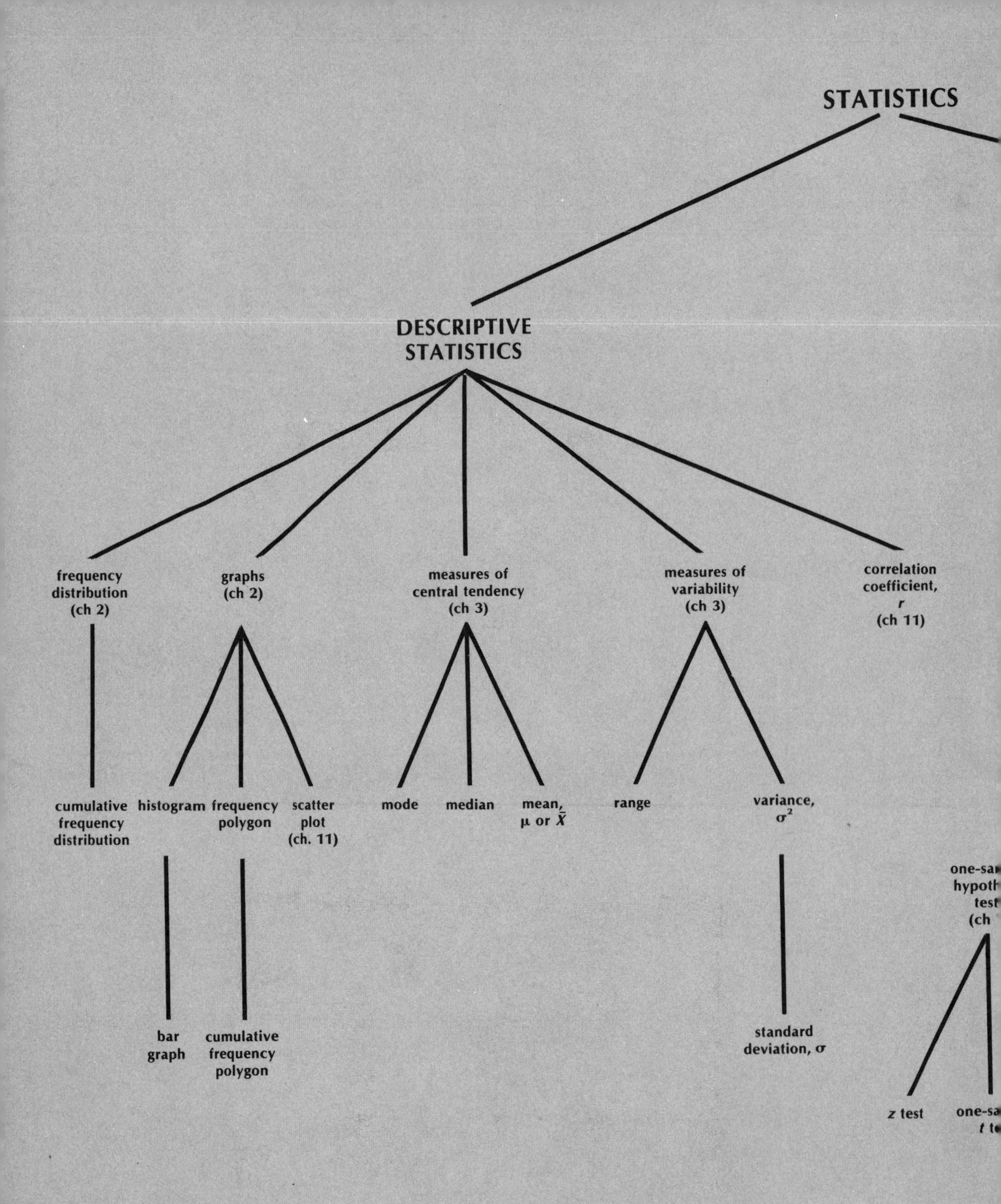

STATISTICS
DESCRIPTIVE STATISTICS
frequency distribution (ch 2)
graphs (ch 2)
measures of central tendency (ch 3)
measures of variability (ch 3)
correlation coefficient, r (ch 11)
cumulative frequency distribution
histogram
frequency polygon
scatter plot (ch. 11)
mode
median
mean, μ or X̄
range
variance, σ²
bar graph
cumulative frequency polygon
standard deviation, σ
z test